NOTABLE SELECTIONS IN

Environmental Studies

Second Edition

About the Editor

THEODORE D. GOLDFARB is a professor of chemistry at the State University of New York at Stony Brook. He earned a B.A. from Cornell University and a Ph.D. from the University of California, Berkeley. He is the author of over 35 research papers and articles on molecular structure, environmental chemistry, and science policy, as well as the book *A Search for Order in the Physical Universe* (W. H. Freeman, 1974). He is also the editor of *Taking Sides: Clashing Views on Controversial Environmental Issues* (Dushkin/McGraw-Hill), now in its eighth edition. Dr. Goldfarb is a recipient of the State University of New York's Chancellor's Award for Excellence in Teaching. In addition to teaching undergraduate and graduate courses in environmental and physical chemistry, he has taught summer institutes and special seminars for college and secondary school teachers and for undergraduate and graduate research students on a variety of topics, including energy policy, integrated waste management strategies, sustainable development, and ethics in science. He is presently directing a program funded by the National Science Foundation that is designed to promote the incorporation of ethics and values issues in the teaching of secondary school science. Dr. Goldfarb has served as a consultant and adviser to citizens' groups, town and city governments, and federal and state agencies on environmental matters. He is an active member of several professional organizations, including the American Chemical Society, the American Association for the Advancement of Science, and the New York Academy of Sciences.

S·O·U·R·C·E·S

NOTABLE SELECTIONS IN

Environmental Studies

Second Edition

EDITED BY

THEODORE D. GOLDFARB
State University of New York at Stony Brook

Dushkin/McGraw-Hill
A Division of The McGraw-Hill Companies

I dedicate this book to my friend and partner, Kate Lehman, who is a constant source of inspiration.

Manufactured in the United States of America

Second Edition

9 10 FGR/FGR 0 5 4

Library of Congress Cataloging-in-Publication Data

Main entry under title:

Sources: notable selections in environmental studies/edited by Theodore D. Goldfarb.—2nd ed.

Includes bibliographical references and index.

1. Environmental protection. 2. Conservation of natural resources. 3. Human ecology. I. Goldfarb, Theodore D., *comp.*

333.7

0-07-303186-0 96-85792

Printed on Recycled Paper

Preface

Over the past three decades, rapidly growing global concerns about environmental issues and problems have given rise to a variety of unconventional concentrations, majors, and minors at colleges and universities. Those who have pioneered this development recognized that an educational program in a traditional disciplinary major supplemented by a series of uncoordinated courses on environmental topics in the sciences, social sciences, and humanities would not prepare students adequately to deal with such an intrinsically interdisciplinary subject as the environment.

Several obstacles had to be surmounted in the creation of these new programs. For example, the curricula in most institutions of higher learning are primarily controlled by semiautonomous disciplinary departments; this inhibits the creation of effective interdisciplinary programs of study. Also, support for the organization and administration of such programs must come from some central academic office, which may not have control of the necessary resources. Another problem is that many academics are skeptical that broad, interdisciplinary majors or minors can provide the depth and substance that are usually characteristic of more narrowly focused degree programs. Furthermore, there is a prejudice against undergraduate programs that are designed to respond to practical, real-world problems rather than to introduce students to the more "basic" set of facts and methods associated with traditional academic disciplines.

Surmounting these obstacles in the context of local realities has produced a diverse set of programs with a variety of structures. Almost all of them require students to complete basic courses in the sciences as well as in humanities and social science disciplines, such as philosophy, politics, economics, and law. Programs called *environmental science* usually require more depth in the physical and biological sciences than those called *environmental studies*.

The vast majority of environmental science/studies programs include integrated courses designed to train students to approach environmental issues from a holistic perspective. One such course generally serves as an entry-level introduction to environmental studies. This book is intended to be used in such a course, either as the principal text or as a source of readings from the original literature to accompany an environmental studies or science text.

This second edition of *Sources: Notable Selections in Environmental Studies* brings together 46 selections of enduring intellectual value—including classic essays, book excerpts, journal articles, and an opinion from a Supreme Court case—that have helped to shape our contemporary understanding of the environment. Deciding which of the numerous selections that fit that description to include has been particularly difficult because there is less agreement about what constitutes the core subject matter of environmental studies than there is for more established academic disciplines such as chemistry, psychology, or philosophy. I am confident that the readings I have chosen to include in this volume can be used to enhance the value of any introductory environmental studies or science course. I sought help in making decisions about the contents of this second edition from several valued colleagues I know who had used the first edition in their courses. I am indebted to Tim Allen, Larry Davis, David Firmage, Katrina Smith Korfmacher, and Harold Ward for their thoughtful responses to that request. In the end I chose to delete 11 selections from the first edition and add 14 new selections.

Several additional sources of information proved helpful in choosing and organizing the selections. The bibliographies in several of the most popular introductory environmental science/studies texts were a rich source of potential readings, and the tables of contents of these texts provided ideas for coherent topical arrangements. Another useful resource was the course syllabi I have collected over the years from colleagues who teach at a wide range of institutions of higher learning, including two-year colleges, small liberal arts colleges, technical colleges, and large universities. Also of great value were the many discussions of environmental curricula that I have participated in at the annual meetings of the Northeastern Environmental Studies (NEES) group. I also received several helpful responses to a message soliciting recommendations of "classic readings" that I sent to several environmental Internet subscriber lists. Finally, my personal experience gained from choosing articles for the eight editions of *Taking Sides: Clashing Views on Controversial Environmental Issues* (Dushkin/McGraw Hill) and from almost 25 years of teaching environmental courses, doing environmental research, and being an environmental activist helped provide the perspective needed to pull it all together.

ORGANIZATION OF THE BOOK The selections are organized topically around major areas of environmental studies. Part 1 provides An Overview of Environmental Studies; Part 2 includes selections on Energy; Part 3, Environmental Degradation; Part 4, Population Issues and the Environment; Part 5, Human Health and the Environment; and Part 6, Environment and Society. Many of the selections included in this book, however, could have been equally well placed in a different chapter. Indeed, the ordering of the selections is not definitive, and the articles can be freely rearranged and read in any order that meets the needs or preferences of any course or instructor. Each selection is preceded by a short introductory headnote that provides biographical information about the author and briefly describes the relevance of the topic and the value of the selection.

ON THE INTERNET Each part in this book is preceded by an *On the Internet* page. This page provides a list of Internet site addresses that are relevant to the part as well as a description of each site.

A WORD TO THE INSTRUCTOR An *Instructor's Manual With Test Questions* (multiple-choice and essay) is available through the publisher for instructors using *Sources: Notable Selections in Environmental Studies*, 2d ed., in the classroom.

ACKNOWLEDGMENTS I wish to thank Mimi Egan, former publisher for the Sources series, for persuading me to take on this project. Theodore Knight, list manager for the Sources series, provided frequent encouragement, valuable feedback, and many helpful suggestions.

I also owe much gratitude to many colleagues who provided thoughtful advice and specific suggestions of notable sources that they have found useful in their own courses.

Comments, criticisms, and especially suggestions of selections to be considered for future editions are welcome from all readers or users of this book.

Theodore D. Goldfarb
State University of New York at Stony Brook

Contents

PART ONE

An Overview of Environmental Studies

On the Internet . . .

Sites appropriate to Part One

The Association for the Study of Literature and Environment is an organization dedicated to promoting an interest in literature that considers the relationship between human beings and the natural world. This is the home page of the association.

`http://www.asle.umn.edu/index.html`

Environment and History is an interdisciplinary journal containing articles by scholars in the humanities and the biological sciences. These articles examine current environmental problems from an historical perspective. This is the home page of the journal.

`http://www.erica.demon.co.uk/EH.html`

The mission of the Environmental Protection Agency is to protect human health and to safeguard the natural environment—air, water, and land—upon which life depends. This home page is a major resource for information on a wide variety of current environmental issues.

`http://www.epa.gov`

EnviroLink is a nonprofit organization dedicated to providing comprehensive, up-to-date information about available environmental resources. This Web site contains information about the organization as well as a searchable library.

`http://www.envirolink.org`

CHAPTER 1 Preservation vs. Conservation

1.1 JOHN MUIR

Hetch Hetchy Valley

John Muir (1838–1914) was, perhaps, America's most influential naturalist. In 1849 he and his family came to the United States from Scotland, where he was born. After abandoning plans to attend medical school, he worked at odd jobs and established a reputation as an inventor. In 1867 he decided to tramp the wilderness as an avocation. He took a 1,000-mile walk from Indiana to Florida, sailed to Cuba and then through the Isthmus of Panama, and arrived in San Francisco, California, in 1868. He eventually settled in the summits of the Sierra Nevada.

Muir referred to his years of wandering across meadows, around lakes, and up mountains as a lifelong education in the "University of the Wilderness." The detailed observations that he recorded daily were the source of both the rapturous, poetic descriptions and the scientific theories that he published. In 1871 Muir proposed that Yosemite Valley had been formed by glacial action. This theory, now widely accepted, was ridiculed by the geologists of the time. The establishment of Yosemite National Park in 1890 was largely a result of Muir's inspirational descriptions of this remarkable wonderland.

Muir was a vociferous proponent of an ecocentric rather than an ethnocentric philosophical perspective. He challenged the prevailing view of most of the conservationists of his time, who tempered their respect for nature with the multiple-use concept that gave primacy to human needs and appetites. The preservationist movement, supported by Muir, saw the need to set aside wilderness areas where no commercial or industrial activity would be permitted. The first such "primitive areas" were established by an administrative fiat of the U.S. Forest Service in the 1920s, a decade after Muir's death, but they were not officially protected by federal law until the passage of the Wilderness Act in 1964.

The most bitter disappointment of Muir's life was his failure to prevent the flooding and submerging of the Hetch Hetchy Valley in Yosemite National Park by a dam built to supply water and electricity to San Francisco. Muir focuses on the Hetch Hetchy Valley in the following selection from his book *The Mountains of California* (Houghton Mifflin, 1916). In it, Muir displays his mastery in conveying an image of the wonders of nature and his animosity toward those who would sacrifice such wonders in the interest of commercialism.

Key Concept: the primacy of nature and the case for wilderness preservation

Yosemite is so wonderful that we are apt to regard it as an exceptional creation, the only valley of its kind in the world; but Nature is not so poor as to have only one of anything. Several other yosemites have been discovered in the Sierra that occupy the same relative positions on the range and were formed by the same forces in the same kind of granite. One of these, the Hetch Hetchy Valley, is in the Yosemite National Park, about twenty miles from Yosemite, and is easily accessible to all sorts of travelers by a road and trail that leaves the Big Oak Flat road at Bronson Meadows a few miles below Crane Flat,and to mountaineers by way of Yosemite Creek basin and the head of the middle fork of the Tuolumne.

It is said to have been discovered by Joseph Screech, a hunter, in 1850, a year before the discovery of the great Yosemite. After my first visit to it in the autumn of 1871, I have always called it the "Tuolumne Yosemite," for it is a wonderfully exact counterpart of the Merced Yosemite, not only in its sublime rocks and waterfalls but in the gardens, groves and meadows of its flowery park-like floor. The floor of Yosemite is about four thousand feet above the sea; the Hetch Hetchy floor about thirty-seven hundred feet. And as the Merced River flows through Yosemite, so does the Tuolumne through Hetch Hetchy. The walls of both are of gray granite, rise abruptly from the floor, are sculptured in the same style and in both every rock is a glacier monument.

Standing boldly out from the south wall is a strikingly picturesque rock called by the Indians, Kolana, the outermost of a group twenty-three hundred feet high, corresponding with the Cathedral Rocks of Yosemite both in relative position and form. On the opposite side of the Valley, facing Kolana, there is a counterpart of El Capitan that rises sheer and plain to a height of eighteen hundred feet, and over its massive brow flows a stream which makes the most graceful fall I have ever seen. From the edge of the cliff to the top of an earthquake talus it is perfectly free in the air for a thousand feet before it is broken into cascades among talus boulders. It is in all its glory in June, when the snow is melting fast, but fades and vanishes toward the end of summer. The only fall I know with which it may fairly be compared is the Yosemite Bridal Veil; but it excels even that favorite fall both in height and airy-fairy beauty and behavior. Lowlanders are apt to suppose that mountain streams in their wild career over cliffs lose control of themselves and tumble in a noisy chaos of mist and spray. On the contrary, on no part of their travels are they more harmonious and self-controlled. Imagine yourself in Hetch Hetchy on a sunny day in June, standing waist-deep in grass and flowers (as I have often stood), while the great pines sway dreamily with scarcely perceptible motion. Looking northward across the Valley you see a plain, gray granite cliff rising abruptly out of the gardens and groves to a height of eighteen hundred feet, and in front of it Tueeulala's silvery scarf burning with irised sun-fire. In the first white outburst at the head there is abundance of visible energy, but it is speedily hushed and concealed in divine repose, and its tranquil progress to the base of the cliff is like that of a downy feather in a still room. Now observe the fineness and marvelous distinctness of the various sun-illumined fabrics into which the water is woven; they sift and float from form to form down the face of that grand gray rock in so leisurely and unconfused a manner that you can examine their texture, and patterns and tones of color as you would a piece of embroidery held in the hand. Toward the top of the fall you see groups of booming, comet-like masses, their solid, white heads separate, their tails like combed silk interlacing among delicate gray and purple shadows, ever forming and dissolving, worn out by friction in their rush through the air. Most of these vanish a few hundred feet below the summit, changing to varied forms of cloud-like drapery. Near the bottom the width of the fall has increased from about twenty-five feet to a hundred feet. Here it is composed of yet finer tissues, and is still without a trace of disorder—air, water and sunlight woven into stuff that spirits might wear.

So fine a fall might well seem sufficient to glorify any valley; but here, as in Yosemite, Nature seems in nowise moderate, for a short distance to the eastward of Tueeulala booms and thunders the great Hetch Hetchy Fall, Wapama, so near that you have both of them in full view from the same standpoint. It is the counterpart of the Yosemite Fall, but has a much greater volume of water, is about seventeen hundred feet in height, and appears to be nearly vertical, though considerably inclined, and is dashed into huge outbounding bosses of foam on projecting shelves and knobs. No two falls could be more unlike—Tueeulala out in the open sunshine descending like thistledown; Wapama in a jagged, shadowy gorge roaring and thundering, pounding its way like an earthquake avalanche.

Besides this glorious pair there is a broad, massive fall on the main river a short distance above the head of the Valley. Its position is something like that of the Vernal in Yosemite, and its roar as it plunges into a surging trout-pool may be heard a long way, though it is only about twenty feet high. On Rancheria Creek, a large stream, corresponding in position with the Yosemite Tenaya Creek, there is a chain of cascades joined here and there with swift flashing plumes like the one between the Vernal and Nevada Falls, making magnificent shows as they go their glacier-sculptured way, sliding, leaping, hurrahing, covered with crisp clashing spray made glorious with sifting sunshine. And besides all these a few small streams come over the walls at wide intervals, leaping from ledge to ledge with bird-like song and watering many a hidden cliff-garden and fernery, but they are too unshowy to be noticed in so grand a place. . . .

It appears . . . that Hetch Hetchy Valley, far from being a plain, common, rockbound meadow, as many who have not seen it seem to suppose, is a grand landscape garden, one of Nature's rarest and most precious mountain temples. As in Yosemite, the sublime rocks of its walls seem to glow with life, whether leaning back in repose or standing erect in thoughtful attitudes, giving welcome to storms and calms alike, their brows in the sky, their feet set in the groves and gay flowery meadows, while birds, bees, and butterflies help the river and waterfalls to stir all the air into music—things frail and fleeting and types of permanence meeting here and blending, just as they do in Yosemite, to draw her lovers into close and confiding communion with her.

Sad to say, this most precious and sublime feature of the Yosemite National Park, one of the greatest of all our natural resources for the uplifting joy and peace and health of the people, is in danger of being dammed and made into a reservoir to help supply San Francisco with water and light, thus flooding it from wall to wall and burying its gardens and groves one or two hundred feet deep. This grossly destructive commercial scheme has long been planned and urged (though water as pure and abundant can be got from sources outside of the people's park, in a dozen different places), because of the comparative cheapness of the dam and of the territory which it is sought to divert from the great uses to which it was dedicated in the Act of 1890 establishing the Yosemite National Park.

The making of gardens and parks goes on with civilization all over the world, and they increase both in size and number as their value is recognized. Everybody needs beauty as well as bread, places to play in and pray in, where Nature may heal and cheer and give strength to body and soul alike. This natural beauty-hunger is made manifest in the little window-sill gardens of the poor, though perhaps only a geranium slip in a broken cup, as well as in the carefully tended rose and lily gardens of the rich, the thousands of spacious city parks and botanical gardens, and in our magnificent national parks—the Yellowstone, Yosemite, Sequoia, etc.—Nature's sublime wonderlands, the admiration and joy of the world. Nevertheless, like anything else worth while, from the very beginning, however well guarded, they have always been subject to attack by despoiling gain-seekers and mischief-makers of every degree from Satan to Senators, eagerly trying to make everything immediately and selfishly commercial, with schemes disguised in smug-smiling philanthropy, industriously, shampiously crying, "Conservation, conservation, panutilization," that man and beast may be fed and the dear Nation made great. Thus long ago a few enterprising merchants utilized the Jerusalem temple as a place of business instead of a place of prayer, changing money, buying and selling cattle and sheep and doves; and earlier still, the first forest reservation, including only one tree, was likewise despoiled. Ever since the establishment of the Yosemite National Park, strife has been going on around its borders and I suppose this will go on as part of the universal battle between right and wrong, however much its boundaries may be shorn, or its wild beauty destroyed.

The first application to the Government by the San Francisco Supervisors for the commercial use of Lake Eleanor and the Hetch Hetchy Valley was made in 1903, and on December 22 of that year it was denied by the Secretary of the Interior, Mr. Hitchcock, who truthfully said:

> Presumably the Yosemite National Park was created such by law because of the natural objects of varying degrees of scenic importance located within its boundaries, inclusive alike of its beautiful small lakes, like Eleanor, and its majestic wonders, like Hetch Hetchy and Yosemite Valley. It is the aggregation of such natural scenic features that makes the Yosemite Park a wonderland which the Congress of the United States sought by law to reserve for all coming time as nearly as practicable in the condition fashioned by the hand of the Creator—a worthy object of national pride and a source of healthful pleasure and rest for the thousands of people who may annually sojourn there during the heated months.

... That any one would try to destroy such a place seems incredible; but sad experience shows that there are people good enough and bad enough for anything. The proponents of the dam scheme bring forward a lot of bad arguments to prove that the only righteous thing to do with the people's parks is to destroy them bit by bit as they are able. Their arguments are curiously like those of the devil, devised for the destruction of the first garden—so much of the very best Eden fruit going to waste; so much of the best Tuolumne water and Tuolumne scenery going to waste. Few of their statements are even partly true, and all are misleading.

Thus, Hetch Hetchy, they say, is a "low-lying meadow." On the contrary, it is a high-lying natural landscape garden, as the photographic illustrations show.

"It is a common minor feature, like thousands of others." On the contrary it is a very uncommon feature; after Yosemite, the rarest and in many ways the most important in the National Park.

"Damming and submerging it one hundred and seventy-five feet deep would enhance its beauty by forming a crystal-clear lake." Landscape gardens, places of recreation and worship, are never made beautiful by destroying and burying them. The beautiful sham lake, forsooth, would be only an eyesore, a dismal blot on the landscape, like many others to be seen in the Sierra. For, instead of keeping it at the same level all the year, allowing Nature centuries of time to make new shores, it would, of course, be full only a month or two in the spring, when the snow is melting fast; then it would be gradually drained, exposing the slimy sides of the basin and shallower parts of the bottom, with the gathered drift and waste, death and decay of the upper basins, caught here instead of being swept on to decent natural burial along the banks of the river or in the sea. Thus the Hetch Hetchy dam-lake would be only a rough imitation of a natural lake for a few of the spring months, an open sepulcher for the others.

"Hetch Hetchy water is the purest of all to be found in the Sierra, unpolluted, and forever unpollutable." On the contrary, excepting that of the Merced below Yosemite, it is less pure than that of most of the other Sierra streams, because of the sewerage of camp-grounds draining into it, especially of the Big Tuolumne Meadows camp-ground, occupied by hundreds of tourists and mountaineers, with their animals, for months every summer, soon to be followed by thousands from all the world.

These temple destroyers, devotees of ravaging commercialism, seem to have a perfect contempt for Nature, and, instead of lifting their eyes to the God of the mountains, lift them to the Almighty Dollar.

Dam Hetch Hetchy! As well dam for watertanks the people's cathedrals and churches, for no holier temple has ever been consecrated by the heart of man.

1.2 GIFFORD PINCHOT

Principles of Conservation

Conservation, or the wise use of natural resources to meet human needs and desires, has always had more public and political support than more restrictive forms of nature preservation. Gifford Pinchot (1865–1946) was one of the founders and one of the most effective leaders of the U.S. conservation movement that began at the end of the nineteenth century. He served as chief of the U.S. Forest Service from 1898 to 1910. In 1902 President Theodore Roosevelt directed Pinchot to lead a campaign to restore the forest and water resources, which had been ravaged by the environmentally destructive practices of large lumbering, mining, and agricultural enterprises. In 1905 Roosevelt established a Bureau of Forestry within the U.S. Department of Agriculture, enhancing Pinchot's ability to defend public lands against the destructive plans of private interests. Pinchot also established Yale University's School of Forestry, where he was a professor from 1903 to 1936. An effective politician, Pinchot helped found the Progressive Party, and he served two terms as governor of Pennsylvania, 1923–1927 and 1931–1935.

In the following selection from his book *The Fight for Conservation* (Doubleday, 1910), Pinchot details the principles that guided his actions. It is clear that his goal was to maximize the value of resources for human exploitation rather than to preserve wilderness.

Key Concept: the conservation of resources for human use

The principles which the word Conservation has come to embody are not many, and they are exceedingly simple. I have had occasion to say a good many times that no other great movement has ever achieved such progress in so short a time, or made itself felt in so many directions with such vigor and effectiveness, as the movement for the conservation of natural resources.

Forestry made good its position in the United States before the conservation movement was born. As a forester I am glad to believe that conservation began with forestry, and that the principles which govern the Forest Service in particular and forestry in general are also the ideas that control conservation.

The first idea of real foresight in connection with natural resources arose in connection with the forest. From it sprang the movement which gathered impetus until it culminated in the great Convention of Governors at Washington in May, 1908. Then came the second official meeting of the National Conservation movement, December, 1908, in Washington. Afterward came the various gatherings of citizens in convention, come together to express their judgment on what ought to be done, and to contribute, as only such meetings can, to the formation of effective public opinion.

The movement so begun and so prosecuted has gathered immense swing and impetus. In 1907 few knew what Conservation meant. Now it has become a household word. While at first Conservation was supposed to apply only to forests, we see now that its sweep extends even beyond the natural resources.

The principles which govern the conservation movement, like all great and effective things, are simple and easily understood. Yet it is often hard to make the simple, easy, and direct facts about a movement of this kind known to the people generally.

The first great fact about conservation is that it stands for development. There has been a fundamental misconception that conservation means nothing but the husbanding of resources for future generations. There could be no more serious mistake. Conservation does mean provision for the future, but it means also and first of all the recognition of the right of the present generation to the fullest necessary use of all the resources with which this country is so abundantly blessed. Conservation demands the welfare of this generation first, and afterward the welfare of the generations to follow.

The first principle of conservation is development, the use of the natural resources now existing on this continent for the benefit of the people who live here now. There may be just as much waste in neglecting the development and use of certain natural resources as there is in their destruction. We have a limited supply of coal, and only a limited supply. Whether it is to last for a hundred or a hundred and fifty or a thousand years, the coal is limited in amount, unless through geological changes which we shall not live to see, there will never be any more of it than there is now. But coal is in a sense the vital essence of our civilization. If it can be preserved, if the life of the mines can be extended, if by preventing waste there can be more coal left in this country after we of this generation have made every needed use of this source of power, then we shall have deserved well of our descendants.

Conservation stands emphatically for the development and use of water-power now, without delay. It stands for the immediate construction of navigable waterways under a broad and comprehensive plan as assistants to the railroads. More coal and more iron are required to move a ton of freight by rail than by water, three to one. In every case and in every direction the conservation movement has development for its first principle, and at the very beginning of its work. The development of our natural resources and the fullest use of them for the present generation is the first duty of this generation. So much for development.

In the second place conservation stands for the prevention of waste. There has come gradually in this country an understanding that waste is not a good thing and that the attack on waste is an industrial necessity. I recall very well indeed how, in the early days of forest fires, they were considered simply and solely as acts of God, against which any opposition was hopeless and any attempt to control them not merely hopeless but childish. It was assumed that they came in the natural order of things, as inevitably as the seasons or the rising and setting of the sun. To-day we understand that forest fires are wholly within the control of men. So we are coming in like manner to understand that the prevention of waste in all other directions is a simple matter of good business. The first duty of the human race is to control the earth it lives upon.

We are in a position more and more completely to say how far the waste and destruction of natural resources are to be allowed to go on and where they are to stop. It is curious that the effort to stop waste, like the effort to stop forest fires, has often been considered as a matter controlled wholly by economic law. I think there could be no greater mistake. Forest fires were allowed to burn long after the people had means to stop them. The idea that men were helpless in the face of them held long after the time had passed when the means of control were fully within our reach. It was the old story that "as a man thinketh, so is he"; we came to see that we could stop forest fires, and we found that the means had long been at hand. When at length we came to see that the control of logging in certain directions was profitable, we found it had long been possible. In all these matters of waste of natural resources, the education of the people to understand that they can stop the leakage comes before the actual stopping and after the means of stopping it have long been ready at our hands.

In addition to the principles of development and preservation of our resources there is a third principle. It is this: The natural resources must be developed and preserved for the benefit of the many, and not merely for the profit of a few. We are coming to understand in this country that public action for public benefit has a very much wider field to cover and a much larger part to play than was the case when there were resources enough for every one, and before certain constitutional provisions had given so tremendously strong a position to vested rights and property in general....

The conservation idea covers a wider range than the field of natural resources alone. Conservation means the greatest good to the greatest number for the longest time. One of its great contributions is just this, that it has added to the worn and well-known phrase, "the greatest good to the greatest number," the additional words "for the longest time," thus recognizing that this nation of ours must be made to endure as the best possible home for all its people.

Conservation advocates the use of foresight, prudence, thrift, and intelligence in dealing with public matters, for the same reasons and in the same way that we each use foresight, prudence, thrift, and intelligence in dealing with our own private affairs. It proclaims the right and duty of the people to act for the benefit of the people. Conservation demands the application of common-sense to the common problems for the common good.

The principles of conservation thus described—development, preservation, the common good—have a general application which is growing rapidly wider. The development of resources and the prevention of waste and loss, the protection of the public interests, by foresight, prudence, and the ordinary business and home-making virtues, all these apply to other things as well as to the natural resources. There is, in fact, no interest of the people to which the principles of conservation do not apply.

The conservation point of view is valuable in the education of our people as well as in forestry; it applies to the body politic as well as to the earth and its minerals. A municipal franchise is as properly within its sphere as a franchise for water-power. The same point of view governs in both. It applies as much to the subject of good roads as to waterways, and the training of our people in citizenship is as germane to it as the productiveness of the earth. The application of common-sense to any problem for the Nation's good will lead directly to national efficiency wherever applied. In other words, and that is the burden of the message, we are coming to see the logical and inevitable outcome that these principles, which arose in forestry and have their bloom in the conservation of natural resources, will have their fruit in the increase and promotion of national efficiency along other lines of national life.

1.3 ALDO LEOPOLD

A Sand County Almanac: And Sketches Here and There

The struggle between those who advocate the conservationist multiple-use concept of environmental protection and proponents of preserving nature for its own sake continues to this day. One of the most influential American preservationists of the first half of the twentieth century was Aldo Leopold (1887–1948). As an officer of the U.S. Forest Service in the 1920s, Leopold became concerned about the failure of management practices in the national forests to adequately protect the natural environment from the effects of commercial activities. Along with fellow officer Robert Marshall, he helped establish 70 "primitive areas" where all development was prohibited. In 1933 Leopold was appointed to chair of game management, a position that was created for him at the University of Wisconsin. In 1935 Leopold and Marshall founded the Wilderness Society, which was to lead the protracted struggle for wilderness preservation. Success was finally achieved—16 years after Leopold's death—with the approval of the Federal Wilderness Act in 1964. Although Leopold was a vociferous advocate of the intrinsic worth of all living things, he supported a strong role for human beings in the management and protection of wild lands.

A Sand County Almanac: And Sketches Here and There (Oxford University Press, 1949), from which the following selection was taken, is a collection of Leopold's lyrical, philosophical writings about nature. In the section under the heading "Thinking Like a Mountain," Leopold reflects on his early enthusiasm for killing wolves as indicative of human ignorance about the ecological interdependence that sustains a mountain ecosystem. Leopold's land ethic, which he discusses in the final section of *A Sand County Almanac,* is his most quoted and influential writing. Leopold held that ethical considerations, which historically had encompassed only the relationships among human beings, must be extended to include interactions of humans with the animate and inanimate components of the natural world.

Key Concept: an ethical relationship between humans and the land

Chapter 1
Preservation vs. Conservation

A deep chesty bawl echoes from rimrock to rimrock, rolls down the mountain, and fades into the far blackness of the night. It is an outburst of wild defiant sorrow, and of contempt for all the adversities of the world.

Every living thing (and perhaps many a dead one as well) pays heed to that call. To the deer it is a reminder of the way of all flesh, to the pine a forecast of midnight scuffles and of blood upon the snow, to the coyote a promise of gleanings to come, to the cowman a threat of red ink at the bank, to the hunter a challenge of fang against bullet. Yet behind these obvious and immediate hopes and fears there lies a deeper meaning, known only to the mountain itself. Only the mountain has lived long enough to listen objectively to the howl of a wolf.

Those unable to decipher the hidden meaning know nevertheless that it is there, for it is felt in all wolf country, and distinguishes that country from all other land. It tingles in the spine of all who hear wolves by night, or who scan their tracks by day. Even without sight or sound of wolf, it is implicit in a hundred small events: the midnight whinny of a pack horse, the rattle of rolling rocks, the bound of a fleeing deer, the way shadows lie under the spruces. Only the ineducable tyro can fail to sense the presence or absence of wolves, or the fact that mountains have a secret opinion about them.

My own conviction on this score dates from the day I saw a wolf die. We were eating lunch on a high rimrock, at the foot of which a turbulent river elbowed its way. We saw what we thought was a doe fording the torrent, her breast awash in white water. When she climbed the bank toward us and shook out her tail, we realized our error: it was a wolf. A half-dozen others, evidently grown pups, sprang from the willows and all joined in a welcoming mêlée of wagging tails and playful maulings. What was literally a pile of wolves writhed and tumbled in the center of an open flat at the foot of our rimrock.

amazing intro

In those days we had never heard of passing up a chance to kill a wolf. In a second we were pumping lead into the pack, but with more excitement than accuracy: how to aim a steep downhill shot is always confusing. When our rifles were empty, the old wolf was down, and a pup was dragging a leg into impassable slide-rocks.

We reached the old wolf in time to watch a fierce green fire dying in her eyes. I realized then, and have known ever since, that there was something new to me in those eyes—something known only to her and to the mountain. I was young then, and full of trigger-itch; I thought that because fewer wolves meant more deer, that no wolves would mean hunters' paradise. But after seeing the green fire die, I sensed that neither the wolf nor the mountain agreed with such a view.

Since then I have lived to see state after state extirpate its wolves. I have watched the face of many a newly wolfless mountain, and seen the south-facing slopes wrinkle with a maze of new deer trails. I have seen every edible bush and seedling browsed, first to anaemic desuetude, and then to death. I have seen every edible tree defoliated to the height of a saddlehorn. Such a mountain looks as if someone had given God a new pruning shears, and forbidden Him all other exercise. In the end the starved bones of the hoped-for deer herd, dead of its own too-much, bleach with the bones of the dead sage, or molder under the high-lined junipers.

I now suspect that just as a deer herd lives in mortal fear of its wolves, so does a mountain live in mortal fear of its deer. And perhaps with better cause, for while a buck pulled down by wolves can be replaced in two or three years, a range pulled down by too many deer may fail of replacement in as many decades.

So also with cows. The cowman who cleans his range of wolves does not realize that he is taking over the wolf's job of trimming the herd to fit the range. He has not learned to think like a mountain. Hence we have dustbowls, and rivers washing the future into the sea.

We all strive for safety, prosperity, comfort, long life, and dullness. The deer strives with his supple legs, the cowman with trap and poison, the statesman with pen, the most of us with machines, votes, and dollars, but it all comes to the same thing: peace in our time. A measure of success in this is all well enough, and perhaps is a requisite to objective thinking, but too much safety seems to yield only danger in the long run. Perhaps this is behind Thoreau's dictum: In wildness is the salvation of the world. Perhaps this is the hidden meaning in the howl of the wolf, long known among mountains, but seldom perceived among men. . . .

THE LAND ETHIC

When god-like Odysseus returned from the wars in Troy, he hanged all on one rope a dozen slave-girls of his household whom he suspected of misbehavior during his absence.

This hanging involved no question of propriety. The girls were property. The disposal of property was then, as now, a matter of expediency, not of right and wrong.

Concepts of right and wrong were not lacking from Odysseus' Greece: witness the fidelity of his wife through the long years before at last his black-prowed galleys clove the wine-dark seas for home. The ethical structure of that day covered wives, but had not yet been extended to human chattels. During the three thousand years which have since elapsed, ethical criteria have been extended to many fields of conduct, with corresponding shrinkages in those judged by expediency only.

The Ethical Sequence

This extension of ethics, so far studied only by philosophers, is actually a process in ecological evolution. Its sequences may be described in ecological as well as in philosophical terms. An ethic, ecologically, is a limitation on freedom of action in the struggle for existence. An ethic, philosophically, is a differentiation of social from anti-social conduct. These are two definitions of one thing. The thing has its origin in the tendency of interdependent individuals or groups to evolve modes of co-operation. The ecologist calls these symbioses. Politics and economics are advanced symbioses in which the original free-for-all competition has been replaced, in part, by co-operative mechanisms with an ethical content.

The complexity of co-operative mechanisms has increased with population density, and with the efficiency of tools. It was simpler, for example, to define the anti-social uses of sticks and stones in the days of the mastodons than of bullets and billboards in the age of motors.

The first ethics dealt with the relation between individuals; the Mosaic Decalogue is an example. Later accretions dealt with the relation between the individual and society. The Golden Rule tries to integrate the individual to society; democracy to integrate social organization to the individual.

There is as yet no ethic dealing with man's relation to land and to the animals and plants which grow upon it. Land, like Odysseus' slave-girls, is still property. The land-relation is still strictly economic, entailing privileges but not obligations.

The extension of ethics to this third element in human environment is, if I read the evidence correctly, an evolutionary possibility and an ecological necessity. It is the third step in a sequence. The first two have already been taken. Individual thinkers since the days of Ezekiel and Isaiah have asserted that the despoliation of land is not only inexpedient but wrong. Society, however, has not yet affirmed their belief. I regard the present conservation movement as the embryo of such an affirmation.

An ethic may be regarded as a mode of guidance for meeting ecological situations so new or intricate, or involving such deferred reactions, that the path of social expediency is not discernible to the average individual. Animal instincts are modes of guidance for the individual in meeting such situations. Ethics are possibly a kind of community instinct in-the-making.

The Community Concept

All ethics so far evolved rest upon a single premise: that the individual is a member of a community of interdependent parts. His instincts prompt him to compete for his place in that community, but his ethics prompt him also to co-operate (perhaps in order that there may be a place to compete for).

The land ethic simply enlarges the boundaries of the community to include soils, waters, plants, and animals, or collectively: the land. . . .

The Outlook

It is inconceivable to me that an ethical relation to land can exist without love, respect, and admiration for land, and a high regard for its value. By value,

I of course mean something far broader than mere economic value; I mean value in the philosophical sense.

Perhaps the most serious obstacle impeding the evolution of a land ethic is the fact that our educational and economic system is headed away from, rather than toward, an intense consciousness of land. Your true modern is separated from the land by many middlemen, and by innumerable physical gadgets. He has no vital relation to it; to him it is the space between cities on which crops grow. Turn him loose for a day on the land, and if the spot does not happen to be a golf links or a 'scenic' area, he is bored stiff. If crops could be raised by hydroponics instead of farming, it would suit him very well. Synthetic substitutes for wood, leather, wool, and other natural land products suit him better than the originals. In short, land is something he has 'outgrown.'

Almost equally serious as an obstacle to a land ethic is the attitude of the farmer for whom the land is still an adversary, or a taskmaster that keeps him in slavery. Theoretically, the mechanization of farming ought to cut the farmer's chains, but whether it really does is debatable.

One of the requisites for an ecological comprehension of land is an understanding of ecology, and this is by no means co-extensive with 'education'; in fact, much higher education seems deliberately to avoid ecological concepts. An understanding of ecology does not necessarily originate in courses bearing ecological labels; it is quite as likely to be labeled geography, botany, agronomy, history, or economics. This is as it should be, but whatever the label, ecological training is scarce.

The case for a land ethic would appear hopeless but for the minority which is in obvious revolt against these 'modern' trends.

The 'key log' which must be moved to release the evolutionary process for an ethic is simply this: quit thinking about decent land-use as solely an economic problem. Examine each question in terms of what is ethically and esthetically right, as well as what is economically expedient. A thing is right when it tends to preserve the integrity, stability, and beauty of the biotic community. It is wrong when it tends otherwise.

It of course goes without saying that economic feasibility limits the tether of what can or cannot be done for land. It always has and it always will. The fallacy the economic determinists have tied around our collective neck, and which we now need to cast off, is the belief that economics determines *all* land-use. This is simply not true. An innumerable host of actions and attitudes, comprising perhaps the bulk of all land relations, is determined by the land-users' tastes and predilections, rather than by his purse. The bulk of all land relations hinges on investments of time, forethought, skill, and faith rather than on investments of cash. As a land-user thinketh, so is he.

I have purposely presented the land ethic as a product of social evolution because nothing so important as an ethic is ever 'written.' Only the most superficial student of history supposes that Moses 'wrote' the Decalogue; it evolved in the minds of a thinking community, and Moses wrote a tentative summary of it for a 'seminar.' I say tentative because evolution never stops.

The evolution of a land ethic is an intellectual as well as emotional process. Conservation is paved with good intentions which prove to be futile, or even dangerous, because they are devoid of critical understanding either of the land, or of economic land-use. I think it is a truism that as the ethical frontier advances from the individual to the community, its intellectual content increases.

The mechanism of operation is the same for any ethic: social approbation for right actions: social disapproval for wrong actions.

By and large, our present problem is one of attitudes and implements. We are remodeling the Alhambra with a steam-shovel, and we are proud of our yardage. We shall hardly relinquish the shovel, which after all has many good points, but we are in need of gentler and more objective criteria for its successful use.

CHAPTER 2 Fundamental Causes of Environmental Problems

2.1 LYNN WHITE, JR.

The Historical Roots of Our Ecological Crisis

Few people would deny that the world's religions have been a major influence in shaping human philosophical attitudes and social actions. No attempt to assess this influence has proven to be as controversial as the lecture delivered by Lynn White, Jr., at the 1966 meeting of the American Association for the Advancement of Science entitled "The Historical Roots of Our Ecological Crisis," from which the following selection has been taken. In it, White, a distinguished professor of medieval and renaissance history at the University of California, Los Angeles, argues that Christianity, as an institution in the Western world, is to blame for the attitudes that have resulted in environmental degradation. In his view, Christian dogma has interpreted the dominion over nature that God granted to man to mean that "nature has no reason for existence save to serve man." According to White, this attitude, coupled with the power unleashed by the development of modern science and technology, has produced ecological crises.

Many of White's critics have based their arguments on biblical texts. They cite passages in Judeo-Christian scripture that suggest that dominion should be interpreted as requiring stewardship rather than wanton exploitation of nature. Such criticism fails to address the central thesis of White's analysis. White acknowledges that Biblical references to nature can be interpreted in many ways. His focus is on the actual preaching of the Christian church, which in his view emphasizes a reading of the Bible that condones the thoughtless conquest of nature but says little or nothing about the need for sensitive stewardship.

Key Concept: Judeo-Christian justification for the exploitation of nature

A conversation with Aldous Huxley not infrequently put one at the receiving end of an unforgettable monologue. About a year before his lamented death he was discoursing on a favorite topic: Man's unnatural treatment of nature and its sad results. To illustrate his point he told how, during the previous summer, he had returned to a little valley in England where he had spent many happy months as a child. Once it had been composed of delightful grassy glades; now it was becoming overgrown with unsightly brush because the rabbits that formerly kept such growth under control had largely succumbed to a disease, myxomatosis, that was deliberately introduced by the local farmers to reduce the rabbits' destruction of crops. Being something of a Philistine, I could be silent no longer, even in the interests of great rhetoric. I interrupted to point out that the rabbit itself had been brought as a domestic animal to England in 1176, presumably to improve the protein diet of the peasantry.

All forms of life modify their contexts. The most spectacular and benign instance is doubtless the coral polyp. By serving its own ends, it has created a vast undersea world favorable to thousands of other kinds of animals and plants. Ever since man became a numerous species he has affected his environment notably. The hypothesis that his fire-drive method of hunting created the world's great grasslands and helped to exterminate the monster mammals of the Pleistocene from much of the globe is plausible, if not proved. For 6 millennia at least, the banks of the lower Nile have been a human artifact rather than the swampy African jungle which nature, apart from man, would have made it. The Aswan Dam, flooding 5000 square miles, is only the latest stage in a long process. In many regions terracing or irrigation, overgrazing, the cutting of forests by Romans to build ships to fight Carthaginians or by Crusaders to solve the logistics problems of their expeditions, have profoundly changed some ecologies. Observation that the French landscape falls into two basic types, the open fields of the north and the *bocage* of the south and west, inspired Marc Bloch to undertake his classic study of medieval agricultural methods. Quite unintentionally, changes in human ways often affect nonhuman nature. It has been noted, for example, that the advent of the automobile eliminated huge flocks of sparrows that once fed on the horse manure littering every street.

The history of ecologic change is still so rudimentary that we know little about what really happened, or what the results were. The extinction of the European aurochs as late as 1627 would seem to have been a simple case of overenthusiastic hunting. On more intricate matters it often is impossible to find solid information. For a thousand years or more the Frisians and Hollanders have been pushing back the North Sea, and the process is culminating in our own time in the reclamation of the Zuider Zee. What, if any, species of animals, birds, fish, shore life, or plants have died out in the process? In their epic combat with Neptune have the Netherlanders overlooked ecological values in such a way that the quality of human life in the Netherlands has suffered? I cannot discover that the questions have ever been asked, much less answered.

People, then, have often been a dynamic element in their own environment, but in the present state of historical scholarship we usually do not know exactly when, where, or with what effects man-induced changes came. As we enter the last third of the 20th century, however, concern for the problem of ecologic backlash is mounting feverishly. Natural science, conceived as the effort to understand the nature of things, had flourished in several eras and among several peoples. Similarly there had been an age-old accumulation of technological skills, sometimes growing rapidly, sometimes slowly. But it was not until about four generations ago that Western Europe and North America arranged a marriage between science and technology, a union of the theoretical and the empirical approaches to our natural environment. The emergence in widespread practice of the Baconian creed that scientific knowledge means technological power over nature can scarcely be dated before about 1850, save in the chemical industries, where it is anticipated in the 18th century. Its acceptance as a normal pattern of action may mark the greatest event in human history since the invention of agriculture, and perhaps in nonhuman terrestrial history as well.

Almost at once the new situation forced the crystallization of the novel concept of ecology; indeed, the word *ecology* first appeared in the English language in 1873. Today, less than a century later, the impact of our race upon the environment has so increased in force that it has changed in essence. When the first cannons were fired, in the early 14th century, they affected ecology by sending workers scrambling to the forests and mountains for more potash, sulfur, iron ore, and charcoal, with some resulting erosion and deforestation. Hydrogen bombs are of a different order: a war fought with them might alter the genetics of all life on this planet. By 1285 London had a smog problem arising from the burning of soft coal, but our present combustion of fossil fuels threatens to change the chemistry of the globe's atmosphere as a whole, with consequences which we are only beginning to guess. With the population explosion, the carcinoma of planless urbanism, the now geological deposits of sewage and garbage, surely no create other than man has ever managed to foul its nest in such short order.

There are many calls to action, but specific proposals, however worthy as individual items, seem too partial, palliative, negative: ban the bomb, tear down the billboards, give the Hindus contraceptives and tell them to eat their sacred cows. The simplest solution to any suspect change is, of course, to stop it, or, better yet, to revert to a romanticized past: make those ugly gasoline stations look like Anne Hathaway's cottage or (in the Far West) like ghost-town saloons. The "wilderness area" mentality invariably advocates deep-freezing an ecology, whether San Gimignano or the High Sierra, as it was before the first Kleenex was dropped. But neither atavism nor prettification will cope with the ecologic crisis of our time.

What shall we do? No one yet knows. Unless we think about fundamentals, our specific measures may produce new backlashes more serious than those they are designed to remedy.

As a beginning we should try to clarify our thinking by looking, in some historical depth, at the presuppositions that underlie modern technology and science. Science was traditionally aristocratic, speculative, intellectual in intent; technology was lower-class, empirical, action-oriented. The quite sudden fusion of these two, towards the middle of the 19th century, is surely related to the slightly prior and contemporary democratic revolutions which, by reducing social barriers, tended to assert a functional unity of brain and hand. Our ecologic crisis is the product of an emerging, entirely novel, democratic culture. The issue is whether a democratized world can survive its own implications. Presumably we cannot unless we rethink our axioms.

THE WESTERN TRADITIONS OF TECHNOLOGY AND SCIENCE

One thing is so certain that it seems stupid to verbalize it: both modern technology and modern science are distinctively *Occidental.* Our technology has absorbed elements from all over the world, notably from China; yet everywhere today, whether in Japan or in Nigeria, successful technology is Western. Our science is the heir to all the sciences of the past, especially perhaps to the work of the great Islamic scientists of the Middle Ages, who so often outdid the ancient Greeks in skill and perspicacity: al Rāzi in medicine, for example; or ibn-al-Haytham in optics; or Omar Khayyám in mathematics. Indeed, not a few works of such geniuses seem to have vanished in the original Arabic and to survive only in medieval Latin translations that helped to lay foundations for later Western developments. Today, around the globe, all significant science is Western in style and method, whatever the pigmentation or language of the scientists.

A second pair of fact is less well recognized because they result from quite recent historical scholarship. The leadership of the West, both in technology and in science, is far older than the so-called Scientific Revolution of the 17th century or the so-called Industrial Revolution of the 18th century. These terms are in fact outmoded and obscure the true nature of what they try to describe—significant stages in two long and separate developments. By A.D. 1000 at the latest—and perhaps, feebly, as much as 200 years earlier—the West began to apply water power to industrial processes other than milling grain. This was followed in the late 12th century by the harnessing of wind power. From simple beginnings, but with remarkable consistency of style, the West rapidly expanded its skills in the development of power machinery, labor-saving devices, and automation. . . .

Since both our technological and our scientific movements got their start, acquired their character, and achieved world dominance in the Middle Ages, it would seem that we cannot understand their nature or their present impact upon ecology without examining fundamental medieval assumptions and developments.

MEDIEVAL VIEW OF MAN AND NATURE

Until recently, agriculture has been the chief occupation even in "advanced" societies; hence, any change of methods of tillage has much importance. Early plows, drawn by two oxen, did not normally turn the sod but merely scratched it. Thus, cross-plowing was needed and fields tended to be squarish. In the fairly light soils and semiarid climates of the Near East and Mediterranean, this worked well. But such a plow was inappropriate to the wet climate and often sticky soils of northern Europe. By the latter part of the 7th century after Christ, however, following obscure beginnings, certain northern peasants were using an entirely new kind of plow, equipped with a vertical knife to cut the line of the furrow, a horizontal share to slice under the sod, and a moldboard to turn it over. The friction of this plow with the soil was so great that it normally required not two but eight oxen. It attacked the land with such violence that cross-plowing was not needed, and fields tended to be shaped in long strips.

In the days of the scratch-plow, fields were distributed generally in units capable of supporting a single family. Subsistence farming was the presupposition. But no peasant owned eight oxen: to use the new and more efficient plow, peasants pooled their oxen to form large plow-teams, originally receiving (it would appear) plowed strips in proportion to their contribution. Thus, distribution of land was based no longer on the needs of a family but, rather, on the capacity of a power machine to till the earth. Man's relation to the soil was profoundly changed. Formerly man had been part of nature; now he was the exploiter of nature. Nowhere else in the world did farmers develop any analogous agricultural implement. Is it coincidence that modern technology, with its ruthlessness toward nature, has so largely been produced by descendants of these peasants of northern Europe?

This same exploitive attitude appears slightly before A.D. 830 in Western illustrated calendars. In older calendars the months were shown as passive personifications. The new Frankish calendars, which set the style for the Middle Ages, are very different: they show men coercing the world around them—plowing, harvesting, chopping trees, butchering pigs. Man and nature are two things, and man is master.

These novelties seem to be in harmony with larger intellectual patterns. What people do about their ecology depends on what they think about themselves in relation to things around them. Human ecology is deeply conditioned by beliefs about our nature and destiny—that is, by religion. To Western eyes this is very evident in, say, India or Ceylon. It is equally true of ourselves and of our medieval ancestors.

The victory of Christianity over paganism was the greatest psychic revolution in the history of our culture. It has become fashionable today to say that, for better or worse, we live in "the post-Christian age." Certainly the forms of our thinking and language have largely ceased to be Christian, but to my eye the substance often remains amazingly akin to that of the past. Our daily habits of action, for example, are dominated by an implicit faith in perpetual progress which was unknown either to Greco-Roman antiquity or to the Orient. It is rooted in, and is indefensible apart from, Judeo-Christian teleology. The fact that Communists share it merely helps to show what can be demonstrated on many other grounds: that Marxism, like Islam, is a Judeo-Christian heresy. We continue today to live, as we have lived for about 1700 years, very largely in a context of Christian axioms.

What did Christianity tell people about their relations with the environment?

While many of the world's mythologies provide stories of creation, Greco-Roman mythology was singularly incoherent in this respect. Like Aristotle, the intellectuals of the ancient West denied that the visible world had had a beginning. Indeed, the idea of a beginning was impossible in the framework of their cyclical notion of time. In sharp contrast, Christianity inherited from Judaism not only a concept of time as nonrepetitive and linear but also a striking story of creation. By gradual stages a loving and all-powerful God had created light and darkness, the heavenly bodies, the earth and all its plants, animals, birds, and fishes. Finally, God had created Adam and, as an afterthought, Eve to keep man from being lonely. Man named all the animals, thus establishing his dominance over them. God planned all of this explicitly for man's benefit and rule: no item in the physical creation had any purpose save to serve man's purposes. And, although man's body is made of clay, he is not simply part of nature: he is made in God's image.

Especially in its Western form, Christianity is the most anthropocentric religion the world has seen. As early as the 2nd century both Tertullian and Saint Irenaeus of Lyons were insisting that when God shaped Adam he was foreshadowing the image of the incarnate Christ, the Second Adam. Man shares, in great measure, God's transcendence of nature. Christianity, in absolute contrast to ancient paganism and Asia's religions (except, perhaps, Zoroastrianism), not only established a dualism of man and nature but also insisted that it is God's will that man exploit nature for his proper ends. . . .

The Christian dogma of creation, which is found in the first clause of all the Creeds, has another meaning for our comprehension of today's ecologic crisis. By revelation, God had given man the Bible, the Book of Scripture. But since God had made nature, nature also must reveal the divine mentality. The religious study of nature for the better understanding of God was known as natural theology. In the early Church, and always in the Greek East, nature was conceived primarily as a symbolic system through which God speaks to men: the ant is a sermon to sluggards; rising flames are the symbol of the soul's aspiration. This view of nature was essentially artistic rather than scientific. While Byzantium preserved and copied great numbers of ancient Greek scientific texts, science as we conceive it could scarcely flourish in such an ambience.

However, in the Latin West by the early 13th century natural theology was following a very different bent. It was ceasing to be the decoding of the physical symbols of God's communication with man and was becoming the effort to understand God's mind by discovering how his creation operates. The rainbow was no longer simply a symbol of hope first sent to Noah after the Deluge: Robert Grosseteste, Friar Roger Bacon, and Theodoric of Freiberg produced startlingly sophisticated work on the optics of the rainbow, but they did it as a venture in religious understanding. From the 13th century onward, up to and including Leibnitz and Newton, every major scientist, in effect, explained his motivations in religious terms. Indeed, if Galileo had not been so expert an amateur theologian he would have got into far less trouble: the professionals resented his intrusion. And Newton seems to have regarded himself more as a theologian than as a scientist. It was not until the late 18th century that the hypothesis of God became unnecessary to many scientists.

It is often hard for the historian to judge, when men explain why they are doing what they want to do, whether they are offering real reasons or merely culturally acceptable reasons. The consistency with which scientists during the long formative centuries of Western science said that the task and the reward of the scientist was "to think God's thoughts after him" leads one to believe that this was their real motivation. If so, then modern Western science was cast in a matrix of Christian theology. The dynamism of religious devotion, shaped by the Judeo-Christian dogma of creation, gave it impetus.

AN ALTERNATIVE CHRISTIAN VIEW

We would seem to be headed toward conclusions unpalatable to many Christians. Since both *science* and *technology* are blessed words in our contemporary vocabulary, some may be happy at the notions, first, that, viewed historically, modern science is an extrapolation of natural theology and, second, that modern technology is at least partly to be explained as an Occidental voluntarist realization of the Christian dogma of man's transcendence of, and rightful mastery over, nature. But, as we now recognize, somewhat over a century ago science and technology—hitherto quite separate activities—joined to give mankind powers which, to judge by many of the ecologic effects, are out of control. If so, Christianity bears a huge burden of guilt.

I personally doubt that disastrous ecologic backlash can be avoided simply by applying to our problems more science and more technology. Our science and technology have grown out of Christian attitudes toward man's relation to nature which are almost universally held not only by Christians and neo-Christians but also by those who fondly regard themselves as post-Christians. Despite Copernicus, all the cosmos rotates around our little globe. Despite Darwin, we are *not*, in our hearts, part of the natural process. We are superior to nature, contemptuous of it, willing to use it for our slightest whim. The newly elected Governor of California, like myself a churchman but less troubled than I, spoke for the Christian tradition when he said (as is alleged), "when you've seen one redwood tree, you've seen them all." To a Christian a tree can be no more than a physical fact. The whole concept of the sacred grove is alien to Christianity and to the ethos of the West. For nearly 2 millennia Christian missionaries have been chopping down sacred groves, which are idolatrous because they assume spirit in nature.

What we do about ecology depends on our ideas of the man-nature relationship. More science and more technology are not going to get us out of the present ecologic crisis until we find a new religion, or rethink our old one. The beatniks, who are the basic revolutionaries of our time, show a sound instinct in their affinity for Zen Buddhism, which conceives of the man-nature relationship as very nearly the mirror image of the Christian view. Zen, however, is as deeply conditioned by Asian history as Christianity is by the experience of the West, and I am dubious of its viability among us.

Possibly we should ponder the greatest radical in Christian history since Christ: Saint Francis of Assisi. The prime miracle of Saint Francis is the fact that he did not end at the stake, as many of his left-wing followers did. He was so clearly heretical that a General of the Franciscan Order, Saint Bonaventura, a great and perceptive Christian, tried to suppress the early account of Franciscanism. The key to an understanding of Francis is his belief in the virtue of humility—not merely for the individual but for man as a species. Francis tried to depose man from his monarchy over creation and set up a democracy of all God's creatures. With him the ant is no longer simply a homily for the lazy, flames a sign of the thrust of the soul toward union with God; now they are Brother Ant and Sister Fire, praising the Creator in their own ways as Brother Man does in his.

Later commentators have said that Francis preached to the birds as a rebuke to men who would not listen. The records do not read so: he urged the little birds to praise God, and in spiritual ecstasy they flapped their wings and chirped rejoicing. Legends of saints, especially the Irish saints, had long told of their dealings with animals but always, I believe, to show their human dominance over creatures. With Francis it is different. The land around Gubbio in the Apennines was being ravaged by a fierce wolf. Saint Francis, says the legend, talked to the wolf and persuaded him of the error of his ways. The wolf repented, died in the odor of sanctity, and was buried in consecrated ground.

What Sir Steven Ruciman calls "the Franciscan doctrine of the animal soul" was quickly stamped out. Quite possible it was in part inspired, consciously or unconsciously, by the belief in reincarnation held by the Cathar heretics who at that time teemed in Italy and southern France, and who presumably had got it originally from India. It is significant that at just the same moment, about 1200, traces of metempsychosis are found also in western Judaism, in the Provençal *Cabbala.* But Francis held neither to transmigration of souls nor to pantheism. His view of nature and of man rested on a unique sort of pan-psychism of all things animate and inanimate, designed for the glorification of their transcendent Creator, who, in the ultimate gesture of cosmic humility, assumed flesh, lay helpless in a manger, and hung dying on a scaffold.

I am not suggesting that many contemporary Americans who are concerned about our ecologic crisis will be either able or willing to counsel with wolves or exhort birds. However, the present increasing disruption of the global environment is the product of a dynamic technology and science which were originating in the Western medieval world against which Saint Francis was rebelling in so original a way. Their growth cannot be understood historically apart from distinctive attitudes toward nature which are deeply grounded in Christian dogma. The fact that most people do not think of these attitudes as Christian is irrelevant. No new set of basic values has been accepted in our society to displace those of Christianity. Hence we shall continue to have a worsening ecologic crisis until we reject the Christian axiom that nature has no reason for existence save to serve man.

The greatest spiritual revolutionary in Western history, Saint Francis, proposed what he thought was an alternative Christian view of nature and man's relation to it: he tried to substitute the idea of the equality of all creatures, including man, for the idea of man's limitless rule of creation. He failed. Both our present science and our present technology are so tinctured with orthodox Christian arrogance toward nature that no solution for our ecologic crisis can be expected from them alone. Since the roots of our trouble are so largely religious, the remedy must also be essentially religious, whether we call it that or not. We must rethink and refeel our nature and destiny. The profoundly religious, but heretical, sense of the primitive Franciscans for the spiritual autonomy of all parts of nature may point a direction. I propose Francis as a patron saint for ecologists.

The Closing Circle: Nature, Man and Technology

For more than 50 years biologist Barry Commoner (b. 1917) has been one of the most provocative and influential of the world's environmental analysts, activists, and educators. As a professor at Washington University in St. Louis, Missouri, Commoner founded that city's Committee for Environmental Information. He went on to create and become director of the university's Center for the Biology of Natural Systems, which has since relocated to Queens College of the City University of New York. Commoner played a large role in the documentation of the hazards of radioactive fallout from atmospheric testing of nuclear weapons and in the movement that led to the international agreement to end such testing.

Commoner's voluminous environmental writings focus on one central theme. In his opinion, it is the contradiction between the entrepreneur's demand for short-term profits and the social needs of the human race that is the root cause of our ecological crises. Although he acknowledges the impact of population growth, Commoner argues that inappropriate technology is more of a factor in the creation of environmental problems. According to Commoner, the failure to base industrial developmental strategies on the short- and long-term impacts of alternative technologies has produced our local and global pollution problems.

In the following selection from his best-seller *The Closing Circle: Nature, Man and Technology* (Bantam Books, 1971), Commoner attributes the failure of human society to demand sound environmental practices to a lifestyle that has separated us from the natural systems that we depend on. Commoner, however, is optimistic that our environmental problems will be solved through informed, collective social action.

Key Concept: inappropriate technology as the cause of environmental problems

THE ENVIRONMENTAL CRISIS

The environment has just been rediscovered by the people who live in it. In the United States the event was celebrated in April 1970, during Earth Week. It was a sudden, noisy awakening. School children cleaned up rubbish; college students organized huge demonstrations; determined citizens recaptured the streets from the automobile, at least for a day. Everyone seemed to be aroused to the environmental danger and eager to do something about it. . . .

Any living thing that hopes to live on the earth must fit into the ecosphere or perish. The environmental crisis is a sign that the finely sculptured fit between life and its surroundings has begun to corrode. As the links between one living thing and another, and between all of them and their surroundings, begin to break down, the dynamic interactions that sustain the whole have begun to falter and, in some places, stop.

Why, after millions of years of harmonious co-existence, have the relationships between living things and their earthly surroundings begun to collapse? Where did the fabric of the ecosphere begin to unravel? How far will the process go? How can we stop it and restore the broken links? . . .

THE ECOSPHERE

To survive on the earth, human beings require the stable, continuing existence of a suitable environment. Yet the evidence is overwhelming that the way in which we now live on the earth is driving its thin, life-supporting skin, and ourselves with it, to destruction. To understand this calamity, we need to begin with a close look at the nature of the environment itself. Most of us find this is a difficult thing to do, for there is a kind of ambiguity in our relation to the environment. Biologically, human beings *participate* in the environmental system as subsidiary parts of the whole. Yet, human society is designed to *exploit* the environment as a whole, to produce wealth. The paradoxical role we play in the natural environment—at once participant and exploiter—distorts our perception of it.

Among primitive people, a person is seen as a dependent part of nature, a frail reed in a harsh world governed by natural laws that must be obeyed if he is to survive. Pressed by this need, primitive peoples can achieve a remarkable knowledge of their environment. The African Bushman lives in one of the most stringent habitats on earth; food and water are scarce, and the weather is extreme. The Bushman survives because he has an incredibly intimate understanding of this environment. A Bushman can, for example, return after many months and miles of travel to find a single underground tuber, noted in his previous wanderings, when he needs it for his water supply in the dry season.

We who call ourselves advanced seem to have escaped from this kind of dependence on the environment. The Bushman must squeeze water from a searched-out tuber; we get ours by the turn of a tap. Instead of trackless terrain, we have the grid of city streets. Instead of seeking the sun's heat when we need it, or shunning it when it is too strong, we warm and cool ourselves with man-made machines. All this leads us to believe that we have made our own environment and no longer depend on the one provided by nature. In the eager search for the benefits of modern science and technology we have become enticed into a nearly fatal illusion: that through our machines we have at last escaped from dependence on the natural environment.

A good place to experience this illusion is a jet airplane. Safely seated on a plastic cushion, carried in a winged aluminum tube, streaking miles above the earth, through air nearly thin enough to boil the blood, at a speed that seems to make the sun stand still, it is easy to believe that we have conquered nature and have escaped from the ancient bondage to air, water, and soil. . . .

THE CLOSING CIRCLE

. . . We live in a time that is dominated by enormous technical power and extreme human need. The power is painfully self-evident in the megawattage of power plants, and in the megotonnage of nuclear bombs. The human need is evident in the sheer numbers of people now and soon to be living, in the deterioration of their habitat, the earth, and in the tragic world-wide epidemic of hunger and want. The gap between brute power and human need continues to grow, as the power fattens on the same faulty technology that intensifies the need.

Everywhere in the world there is evidence of a deep-seated failure in the effort to use the competence, the wealth, the power at human disposal for the maximum good of human beings. The environmental crisis is a major example of this failure. For we are in an environmental crisis because the means by which we use the ecosphere to produce wealth are destructive of the ecosphere itself. The present system of production is self-destructive; the present course of human civilization is suicidal.

The environmental crisis is somber evidence of an insidious fraud hidden in the vaunted productivity and wealth of modern, technology-based society. This wealth has been gained by rapid short-term exploitation of the environmental system, but it has blindly accumulated a debt to nature (in the form of environmental destruction in developed countries and of population pressure in developing ones)—a debt so large and so pervasive that in the next generation it may, if unpaid, wipe out most of the wealth it has gained us. In effect, the account books of modern society are drastically out of balance, so that, largely unconsciously, a huge fraud has been perpetrated on the people of the world. The rapidly worsening course of environmental pollution is a warning that the bubble is about to burst, that the demand to pay the global debt may find the world bankrupt.

This does *not* necessarily mean that to survive the environmental crisis, the people of industrialized nations will need to give up their "affluent" way of life. For as shown earlier, this "affluence," as judged by conventional measures —such as GNP, power consumption, and production of metals—is itself an illusion. To a considerable extent it reflects ecologically faulty, socially wasteful types of production rather than the actual welfare of individual human beings. Therefore, the needed productive reforms can be carried out without seriously reducing the present level of *useful* goods available to the individual; and, at the same time, by controlling pollution the quality of life can be improved significantly.

There are, however, certain luxuries which the environmental crisis, and the approaching bankruptcy that it signifies, will, I believe, force us to give up. These are the *political* luxuries which have so long been enjoyed by those who can benefit from them: the luxury of allowing the wealth of the nation to serve preferentially the interests of so few of its citizens; of failing fully to inform citizens of what they need to know in order to exercise their right of political governance; of condemning as anathema any suggestion which re-examines basic economic values; of burying the issues revealed by logic in a morass of self-serving propaganda.

To resolve the environmental crisis, we shall need to forego, at last, the luxury of tolerating poverty, racial discrimination, and war. In our unwitting march toward ecological suicide we have run out of options. Now that the bill for the environmental debt has been presented, our options have become reduced to two: either the rational, social organization of the use and distribution of the earth's resources, or a new barbarism.

This iron logic has recently been made explicit by one of the most insistent proponents of population control, Garrett Hardin. Over recent years he has expounded on the "tragedy of the commons"—the view that the world ecosystem is like a common pasture where each individual, guided by a desire for personal gain, increases his herd until the pasture is ruined for all. Until recently, Hardin drew two rather general conclusions from this analogy: first, that "freedom in a commons brings ruin to all," and second, that the freedom which must be constrained if ruin is to be avoided is not the derivation of private gain from a social good (the commons), but rather "the freedom to breed."

Hardin's logic is clear, and follows the course outlined earlier: if we accept as unchangeable the present governance of a social good (the commons, or the ecosphere) by private need, then survival requires the immediate, drastic limitation of population. Very recently, Hardin has carried this course of reasoning to its logical conclusion; in an editorial in *Science*, he asserts:

> Every day we [i.e., Americans] are a smaller minority. We are increasing at only one per cent a year; the rest of the world increases twice as fast. By the year 2000, one person in twenty-four will be an American; in one hundred years only one in forty-six. . . . If the world is one great commons, in which all food is shred equally, then we are lost. Those who breed faster will replace the rest. . . . In the absence of breeding control a policy of "one mouth one meal" ultimately produces one totally miserable world. In a less than perfect world, the allocation of rights based on territory must be defended if a ruinous breeding race is to be avoided. It is unlikely that civilization and dignity can survive everywhere; but better in a few places than in none. Fortunate minorities must act as the trustees of a civilization that is threatened by uninformed good intentions.

Here, only faintly masked, is barbarism. It denies the equal right of all the human inhabitants of the earth to a humane life. It would condemn most of the people of the world to the material level of the barbarian, and the rest, the "fortunate minorities," to the moral level of the barbarian. Neither within Hardin's tiny enclaves of "civilization," nor in the larger world around them, would anything that we seek to preserve—the dignity and the humaneness of man, the grace of civilization—survive.

In the narrow options that are possible in a world gripped by environmental crisis, there is no apparent alternative between barbarism and the acceptance of the economic consequence of the ecological imperative—that the social, global nature of the ecosphere must determine a corresponding organization of the productive enterprises that depend on it.

One of the common responses to a recitation of the world's environmental ills is a deep pessimism, which is perhaps the natural aftermath to the shock of recognizing that the vaunted "progress" of modern civilization is only a thin cloak for global catastrophe. I am convinced, however, that once we pass beyond the mere awareness of impending disaster and begin to understand *why* we have come to the present predicament, and where the alternative paths ahead can lead, there is reason to find in the very depths of the environmental crisis itself a source of optimism.

There is, for example, cause for optimism in the very complexity of the issues generated by the environmental crisis; once the links between the separate parts of the problem are perceived, it becomes possible to see new means of solving the whole. Thus, confronted separately, the need of developing nations for new productive enterprises, and the need of industrialized countries to reorganize theirs along ecologically sound lines, may seem hopelessly difficult. However, when the link between the two—the ecological significance of the introduction of synthetic substitutes for natural products—is recognized, ways of solving both can be seen. In the same way, we despair over releasing the grip of the United States on so much of the world's resources until it becomes clear how much of this "affluence" stresses the environment rather than contributes to human welfare. Then the very magnitude of the present United States share of the world's resources is a source of hope—for its reduction through ecological reform can then have a large and favorable impact on the desperate needs of the developing nations.

I find another source of optimism in the very nature of the environmental crisis. It is not the product of man's *biological* capabilities, which could not change in time to save us, but of his *social* actions—which are subject to much more rapid change. Since the environmental crisis is the result of the social mismanagement of the world's resources, then it can be resolved and man can survive in a humane condition when the social organization of man is brought into harmony with the ecosphere.

Here we can learn a basic lesson from nature: that nothing can survive on the planet unless it is a cooperative part of a larger, global whole. Life itself learned that lesson on the primitive earth. For it will be recalled that the earth's first living things, like modern man, consumed their nutritive base as they grew, converting the geochemical store of organic matter into wastes which could no longer serve their needs. Life, as it first appeared on the earth, was embarked on a linear, self-destructive course.

What saved life from extinction was the invention, in the course of evolution, of a new life-form which reconverted the waste of the primitive organisms into fresh, organic matter. The first photosynthetic organisms transformed the rapacious, linear course of life into the earth's first great ecological cycle. By closing the circle, they achieved what no living organism, alone, can accomplish—survival.

Human beings have broken out of the circle of life, driven not by biological need, but by the social organization which they have devised to "conquer" nature: means of gaining wealth that are governed by requirements conflicting with those which govern nature. The end result is the environmental crisis, a crisis of survival. Once more, to survive, we must close the circle. We must learn how to restore to nature the wealth that we borrow from it.

33

Barry Commoner

In our progress-minded society, anyone who presumes to explain a serious problem is expected to offer to solve it as well. But none of us—singly or sitting in committee—can possibly blueprint a specific "plan" for resolving the environmental crisis. To pretend otherwise is only to evade the real meaning of the environmental crisis: that the world is being carried to the brink of ecological disaster not by a singular fault, which some clever scheme can correct, but by the phalanx of powerful economic, political, and social forces that constitute the march of history. Anyone who proposes to cure the environmental crisis undertakes thereby to change the course of history.

But this is a competence reserved to history itself, for sweeping social change can be designed only in the workshop of rational, informed, collective social action. That we must act is now clear. The question which we face is how.

2.3 PAUL R. EHRLICH

Too Many People

According to Paul Ehrlich (b. 1932), the root cause of environmental degradation is worldwide population growth. In most developing countries, he believes, the rate of population growth is far exceeding the expansion of the food supply, and overcrowding is overwhelming the already inadequate sanitation and waste disposal systems. In the developed countries, contends Ehrlich, a more moderate rate of population increase coupled with a disproportionate use of natural resources is causing severe pollution problems.

Ehrlich, a population biologist and the author of many scientific papers in the fields of ecology, evolution, and behavior, is the Bing Professor of Population Studies at Stanford University. He has been an active campaigner for population control and ecological education. He is a founder of Zero Population Growth, an international organization that advocates governmental support for birth control and a voluntary two-child limit for families. *The Population Bomb,* first published in 1968, was the first and perhaps the most influential of Ehrlich's popular books describing the inevitable consequences of unchecked population growth. It conveyed a dramatic message to its wide readership about the severity of what it described as a population explosion. His other, more recent books, coauthored with his wife, Anne, who is a researcher in biological science at Stanford, update and expand the dramatic, well-documented arguments linking resource depletion, ecosystem destruction, and human starvation to the population crisis. The following selection is from the revised edition of *The Population Bomb* (Ballantine Books, 1971).

Key Concept: starvation and ecological crises as inevitable consequences of a population explosion

Americans are beginning to realize that the underdeveloped countries of the world face an inevitable population-food crisis. Each year food production in these countries falls a bit further behind burgeoning population growth, and people go to bed a little bit hungrier. While there are temporary or local reversals of this trend, it now seems inevitable that it will continue to its logical conclusion: mass starvation. The rich may continue to get richer, but the more numerous poor are going to get poorer. Of these poor, a *minimum* of ten million people, most of them children, will starve to death during each year of the 1970s. But this is a mere handful compared to the numbers that will be starving before the end of the century. And it is now too late to take action to save many of those people.

However, most Americans are not aware that the U.S. and other developed countries also have a problem with overpopulation. Rather than suffering from food shortages, these countries show symptoms in the form of environmental deterioration and increased difficulty in obtaining resources to support their affluence. . . .

Of course, population growth is not occurring uniformly over the face of the Earth. Indeed, countries are divided rather neatly into two groups: those with rapid growth rates, and those with relatively slow growth rates. The first group, making up about two-thirds of the world population, coincides closely with what are known as the "underdeveloped countries" (UDCs). The UDCs are not industrialized, tend to have inefficient agriculture, very small gross national products, high illiteracy rates and related problems. That's what UDCs are technically, but a short definition of underdeveloped is "hungry." Most Latin American, African, and Asian countries fall into this category. The second group consists of the "overdeveloped countries" (ODCs). ODCs are modern industrial nations, such as the United States, Canada, most European countries, Israel, the USSR, Japan, and Australia. They consume a disproportionate amount of the world's resources and are the major polluters. Most, but by no means all, people in these countries are adequately nourished.

Doubling times in the UDCs range around 20 to 35 years. Examples of these times (from the 1970 figures released by the Population Reference Bureau) are: Kenya, 23 years; Nigeria, 27; Turkey, 26; Indonesia, 24; Philippines, 21; Brazil, 25; Costa Rica, 19; and El Salvador, 21. Think of what it means for the population of a country to double in 25 years. In order just to keep living standards at the present inadequate level, the food available for the people must be doubled. Every structure and road must be duplicated. The amount of power must be doubled. The capacity of the transport system must be doubled. The number of trained doctors, nurses, teachers, and administrators must be doubled. This would be a fantastically difficult job in the United States—a rich country with a fine agricultural system, immense industries, and access to abundant resources. Think of what it means to a country with none of these.

Remember also that in virtually all UDCs, people have gotten the word about the better life it is possible to have. They have seen colored pictures in magazines of the miracles of Western technology. They have seen automobiles and airplanes. They have seen American and European movies. Many have seen refrigerators, tractors, and even TV sets. Almost all have heard transistor radios. They *know* that a better life is possible. They have what we like to call "rising expectations." If twice as many people are to be happy, the miracle of doubling what they now have will not be enough. It will only maintain today's standard of living. There will have to be a tripling or better. Needless to say, they are not going to be happy.

Doubling times for the populations of the ODCs tend to be in the 50-to-200-year range. Examples of 1970 doubling times are the United States, 70 years; Austria, 175; Denmark, 88; Norway, 78; United Kingdom, 140; Poland, 78; Russia, 70; Italy, 88; Spain, 70; and Japan, 63. These are industrialized countries that have undergone the so-called demographic transition—a transition from high to low growth rates. As industrialization progressed, children became less important to parents as extra hands to work on the farm and as support in old age. At the same time they became a financial drag—expensive to raise and educate. presumably these were the reasons for a slowing of population growth after industrialization. They boil down to a simple fact—people just wanted to have fewer children.

It is important to emphasize, however, that the demographic transition does not result in zero population growth, but in a growth rate which in many of the most important ODCs results in populations doubling every seventy years or so. This means, for instance, that even if most UDCs were to undergo a demographic transition (of which there is no sign) the world would still be faced by catastrophic population growth. *No growth rate can be sustained in the long run.*

Saying that the ODCs have undergone a demographic transition thus does not mean that they have no population problems. First of all, most of them are already overpopulated. They are overpopulated by the simple criterion that they are not able to produce enough food to feed their populations. It is true that they have the money to buy food, but when food is no longer available for sale they will find the money rather indigestible. Similarly, ODCs are overpopulated because they do not themselves have the resources to support their affluent societies; they must coopt much more than their fair share of the world's wealth of minerals and energy. And they are overpopulated because they have exceeded the capacity of their environments to dispose of their wastes. Remember, overpopulation does not normally mean too many people for the area of a country, but too many people in relation to the necessities and amenities of life. *Overpopulation occurs when numbers threaten values.*

ODCs also share with the UDCs serious problems of population distribution. Their urban centers are getting more and more crowded relative to the countryside. This problem is not as severe in ODCs as it is in the UDCs (if current trends should continue, which they cannot, Calcutta would have 66 million inhabitants in the year 2000), but they are very serious and speedily worsening. In the United States, one of the more rapidly growing ODCs, we hear constantly of the headaches related to growing cities: not just garbage in our environment, but overcrowded highways, burgeoning slums, deteriorating school systems, rising tax and crime rates, riots, and other social disorders. Indeed, social and environmental problems not only increase with growing population and urbanization, they tend to increase at an even faster rate. Adding more people to an area increases the damage done by each individual. Doubling the population normally much more than doubles environmental deterioration.

Demographically, the whole problem is quite simple. A population will continue to grow as long as the birth rate exceeds the death rate—if immigration and emigration are not occurring. It is, of course, the balance between birth rate and death rate that is critical. The birth rate is the number of births per thousand people per year in the population. The death rate is the number of deaths per thousand people per year. Subtracting the death rate from the birth rate, ignoring migration, gives the rate of increase. If the birth rate is 30 per thousand per year, and the death rate is 10 per thousand per year, then the rate of increase is 20 per thousand per year ($30 - 10 = 20$). Expressed as a percent (rate per hundred people), the rate of 20 per thousand becomes 2%. If the rate of increase is 2%, then the doubling time will be 35 years. Note that if you simply added 20 people per thousand per year to the population, it would take 50 years to add a second thousand people (20 × 50 = 1,000). But the doubling time is actually much less because populations grow at compound interest rates. Just as interest dollars themselves earn interest, so people added to population produce more people. It's growing at compound interest that makes populations double so much more rapidly than seems possible. Look at the relationship between the annual percent increase (interest rate) and the doubling time of the population (time for your money to double):

Annual percent increase	*Doubling time*
1.0	70
2.0	35
3.0	24
4.0	17

... There are some professional optimists around who like to greet every sign of dropping birth rates with wild pronouncements about the end of the population explosion. They are a little like a person who, after a low temperature of five below zero on December 21, interprets a low of only three below on December 22 as a cheery sign of approaching spring. First of all, birth rates, along with all demographic statistics, show short-term fluctuations caused by many factors. For instance, the birth rate depends rather heavily on the number of women at reproductive age. In the United States the low birth rates of the late 1960's are being replaced by higher rates as more post World War II "baby boom" children move into their reproductive years. In Japan, 1966, the Year of the Fire Horse, was a year of very low birth rates. There is widespread belief that girls born in the Year of the Fire Horse make poor wives, and Japanese couples try to avoid giving birth in that year because they are afraid of having daughters.

But, I repeat, it is the relationship between birth rate and death rate that is most critical. Indonesia, Laos, and Haiti all had birth rates around 46 per thousand in 1966. Costa Rica's birth rate was 41 per thousand. Good for Costa Rica? Unfortunately, not very. Costa Rica's death rate was less than nine per thousand, while the other countries all had death rates above 20 per thousand. The population of Costa Rica in 1966 was doubling every 17 years, while the doubling times of Indonesia, Laos, and Haiti were all above 30 years. Ah, but, you say, it was good for Costa Rica—fewer people per thousand were dying each year. Fine for a few years perhaps, but what then? Some 50% of the people in Costa Rica are under 15 years old. As they get older, they will need more and more food in a world with less and less. In 1983 they will have twice as many mouths to feed as they had in 1966, if the 1966 trend continues. Where will the food come from? Today the death rate in Costa Rica is low in part because they have a large number of physicians in proportion to their population. How do you suppose those physicians will keep the death rate down when there's not enough food to keep people alive?

One of the most ominous facts of the current situation is that over 40% of the population of the underdeveloped world is made up of people *under 15 years old.* As that mass of young people moves into its reproductive years during the next decade, we're going to see the greatest baby boom of all time. Those youngsters are the reason for all the ominous predictions for the year 2000. They are the gunpowder for the population explosion. . . .

It is, of course, socially very acceptable to reduce the death rate. Billions of years of evolution have given us all a powerful will to live. Intervening in the birth rate goes against our evolutionary values. During all those centureis of our evolutionary past, the individuals who had the most children passed on their genetic endowment in greater quantities than those who reproduced less. Their genes dominate our heredity today. All our biological urges are for more reproduction, and they are all too often reinforced by our culture. In brief, death control goes with the grain, birth control against it.

In summary, the world's population will continue to grow as long as the birth rate exceeds the death rate; it's as simple as that. When it stops growing or starts to shrink, it will mean that either the birth rate has gone down or the death rate has gone up or a combination of the two. Basically, then, there are only two kinds of solutions to the population problem. One is a "birth rate solution," in which we find ways to lower the birth rate. The other is a "death rate solution," in which ways to raise the death rate—war, famine, pestilence—*find us.* The problem could have been avoided by *population control,* in which mankind consciously adjusted the birth rate so that a "death rate solution" did not have to occur.

2.4 GARRETT HARDIN

The Tragedy of the Commons

Since 1946 Garrett Hardin has been a professor of biology and human ecology at the University of California, Santa Barbara. The forthright, controversial views on population, evolution, and birth control he has expressed in his numerous articles, books, and speeches have earned him an international reputation as an influential, outspoken critic of the ecological and environmental community.

The best known and most often quoted of Hardin's writings on the social and ethical issues raised by a world of limited resources and increasing numbers of people is "The Tragedy of the Commons," which was printed in the December 1968 issue of *Science* and which is excerpted in the following selection. In this essay, Hardin relates a nineteenth-century tale about a common pasture becoming overgrazed and destroyed because each of the herdsmen whose animals grazed on it considered only the advantage to his own family of increasing his herd. From this parable, Hardin draws a general conclusion: that all resources, such as the oceans, which are held in common and are therefore not anyone's private property, will be overused and ultimately degraded. Among the policy implications he derives from this assessment is that programs that attempt to deal with hunger by providing free food to people are counterproductive. Indeed, Hardin decided to terminate his own research on the culture of algae as a potential major food source because he believes that more food simply encourages further increases in population, which will ultimately produce even greater starvation.

In subsequent writings, Hardin goes on to develop his "lifeboat ethics" theory. He proposes a world model in which the developed, affluent nations that control and use most of the world's resources are in a lifeboat while the struggling developing nations are floundering in the surrounding ocean. He concludes that it is folly to try to rescue all the swimmers and suggests that the ethically appropriate strategy is one of triage, by which the "haves" permit the poorest and least developed of the "have nots" to drown in order to prevent the entire boat from sinking. While this harsh analysis has won praise from many environmentalists who share Hardin's predilection for "pragmatic" decisions based on a competitive "marketplace" model, it has been rejected by others who advocate a more egalitarian approach based on informed social planning.

Key Concept: commonly owned resources are doomed to destruction

An implicit and almost universal assumption of discussions published in professional and semipopular scientific journals is that the problem under discussion has a technical solution. A technical solution may be defined as one that requires a change only in the techniques of the natural sciences, demanding little or nothing in the way of change in human values or ideas of morality.

In our day (though not in earlier times) technical solutions are always welcome. Because of previous failures in prophecy, it takes courage to assert that a desired technical solution is not possible.... [T]he concern here is with the important concept of a class of human problems which can be called "no technical solution problems," and, more specifically, with the identification and discussion of one of these.

It is easy to show that the class is not a null class. Recall the game of tick-tack-toe. Consider the problem, "How can I win the game of tick-tack-toe?" It is well known that I cannot, if I assume (in keeping with the conventions of game theory) that my opponent understands the game perfectly. Put another way, there is no "technical solution" to the problem. I can win only by giving a radical meaning to the word "win." I can hit my opponent over the head; or I can drug him; or I can falsify the records. Every way in which I "win" involves, in some sense, an abandonment of the game, as we intuitively understand it. (I can also, of course, openly abandon the game—refuse to play it. This is what most adults do.)

The class of "No technical solution problems" has members. My thesis is that the "population problem," as conventionally conceived, is a member of this class. How it is conventionally conceived needs some comment. It is fair to say that most people who anguish over the population problem are trying to find a way to avoid the evils of overpopulation without relinquishing any of the privileges they now enjoy. They think that farming the seas or developing new strains of wheat will solve the problem—technologically. I try to show here that the solution they seek cannot be found. The population problem cannot be solved in a technical way, any more than can the problem of winning the game of tick-tack-toe....

We can make little progress in working toward optimum population size until we explicitly exorcize the spirit of Adam Smith in the field of practical demography. In economic affairs, *The Wealth of Nations* (1776) popularized the "invisible hand," the idea that an individual who "intends only his own gain," is, as it were, "led by an invisible hand to promote... the public interest." Adam Smith did not assert that this was invariably true, and perhaps neither did any of his followers. But he contributed to a dominant tendency of thought that has ever since interfered with positive action based on rational analysis, namely, the tendency to assume that decisions reached individually will, in fact, be the best decisions for an entire society. If this assumption is correct it justifies the continuance of our present policy of laissez-faire in reproduction. If it is correct we can assume that men will control their individual fecundity so as to produce the optimum population. If the assumption is not correct, we need to reexamine our individual freedoms to see which ones are defensible.

TRAGEDY OF FREEDOM IN A COMMONS

The rebuttal to the invisible hand in population control is to be found in a scenario first sketched in a little-known pamphlet in 1833 by a mathematical amateur named William Forster Lloyd (1794–1852). We may well call it "the tragedy of the commons," using the word "tragedy" as the philosopher Whitehead used it: "The essence of dramatic tragedy is not unhappiness. It resides in the solemnity of the remorseless working of things." He then goes on to say, "This inevitableness of destiny can only be illustrated in terms of human life by incidents which in fact involve unhappiness. For it is only by them that the futility of escape can be made evident in the drama."

The tragedy of the commons develops in this way. Picture a pasture open to all. It is to be expected that each herdsman will try to keep as many cattle as possible on the commons. Such an arrangement may work reasonably satisfactorily for centuries because tribal wars, poaching, and disease keep the numbers of both man and beast well below the carrying capacity of the land. Finally, however, comes the day of reckoning, that is, the day when the long-desired goal of social stability becomes a reality. At this point, the inherent logic of the commons remorselessly generates tragedy.

As a rational being, each herdsman seeks to maximize his gain. Explicitly or implicitly, more or less consciously, he asks, "What is the utility *to me* of adding one more animal to my herd?" This utility has one negative and one positive component.

1. The positive component is a function of the increment of one animal. Since the herdsman receives all the proceeds from the sale of the additional animal, the positive utility is nearly +1.
2. The negative component is a function of the additional overgrazing created by one more animal. Since, however, the effects of overgrazing are shared by all the herdsmen, the negative utility for any particular decision-making herdsman is only a fraction of −1.

Adding together the component partial utilities, the rational herdsman concludes that the only sensible course for him to pursue is to add another animal to his herd. And another; and another.... But this is the conclusion reached by each and every rational herdsman sharing a commons. Therein is the tragedy. Each man is locked into a system that compels him to increase his herd without limit—in a world that is limited. Ruin is the destination toward which all men rush, each pursuing his own best interest in a society that believes in the freedom of the commons. Freedom in a commons brings ruin to all.

Some would say that this is a platitude. Would that it were! In a sense, it was learned thousands of years ago, but natural selection favors the forces of psychological denial. The individual benefits as an individual from his ability to deny the truth even though society as a whole, of which he is a part, suffers. Education can counteract the natural tendency to do the wrong thing, but the inexorable succession of generations requires that the basis for this knowledge be constantly refreshed.

A simple incident that occurred a few years ago in Leominster, Massachusetts, shows how perishable the knowledge is. During the Christmas shopping season the parking meters downtown were covered with plastic bags that bore tags reading: "Do not open until after Christmas. Free parking courtesy of the mayor and city council." In other words, facing the prospect of an increased demand for already scarce space, the city fathers reinstituted the system of the commons. (Cynically, we suspect that they gained more votes than they lost by this retrogressive act.)

In an approximate way, the logic of the commons has been understood for a long time, perhaps since the discovery of agriculture or the invention of private property in real estate. But it is understood mostly only in special cases which are not sufficiently generalized. Even at this late date, cattlemen leasing national land on the western ranges demonstrate no more than an ambivalent understanding, in constantly pressuring federal authorities to increase the head count to the point where overgrazing produces erosion and weed-dominance. Likewise, the oceans of the world continue to suffer from the survival of the philosophy of the commons. Maritime nations still respond automatically to the shibboleth of the "freedom of the seas." Professing to believe in the "inexhaustible resources of the oceans," they bring species after species of fish and whales closer to extinction.

The National Parks present another instance of the working out of the tragedy of the commons. At present, they are open to all, without limit. The parks themselves are limited in extent—there is only one Yosemite Valley—whereas population seems to grow without limit. The values that visitors seek in the parks are steadily eroded. Plainly, we must soon cease to treat the parks as commons or they will be of no value to anyone.

What shall we do? We have several options. We might sell them off as private property. We might keep them as public property, but allocate the right to enter them. The allocation might be on the basis of wealth, by the use of an auction system. It might be on the basis of merit, as defined by some agreed-upon standards. It might be by lottery. Or it might be on a first-come, first-served basis, administered to long queues. These, I think, are all the reasonable possibilities. They are all objectionable. But we must choose—or acquiesce in the destruction of the commons that we call our national parks.

POLLUTION

In a reverse way, the tragedy of the commons reappears in problems of pollution. Here it is not a question of taking something out of the commons, but of putting something in—sewage, or chemical, radioactive, and heat wastes into water; noxious and dangerous fumes into the air; and distracting and unpleasant advertising signs into the line of sight. The calculations of utility are much the same as before. The rational man finds that his share of the cost of the wastes he discharges into the commons is less than the cost of purifying his wastes before releasing them. Since this is true for everyone, we are locked into a system of "fouling our own nest," so long as we behave only as independent, rational, free-enterprisers.

The tragedy of the commons as a food basket is averted by private property, or something formally like it. But the air and waters surrounding us cannot readily be fenced, and so the tragedy of the commons as a cesspool must be prevented by different means, by coercive laws or taxing devices that make it cheaper for the polluter to treat his pollutants than to discharge them untreated. We have not progressed as far with the solution of this problem as we have with the first. Indeed, our particular concept of private property, which deters us from exhausting the positive resources of the earth, favors pollution. The owner of a factory on the bank of a stream—whose property extends to the middle of the stream—often has difficulty seeing why it is not his natural right to muddy the waters flowing past his door. The law, always behind the times, requires elaborate stitching and fitting to adapt it to this newly perceived aspect of the commons.

The pollution problem is a consequence of population. It did not much matter how a lonely American frontiersman disposed of his waste. "Flowing water purifies itself every 10 miles," my grandfather used to say, and the myth was near enough to the truth when he was a boy, for there were not too many people. But as population became denser, the natural chemical and biological recycling processes became overloaded, calling for a redefinition of property rights. . . .

FREEDOM TO BREED IS INTOLERABLE

The tragedy of the commons is involved in population problems in another way. In a world governed solely by the principle of "dog eat dog"—if indeed there ever was such a world—how many children a family had would not be a matter of public concern. Parents who bred too exuberantly would leave fewer descendants, not more, because they would be unable to care adequately for their children. David Lack and others have found that such a negative feedback demonstrably controls the fecundity of birds. But men are not birds, and have not acted like them for millenniums, at least.

If each human family were dependent only on its own resources; *if* the children of improvident parents starved to death; *if*, thus, overbreeding brought its own "punishment" to the germ line—*then* there would be no public interest in controlling the breeding of families. But our society is deeply committed to the welfare state, and hence is confronted with another aspect of the tragedy of the commons.

In a welfare state, how shall we deal with the family, the religion, the race, or the class (or indeed any distinguishable and cohesive group) that adopts overbreeding as a policy to secure its own aggrandizement? To couple the concept of freedom to breed with the belief that everyone born has an equal right to the commons is to lock the world into a tragic course of action.

Unfortunately this is just the course of action that is being pursued by the United Nations. In late 1967, some 30 nations agreed to the following:

> The Universal Declaration of Human Rights describes the family as the natural and fundamental unit of society. It follows that any choice and decision with regard to the size of the family must irrevocably rest with the family itself, and cannot be made by anyone else.

It is painful to have to deny categorically the validity of this right; denying it, one feels as uncomfortable as a resident of Salem, Massachusetts, who denied the reality of witches in the 17th century. At the present time, in liberal quarters, something like a taboo acts to inhibit criticism of the United Nations. There is a feeling that the United Nations is "our last and best hope," that we shouldn't find fault with it; we shouldn't play into the hands of the archconservatives. However, let us not forget what Robert Louis Stevenson said: "The truth that is suppressed by friends is the readiest weapon of the enemy." If we love the truth we must openly deny the validity of the Universal Declaration of Human Rights, even though it is promoted by the United Nations. We should also join with Kingsley Davis in attempting to get Planned Parenthood-World Population to see the error of its ways in embracing the same tragic ideal.

CONSCIENCE IS SELF-ELIMINATING

It is a mistake to think that we can control the breeding of mankind in the long run by an appeal to conscience. Charles Galton Darwin made this point when he spoke on the centennial of the publication of his grandfather's great book. The argument is straightforward and Darwinian.

People vary. Confronted with appeals to limit breeding, some people will undoubtedly respond to the plea more than others. Those who have more children will produce a larger fraction of the next generation than those with more susceptible consciences. The difference will be accentuated, generation by generation.

In C. G. Darwin's words: "It may well be that it would take hundreds of generations for the progenitive instinct to develop in this way, but if it should do so, nature would have taken her revenge, and the variety *Homo contracipiens* would become extinct and would be replaced by the variety *Homo progenitivus*."

The argument assumes that conscience or the desire for children (no matter which) is hereditary—but hereditary only in the most general formal sense.

The result will be the same whether the attitude is transmitted through germ cells, or exosomatically, to use A. J. Lotka's term. (If one denies the latter possibility as well as the former, then what's the point of education?) The argument has here been stated in the context of the population problem, but it applies equally well to any instance in which society appeals to an individual exploiting a commons to restrain himself for the general good—by means of his conscience. To make such an appeal is to set up a selective system that works toward the elimination of conscience from the race. . . .

MUTUAL COERCION MUTUALLY AGREED UPON

The social arrangements that produce responsibility are arrangements that create coercion, of some sort. Consider bank-robbing. The man who takes money from a bank acts as if the bank were a commons. How do we prevent such action? Certainly not by trying to control his behavior solely by a verbal appeal to his sense of responsibility. Rather than rely on propaganda we follow [Charles] Frankel's lead and insist that a bank is not a commons; we seek the definitive social arrangements that will keep it from becoming a commons. That we thereby infringe on the freedom of would-be robbers we neither deny nor regret.

The morality of bank-robbing is particularly easy to understand because we accept complete prohibition of this activity. We are willing to say "Thou shalt not rob banks," without providing for exceptions. But temperance also can be created by coercion. Taxing is a good coercive device. To keep downtown shoppers temperate in their use of parking space we introduce parking meters for short periods, and traffic fines for longer ones. We need not actually forbid a citizen to park as long as he wants to; we need merely make it increasingly expensive for him to do so. Not prohibition, but carefully biased options are what we offer him. A Madison Avenue man might call this persuasion; I prefer the greater candor of the word coercion.

Coercion is a dirty word to most liberals now, but it need not forever be so. As with the four-letter words, its dirtiness can be cleansed away by exposure to the light, by saying it over and over without apology or embarrassment. To many, the word coercion implies arbitrary decisions of distant and irresponsible bureaucrats; but this is not a necessary part of its meaning. The only kind of coercion I recommend is mutual coercion, mutually agreed upon by the majority of the people affected.

To say that we mutually agree to coercion is not to say that we are required to enjoy it, or even to pretend we enjoy it. Who enjoys taxes? We all grumble about them. But we accept compulsory taxes because we recognize that voluntary taxes would favor the conscienceless. We institute and (grumblingly) support taxes and other coercive devices to escape the horror of the commons.

An alternative to the commons need not be perfectly just to be preferable. With real estate and other material goods, the alternative we have chosen is the institution of private property coupled with legal inheritance. Is this system perfectly just? As a genetically trained biologist I deny that it is. It seems to me that, if there are to be differences in individual inheritance, legal possession should be perfectly correlated with biological inheritance—that those who are biologically more fit to be the custodians of property and power should legally inherit more. But genetic recombination continually makes a mockery of the doctrine of "like father, like son" implicit in our laws of legal inheritance. An idiot can inherit millions, and a trust fund can keep his estate intact. We must admit that our legal system of private property plus inheritance is unjust—but we put up with it because we are not convinced, at the moment, that anyone has invented a better system. The alternative of the commons is too horrifying to contemplate. Injustice is preferable to total ruin.

It is one of the peculiarities of the warfare between reform and the status quo that it is thoughtlessly governed by a double standard. Whenever a reform measure is proposed it is often defeated when its opponents triumphantly discover a flaw in it. As Kingsley Davis has pointed out, worshippers of the status quo sometimes imply that no reform is possible without unanimous agreement, an implication contrary to historical fact. As nearly as I can make out, automatic rejection of proposed reforms is based on one of two unconscious assumptions: (i) that the status quo is perfect; or (ii) that the choice we face is between reform and no action; if the proposed reform is imperfect, we presumably should take no action at all, while we wait for a perfect proposal.

But we can never do nothing. That which we have done for thousands of years is also action. It also produces evils. Once we are aware that the status quo is action, we can then compare its discoverable advantages and disadvantages with the predicted advantages and disadvantages of the proposed reform, discounting as best we can for our lack of experience. On the basis of such a comparison, we can make a rational decision which will not involve the unworkable assumption that only perfect systems are tolerable.

RECOGNITION OF NECESSITY

Perhaps the simplest summary of this analysis of man's population problems is this: the commons, if justifiable at all, is justifiable only under conditions of low-population density. As the human population has increased, the commons has had to be abandoned in one aspect after another.

First we abandoned the commons in food gathering, enclosing farm land and restricting pastures and hunting and fishing areas. These restrictions are still not complete throughout the world.

Somewhat later we saw that the commons as a place for waste disposal would also have to be abandoned. Restrictions on the disposal of domestic sewage are widely accepted in the Western world; we are still struggling to close the commons to pollution by automobiles, factories, insecticide sprayers, fertilizing operations, and atomic energy installations.

In a still more embryonic state is our recognition of the evils of the commons in matters of pleasure. There is almost no restriction on the propagation of sound waves in the public medium. The shopping public is assaulted with mindless music, without its consent. Our government is paying out billions of

dollars to create supersonic transport which will disturb 50,000 people for every one person who is whisked from coast to coast 3 hours faster. Advertisers muddy the airwaves of radio and television and pollute the view of travelers. We are a long way from outlawing the commons in matters of pleasure. Is this because our Puritan inheritance makes us view pleasure as something of a sin, and pain (that is, the pollution of advertising) as the sign of virtue?

Every new enclosure of the commons involves the infringement of somebody's personal liberty. Infringements made in the distant past are accepted because no contemporary complains of a loss. It is the newly proposed infringements that we vigorously oppose; cries of "rights" and "freedom" fill the air. But what does "freedom" mean? When men mutually agreed to pass laws against robbing, mankind became more free, not less so. Individuals locked into the logic of the commons are free only to bring on universal ruin; once they see the necessity of mutual coercion, they become free to pursue other goals. I believe it was Hegel who said, "Freedom is the recognition of necessity."

The most important aspect of necessity that we must now recognize, is the necessity of abandoning the commons in breeding. No technical solution can rescue us from the misery of overpopulation. Freedom to breed will bring ruin to all. At the moment, to avoid hard decisions many of us are tempted to propagandize for conscience and responsible parenthood. The temptation must be resisted, because an appeal to independently acting consciences selects for the disappearance of all conscience in the long run, and an increase in anxiety in the short.

The only way we can preserve and nurture other and more precious freedoms is by relinquishing the freedom to breed, and that very soon. "Freedom is the recognition of necessity"—and it is the role of education to reveal to all the necessity of abandoning the freedom to breed. Only so, can we put an end to this aspect of the tragedy of the commons.

2.5 DONELLA H. MEADOWS ET AL.

The Limits to Growth

The following selection is from *The Limits to Growth: A Report for the Club of Rome's Project on the Predicament of Mankind* (Universe Books, 1972), in which Donella H. Meadows and her colleagues reported the results of the first phase of a study done in the early 1970s as part of the ambitious Project on the Predicament of Mankind. This project was an undertaking of the Club of Rome, an independent, international group of scientists, economists, humanists, and industrialists who shared the view that traditional institutions and policies could no longer properly evaluate the complex, major problems that the world was facing. The goal of the project was to examine such problems as the contradiction of the simultaneously growing wealth and increasing poverty, environmental degradation, and the loss of faith in existing institutions. Phase One was a study of five factors that might limit global growth and development: population, agricultural production, natural resources, industrial production, and pollution.

Using a computer model called World3 (which is primitive by current standards), a research team that included the authors of the report and that was directed by social policy analyst Dennis L. Meadows reached some startling conclusions. These results, which are summarized in the selection, indicated that a conscious effort to strive for a state of developmental equilibrium was necessary in order to avoid a disastrous decline in population and industrial productivity, which would result from resource depletion and widespread pollution within the following 100 years if current policies of unrestrained growth were allowed to continue.

Critiques of the Club of Rome's study emerged from all sectors of the political spectrum: Conservatives asserted that the marketplace would work to prevent disaster with no need for the imposition of international plans or controls. Liberals argued that restraints on growth would hurt the poor more than the affluent. And radicals contended that the results were only applicable to the type of profit-motivated growth that occurs under capitalism. Subsequent studies using more sophisticated models have softened some of the projected consequences in the original report, but they have also reinforced the basic conclusion that growth without limits will ultimately result in disaster. The 1980 *Global 2000 Report,* which reiterated the need for developmental controls; the 1992 United Nations Conference in Environment and Development in Rio de Janeiro; and the report of the World Commission on Environment and Development, which focused attention on the need for sustainable development, all have roots that extend back to *The Limits to Growth* report.

Key Concept: unrestrained growth will lead to ecological disaster

Donella H. Meadows et al.

I do not wish to seem overdramatic, but I can only conclude from the information that is available to me as Secretary-General, that the Members of the United Nations have perhaps ten years left in which to subordinate their ancient quarrels and launch a global partnership to curb the arms race, to improve the human environment, to defuse the population explosion, and to supply the required momentum to development efforts. If such a global partnership is not forged within the next decade, then I very much fear that the problems I have mentioned will have reached such staggering proportions that they will be beyond our capacity to control.

—U Thant, 1969

The problems U Thant mentions—the arms race, environmental deterioration, the population explosion, and economic stagnation—are often cited as the central, long-term problems of modern man. Many people believe that the future course of human society, perhaps even the survival of human society, depends on the speed and effectiveness with which the world responds to these issues. And yet only a small fraction of the world's population is actively concerned with understanding these problems or seeking their solutions.

HUMAN PERSPECTIVES

Every person in the world faces a series of pressures and problems that require his attention and action. These problems affect him at many different levels. He may spend much of his time trying to find tomorrow's food for himself and his family. He may be concerned about personal power or the power of the nation in which he lives. He may worry about a world war during his lifetime, or a war next week with a rival clan in his neighborhood.

These very different levels of human concern can be represented on a graph like that in figure 1. The graph has two dimensions, space and time. Every human concern can be located at some point on the graph, depending on how much geographical space it includes and how far it extends in time. Most people's worries are concentrated in the lower left-hand corner of the graph. Life for these people is difficult, and they must devote nearly all of their efforts to providing for themselves and their families, day by day. Other people think about and act on problems farther out on the space or time axes. The pressures they perceive involve not only themselves, but the community with which they identify. The actions they take extend not only days, but weeks or years into the future.

A person's time and space perspectives depend on his culture, his past experience, and the immediacy of the problems confronting him on each level. Most people must have successfully solved the problems in a smaller area before they move their concerns to a larger one. In general the larger the space and the longer the time associated with a problem, the smaller the number of people who are actually concerned with its solution.

There can be disappointments and dangers in limiting one's view to an area that is too small. There are many examples of a person striving with all his

FIGURE 1

Human Perspectives

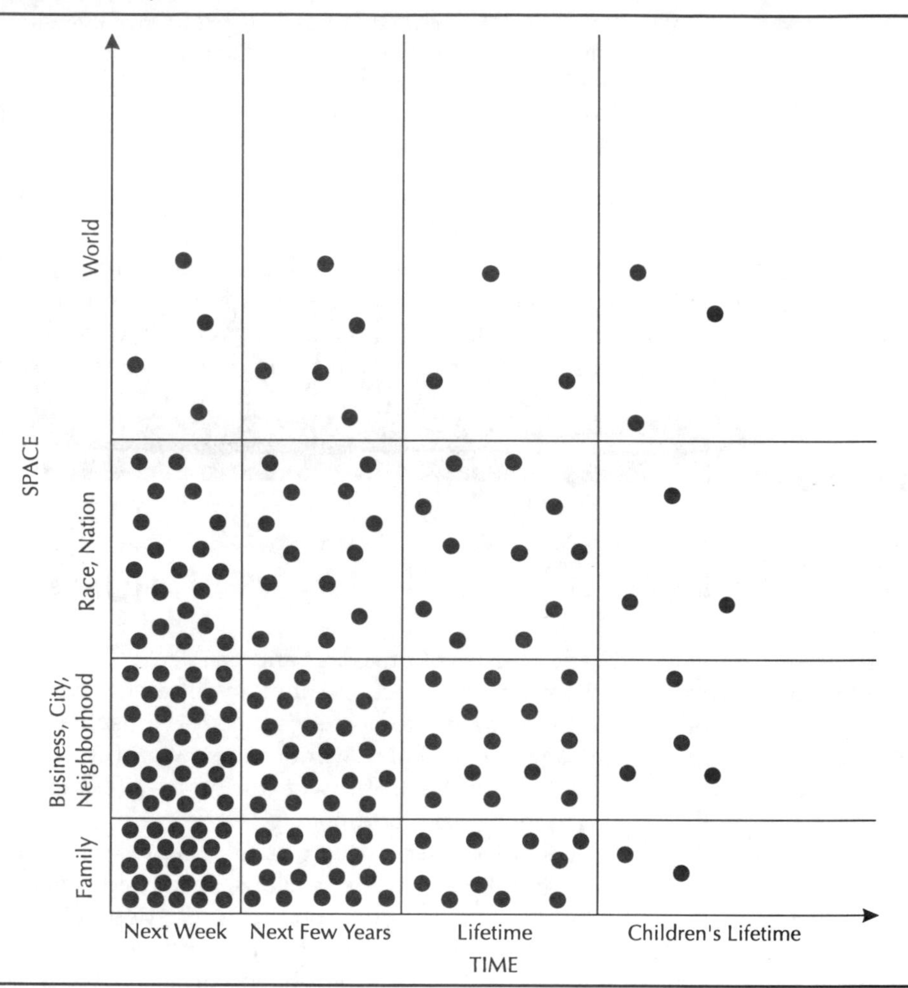

Note: Although the perspectives of the world's people vary in space and in time, every human concern falls somewhere on the space-time graph. The majority of the world's people are concerned with matters that affect only family or friends over a short period of time. Others look further ahead in time or over a larger area—a city or a nation. Only a very vew people have a global perspective that extends far into the future.

might to solve some immediate, local problem, only to find his efforts defeated by events occurring in a larger context. A farmer's carefully maintained fields can be destroyed by an international war. Local officials' plans can be overturned by a national policy. A country's economic development can be thwarted by a lack of world demand for its products. Indeed there is increasing concern today that most personal and national objectives may ultimately be frustrated by long-term, global trends such as those mentioned by U Thant.

Are the implications of these global trends actually so threatening that their resolution should take precedence over local, short-term concerns?

Is it true, as U Thant suggested, that there remains less than a decade to bring these trends under control?

If they are not brought under control; what will the consequences be?

What methods does mankind have for solving global problems, and what will be the results and the costs of employing each of them?

These are the questions that we have been investigating in the first phase of The Club of Rome's Project on the Predicament of Mankind. Our concerns thus fall in the upper right-hand corner of the space-time graph.

PROBLEMS AND MODELS

Every person approaches his problems, wherever they occur on the space-time graph, with the help of models. A model is simply an ordered set of assumptions about a complex system. It is an attempt to understand some aspect of the infinitely varied world by selecting from perceptions and past experience a set of general observations applicable to the problem at hand. A farmer uses a mental model of his land, his assets, market prospects, and past weather conditions to decide which crops to plant each year. A surveyor constructs a physical model—a map—to help in planning a road. An economist uses mathematical models to understand and predict the flow of international trade.

Decision-makers at every level unconsciously use mental models to choose among policies that will shape our future world. These mental models are, of necessity, very simple when compared with the reality from which they are abstracted. The human brain, remarkable as it is, can only keep track of a limited number of the complicated, simultaneous interactions that determine the nature of the real world.

We, too, have used a model. Ours is a formal, written model of the world.[1] It constitutes a preliminary attempt to improve our mental models of long-term, global problems by combining the large amount of information that is already in human minds and in written records with the new information-processing tools that mankind's increasing knowledge has produced—the scientific method, systems analysis, and the modern computer.

Our world model was built specifically to investigate five major trends of global concern—accelerating industrialization, rapid population growth, widespread malnutrition, depletion of nonrenewable resources, and a deteriorating environment. These trends are all interconnected in many ways, and their development is measured in decades or centuries, rather than in months or years. With the model we are seeking to understand the causes of these trends, their interrelationships, and their implications as much as one hundred years in the future.

The model we have constructed is, like every other model, imperfect, oversimplified, and unfinished. We are well aware of its shortcomings, but we believe that it is the most useful model now available for dealing with problems far out on the space-time graph. To our knowledge it is the only formal model in existence that is truly global in scope, that has a time horizon longer than thirty years, and that includes important variables such as population, food production, and pollution, not as independent entities, but as dynamically interacting elements, as they are in the real world.

Since ours is a formal, or mathematical, model it also has two important advantages over mental models. First, every assumption we make is written in a precise form so that it is open to inspection and criticism by all. Second, after the assumptions have been scrutinized, discussed, and revised to agree with our best current knowledge, their implications for the future behavior of the world system can be traced without error by a computer, no matter how complicated they become.

We feel that the advantages listed above make this model unique among all mathematical and mental world models available to us today. But there is no reason to be satisfied with it in its present form. We intend to alter, expand, and improve it as our own knowledge and the world data base gradually improve.

In spite of the preliminary state of our work, we believe it is important to publish the model and our findings now. Decisions are being made every day, in every part of the world, that will affect the physical, economic, and social conditions of the world system for decades to come. These decisions cannot wait for perfect models and total understanding. They will be made on the basis of some model, mental or written, in any case. We feel that the model described here is already sufficiently developed to be of some use to decision-makers. Furthermore, the basic behavior modes we have already observed in this model appear to be so fundamental and general that we do not expect our broad conclusions to be substantially altered by further revisions. . . .

The following conclusions have emerged from our work so far. We are by no means the first group to have stated them. For the past several decades, people who have looked at the world with a global, long-term perspective have reached similar conclusions. Nevertheless, the vast majority of policymakers seems to be actively pursuing goals that are inconsistent with these results.

Our conclusions are:

1. If the present growth trends in world population, industrialization, pollution, food production, and resource depletion continue unchanged, the limits to growth on this planet will be reached sometime within the next one hundred years. The most probable result will be a rather sudden and uncontrollable decline in both population and industrial capacity.
2. It is possible to alter these growth trends and to establish a condition of ecological and economic stability that is sustainable far into the future. The state of global equilibrium could be designed so that the basic material needs of each person on earth are satisfied and each person has an equal opportunity to realize his individual human potential.
3. If the world's people decide to strive for this second outcome rather than the first, the sooner they begin working to attain it, the greater will be their chances of success.

These conclusions are so far-reaching and raise so many questions for further study that we are quite frankly overwhelmed by the enormity of the job that must be done. We hope that this [selection] will serve to interest other people, in many fields of study and in many countries of the world, to raise the space and time horizons of their concerns and to join us in understanding and preparing for a period of great transition—the transition from growth to global equilibrium.

NOTES

1. The prototype model on which we have based our work was designed by Professor Jay W. Forrester of the Massachusetts Institute of Technology. A description of that model has been published in his book *World Dynamics* (Cambridge, Mass.: Wright-Allen Press, 1971.)

Human Domination of Earth's Ecosystems

Political decisions that take into account the present and future impacts of human activities on the environment are necessary to preserve the long-term viability of the human species. This was not always true. For most of human history, the by-products of civilization and the ecological impacts of agricultural and industrial activities were relatively small-scale and local. The industrial revolution brought about the potential to cause greater environmental devastation. Humans began to clear-cut forests, pollute the air, and degrade the water of major rivers and lakes. However, at the beginning of the twentieth century these effects were still primarily local and regional, rather than truly global.

In "Human Domination of Earth's Ecosystems," *Science* (July 25, 1997), from which the following selection is taken, Stanford University biologists Peter M. Vitousek and Harold Mooney, Oregon State University zoologist Jane Lubchenko, and Jerry M. Mellilo of the U.S. Office of Science and Technology Policy present a well-documented overview that illustrates the extent to which human activity has exerted a global impact on the Earth's ecosystems during the past century. They consider the consequences of land transformation, alterations of marine ecosystems, modifications of major biogeochemical cycles, and biotic disruptions. They report that almost 50% of the Earth's land surface has been transformed by human endeavors. The concentration of carbon dioxide in the atmosphere has been increased by 30%, more than half of all available fresh water is presently being used, and approximately 25% of the planet's bird species have been driven to extinction. Furthermore, they point out that activities that have resulted in these impacts are continuing, and in most cases, accelerating.

In the face of this grim assessment Vitousek et al. point out that humans have considerable power to control and reduce the negative impacts of our expanding environmental influence. They end their selection with several recommendations about how humankind should ensure responsible management of the planet in the future.

Key Concept: the impacts of accelerating human dominance of global ecosystems

All organisms modify their environment, and humans are no exception. As the human population has grown and the power of technology has expanded, the scope and nature of this modification has changed drastically. Until recently, the term "human-dominated ecosystems" would have elicited images of agricultural fields, pastures, or urban landscapes; now it applies with greater or lesser force to all of Earth. Many ecosystems are dominated directly by humanity, and no ecosystem on Earth's surface is free of pervasive human influence,

This [selection] provides an overview of human effects on Earth's ecosystems. It is not intended as a litany of environmental disaster, though some disastrous situations are described; nor is it intended either to downplay or to celebrate environmental successes, of which there have been many. Rather, we explore how large humanity looms as a presence on the globe—how, even on the grandest scale, most aspects of the structure and functioning of Earth's ecosystems cannot be understood without accounting for the strong, often dominant influence of humanity.

We view human alterations to the Earth system as operating through the interacting processes summarized in Fig. 1. The growth of the human population, and growth in the resource base used by humanity, is maintained by a suite of human enterprises such as agriculture, industry, fishing, and international commerce. These enterprises transform the land surface (through cropping, forestry, and urbanization), alter the major biogeochemical cycles, and add or remove species and genetically distinct populations in most of Earth's ecosystems. Many of these changes are substantial and reasonably well quantified; all are ongoing. These relatively well-documented changes in turn entrain further alterations to the functioning of the Earth system, most notably by driving global climatic change (*1*) and causing irreversible losses of biological diversity (2).

LAND TRANSFORMATION

The use of land to yield goods and services represents the most substantial human alteration of the Earth system. Human use of land alters the structure and functioning of ecosystems, and it alters how ecosystems interact with the atmosphere, with aquatic systems, and with surrounding land. Moreover, land transformation interacts strongly with most other components of global environmental change.

The measurement of land transformation on a global scale is challenging; changes can be measured more or less straightforwardly at a given site, but it is difficult to aggregate these changes regionally and globally. In contrast to analyses of human alteration of the global carbon cycle, we cannot install instruments on a tropical mountain to collect evidence of land transformation. Remote sensing is a most useful technique, but only recently has there been a

FIGURE 1

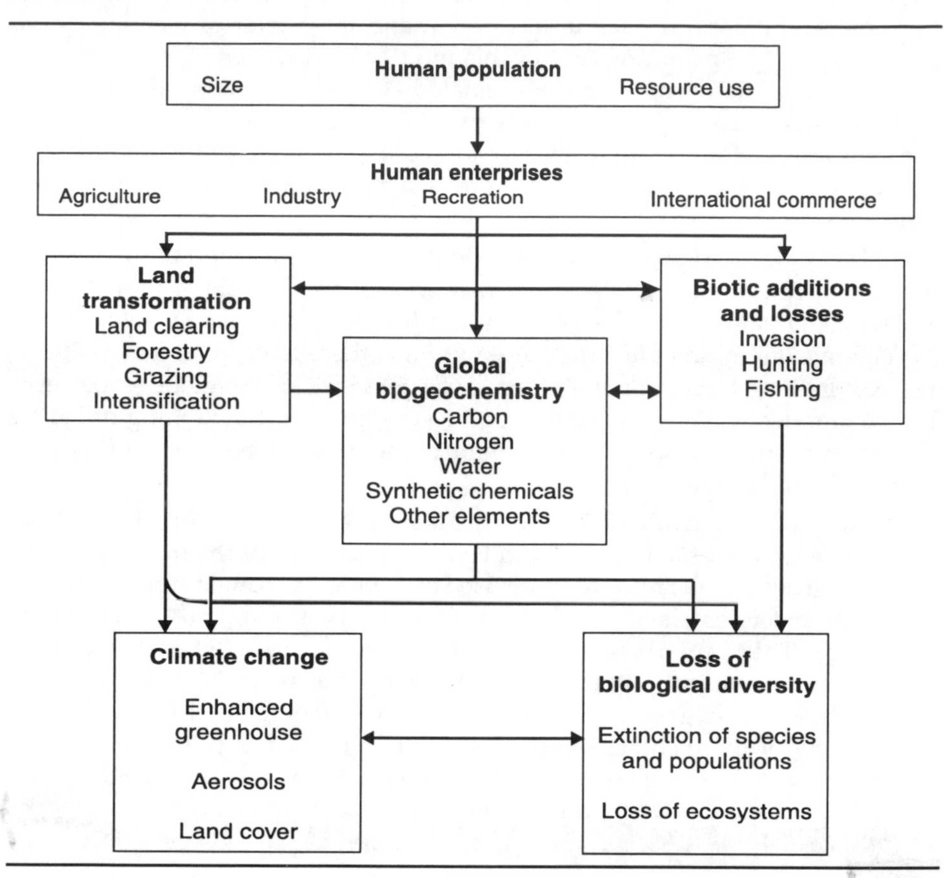

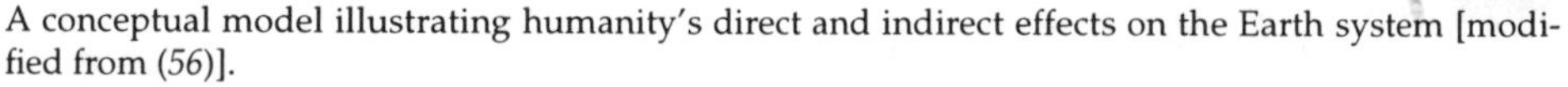
A conceptual model illustrating humanity's direct and indirect effects on the Earth system [modified from (56)].

serious scientific effort to use high-resolution civilian satellite imagery to evaluate even the more visible forms of land transformation, such as deforestation, on continental to global scales (3).

Land transformation encompasses a wide variety of activities that vary substantially in their intensity and consequences. At one extreme, 10 to 15% of Earth's land surface is occupied by row-crop agriculture or by urban-industrial areas, and another 6 to 8% has been converted to pastureland (4); these systems are wholly changed by human activity. At the other extreme, every terrestrial ecosystem is affected by increased atmospheric carbon dioxide (CO_2), and most ecosystems have a history of hunting and other low-intensity resource extraction. Between these extremes lie grassland and semiarid ecosystems that are grazed (and sometimes degraded) by domestic animals, and forests and woodlands from which wood products have been harvested; together, these represent the majority of Earth's vegetated surface.

The variety of human effects on land makes any attempt to summarize land transformations globally a matter of semantics as well as substantial uncer-

FIGURE 2

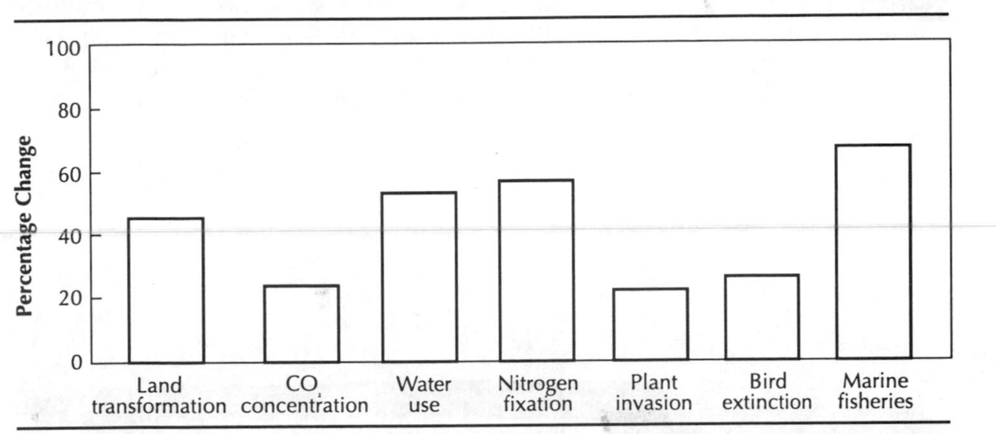

Human dominance or alteration of several major components of the Earth system, expressed as (from left to right) percentage of the land surface transformed (*5*); percentage of the current atmospheric CO_2 concentration that results from human action (*17*); percentage of accessible surface fresh water used (*20*); percentage of terrestrial N fixation that is human-caused (*28*); percentage of plant species in Canada that humanity has introduced from elsewhere (*48*); percentage of bird species on Earth that have become extinct in the past two millennia, almost all of them as a consequence of human activity (*42*); and percentage of major marine fisheries that are fully exploited, overexploited, or depleted (*14*).

tainty. Estimates of the fraction of land transformed or degraded by humanity (or its corollary, the fraction of the land's biological production that is used or dominated) fall in the range of 39 to 50% (*5*) (Fig. 2). These numbers have large uncertainties, but the fact that they are large is not at all uncertain. Moreover, if anything these estimates understate the global impact of land transformation, in that land that has not been transformed often has been divided into fragments by human alteration of the surrounding areas. This fragmentation affects the species composition and functioning of otherwise little modified ecosystems (*6*).

Overall, land transformation represents the primary driving force in the loss of biological diversity worldwide. Moreover, the effects of land transformation extend far beyond the boundaries of transformed lands. Land transformation can affect climate directly at local and even regional scales. It contributes 20% to current anthropogenic CO_2 emissions, and more substantially to the increasing concentrations of the greenhouse gases methane and nitrous oxide; fires associated with it alter the reactive chemistry of the troposphere, bringing elevated carbon monoxide concentrations and episodes of urban-like photochemical air pollution to remote tropical areas of Africa and South America; and it causes runoff of sediment and nutrients that drive substantial changes in stream, lake, estuarine, and coral reef ecosystems (*7–10*).

The central importance of land transformation is well recognized within the community of researchers concerned with global environmental change. Several research programs are focused on aspects of it (*9, 11*); recent and substantial progress toward understanding these aspects has been made (*3*), and much more progress can be anticipated. Understanding land transformation is a difficult challenge; it requires integrating the social, economic, and cultural causes of land transformation with evaluations of its biophysical nature and consequences. This interdisciplinary approach is essential to predicting the course, and to any hope of affecting the consequences, of human-caused land transformation.

OCEANS

Human alterations of marine ecosystems are more difficult to quantify than those of terrestrial ecosystems, but several kinds of information suggest that they are substantial. The human population is concentrated near coasts—about 60% within 100 km—and the oceans' productive coastal margins have been affected strongly by humanity. Coastal wetlands that mediate interactions between land and sea have been altered over large areas; for example, approximately 50% of mangrove ecosystems globally have been transformed or destroyed by human activity (*12*). Moreover, a recent analysis suggested that although humans use about 8% of the primary production of the oceans, that fraction grows to more than 25% for upwelling areas and to 35% for temperate continental shelf systems (*13*).

Many of the fisheries that capture marine productivity are focused on top predators, whose removal can alter marine ecosystems out of proportion to their abundance. Moreover, many such fisheries have proved to be unsustainable, at least at our present level of knowledge and control. As of 1995, 22% of recognized marine fisheries were overexploited or already depleted, and 44% more were at their limit of exploitation (*14*) (Figs. 2 and 3). The consequences of fisheries are not restricted to their target organisms; commercial marine fisheries around the world discard 27 million tons of nontarget animals annually, a quantity nearly one-third as large as total landings (*15*). Moreover, the dredges and trawls used in some fisheries damage habitats substantially as they are dragged along the sea floor.

A recent increase in the frequency, extent, and duration of harmful algal blooms in coastal areas (*16*) suggests that human activity has affected the base as well as the top of marine food chains. Harmful algal blooms are sudden increases in the abundance of marine phytoplankton that produce harmful structures or chemicals. Some but not all of these phytoplankton are strongly pigmented (red or brown tides). Algal blooms usually are correlated with changes in temperature, nutrients, or salinity; nutrients in coastal waters, in particular, are much modified by human activity. Algal blooms can cause extensive fish kills through toxins and by causing anoxia; they also lead to paralytic shellfish poisoning and amnesic shellfish poisoning in humans. Although the existence of harmful algal blooms has long been recognized, they have spread widely in the past two decades (*16*).

FIGURE 3

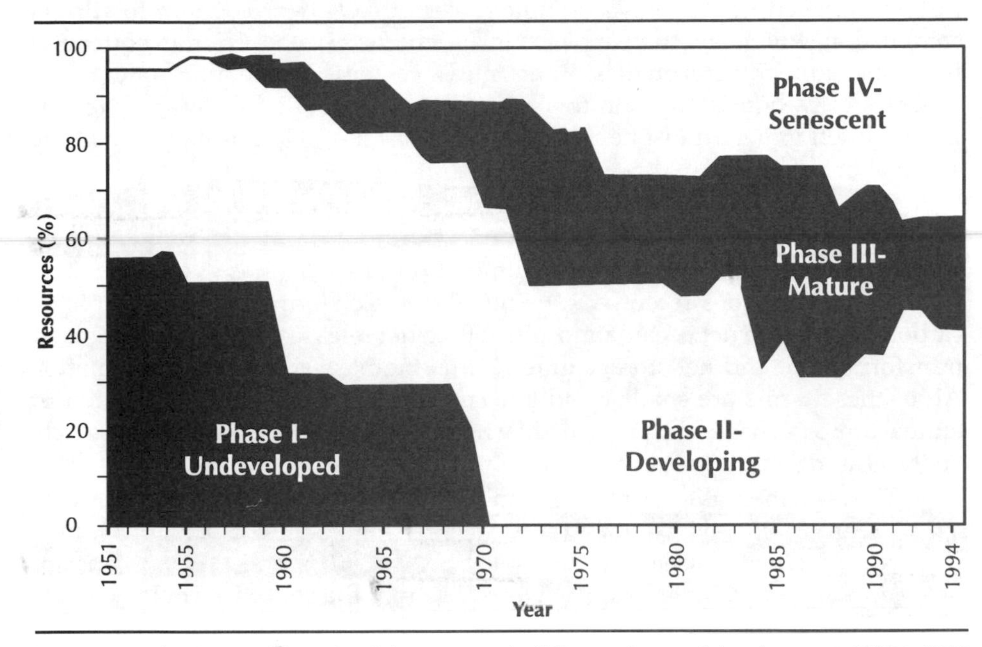

Percentage of major world marine fish resources in different phases of development, 1951 to 1994 (from (57)]. Undeveloped = a low and relatively constant level of catches; developing = rapidly increasing catches; mature = a high and plateauing level of catches; senescent = catches declining from higher levels.

ALTERATIONS OF THE BIOGEOCHEMICAL CYCLES

Carbon. Life on Earth is based on carbon, and the CO_2 in the atmosphere is the primary resource for photosynthesis. Humanity adds CO_2 to the atmosphere by mining and burning fossil fuels, the residue of life from the distant past, and by converting forests and grasslands to agricultural and other low-biomass ecosystems. The net result of both activities is that organic carbon from rocks, organisms, and soils is released into the atmosphere as CO_2.

The modern increase in CO_2 represents the clearest and best documented signal of human alteration of the Earth system. Thanks to the foresight of Roger Revelle, Charles Keeling, and others who initiated careful and systematic measurements of atmospheric CO_2 in 1957 and sustained them through budget crises and changes in scientific fashions, we have observed the concentration of CO_2 as it has increased steadily from 315 ppm to 362 ppm. Analysis of air bubbles extracted from the Antarctic and Greenland ice caps extends the record back much further; the CO_2 concentration was more or less stable near 280 ppm for thousands of years until about 1800, and has increased exponentially since then (*17*).

There is no doubt that this increase has been driven by human activity, today primarily by fossil fuel combustion. The sources of CO_2 can be traced isotopically; before the period of extensive nuclear testing in the atmosphere, carbon depleted in ^{14}C was a specific tracer of CO_2 derived from fossil fuel combustion, whereas carbon depleted in ^{13}C characterized CO_2 from both fossil fuels and land transformation. Direct measurements in the atmosphere, and analyses of carbon isotopes in tree rings, show that both ^{13}C and ^{14}C in CO_2 were diluted in the atmosphere relative to ^{12}C as the CO_2 concentration in the atmosphere increased.

Fossil fuel combustion now adds 5.5 $\pm$ 0.5 billion metric tons of $C0_2$-C to the atmosphere annually, mostly in economically developed regions of the temperate zone (*18*). The annual accumulation of $C0_2$-C has averaged 3.2 $\pm$ 0.2 billion metric tons recently (*17*). The other major terms in the atmospheric carbon balance are net ocean-atmosphere flux, net release of carbon during land transformation, and net storage in terrestrial biomass and soil organic matter. All of these terms are smaller and less certain than fossil fuel combustion or annual atmospheric accumulation; they represent rich areas of current research, analysis, and sometimes contention.

The human-caused increase in atmospheric CO_2 already represents nearly a 30% change relative to the pre-industrial era (Fig. 2), and CO_2 will continue to increase for the foreseeable future. Increased CO_2 represents the most important human enhancement to the greenhouse effect; the consensus of the climate research community is that it probably already affects climate detectably and will drive substantial climate change in the next century (*1*). The direct effects of increased CO_2 on plants and ecosystems may be even more important. The growth of most plants is enhanced by elevated CO_2, but to very different extents; the tissue chemistry of plants that respond to CO_2 is altered in ways that decrease food quality for animals and microbes; and the water use efficiency of plants and ecosystems generally is increased. The fact that increased CO_2 affects species differentially means that it is likely to drive substantial changes in the species composition and dynamics of all terrestrial ecoystems (*19*).

Water. Water is essential to all life. Its movement by gravity, and through evaporation and condensation, contributes to driving Earth's biogeochemical cycles and to controlling its climate. Very little of the water on Earth is directly usable by humans; most is either saline or frozen. Globally, humanity now uses more than half of the runoff water that is fresh and reasonably accessible, with about 70% of this use in agriculture (*20*) (Fig. 2). To meet increasing demands for the limited supply of fresh water, humanity has extensively altered river systems through diversions and impoundments. In the United States only 2% of the rivers run unimpeded, and by the end of this century the flow of about two-thirds of all of Earth's rivers will be regulated (*21*). At present, as much as 6% of Earth's river runoff is evaporated as a consequence of human manipulations (*22*). Major rivers, including the Colorado, the Nile, and the Ganges, are used so extensively that little water reaches the sea. Massive inland water bodies, including the Aral Sea and Lake Chad, have been greatly reduced in extent by water diversions for agriculture. Reduction in the volume of the Aral Sea resulted in the demise of native fishes and the loss of other biota; the loss of a major fishery; exposure of the salt-laden sea bottom, thereby providing a major source of windblown dust; the production of a drier and more continental local climate and a decrease in water quality in the general region; and an increase in human diseases (*23*).

Impounding and impeding the flow of rivers provides reservoirs of water that can be used for energy generation as well as for agriculture. Waterways also are managed for transport, for flood control, and for the dilution of chemical wastes. Together, these activities have altered Earth's freshwater ecosystems profoundly, to a greater extent than terrestrial ecosystems have been altered. The construction of dams affects biotic habitats indirectly as well; the damming of the Danube River, for example, has altered the silica chemistry of the entire Black Sea. The large number of operational dams (36,000) in the world, in conjunction with the many that are planned, ensure that humanity's effects on aquatic biological systems will continue (*24*). Where surface water is sparse or overexploited, humans use groundwater—and in many areas the groundwater that is drawn upon is nonrenewable, or fossil, water (*25*). For example, three-quarters of the water supply of Saudi Arabia currently comes from fossil water (*26*).

Alterations to the hydrological cycle can affect regional climate. Irrigation increases atmospheric humidity in semiarid areas, often increasing precipitation and thunderstorm frequency (*27*). In contrast, land transformation from forest to agriculture or pasture increases albedo and decreases surface roughness; simulations suggest that the net effect of this transformation is to increase temperature and decrease precipitation regionally (*7, 26*).

Conflicts arising from the global use of water will be exacerbated in the years ahead, with a growing human population and with the stresses that global changes will impose on water quality and availability. Of all of the environmental security issues facing nations, an adequate supply of clean water will be the most important.

Nitrogen. Nitrogen (N) is unique among the major elements required for life, in that its cycle includes a vast atmospheric reservoir (N_2) that must be fixed (combined with carbon, hydrogen, or oxygen) before it can be used by most organisms. The supply of this fixed N controls (at least in part) the productivity, carbon storage, and species composition of many ecosystems. Before the extensive human alteration of the N cycle, 90 to 130 million metric tons of N (Tg N) were fixed biologically on land each year; rates of biological fixation in marine systems are less certain, but perhaps as much was fixed there (*28*).

Human activity has altered the global cycle of N substantially by fixing N_2—deliberately for fertilizer and inadvertently during fossil fuel combustion. Industrial fixation of N fertilizer increased from <10 Tg/year in 1950 to 80 Tg/year in 1990; after a brief dip caused by economic dislocations in the former Soviet Union, it is expected to increase to >135 Tg/year by 2030 (*29*). Cultivation of soybeans, alfalfa, and other legume crops that fix N symbiotically enhances fixation by another ~40 Tg/year, and fossil fuel combustion puts >20 Tg/year of reactive N into the atmosphere globally—some by fixing N_2, more from the mobilization of N in the fuel. Overall, human activity adds at least as much fixed N to terrestrial ecosystems as do all natural sources combined (Fig. 2), and it mobilizes >50 Tg/year more during land transformation (*28, 30*).

Alteration of the N cycle has multiple consequences. In the atmosphere, these include (i) an increasing concentration of the greenhouse gas nitrous oxide globally; (ii) substantial increases in fluxes of reactive N gases (two-thirds or more of both nitric oxide and ammonia emissions globally are human-caused); and (iii) a substantial contribution to acid rain and to the photochemical smog that afflicts urban and agricultural areas throughout the world (*31*). Reactive N that is emitted to the atmosphere is deposited downwind, where it can influence the dynamics of recipient ecosystems. In regions where fixed N was in short supply, added N generally increases productivity and C storage within ecosystems, and ultimately increases losses of N and cations from soils, in a set of processes termed "N saturation" (*32*). Where added N increases the productivity of ecosystems, usually it also decreases their biological diversity (*33*).

Human-fixed N also can move from agriculture, from sewage systems, and from N-saturated terrestrial systems to streams, rivers, groundwater, and ultimately the oceans. Fluxes of N through streams and rivers have increased markedly as human alteration of the N cycle has accelerated; river nitrate is highly correlated with the human population of river basins and with the sum of human-caused N inputs to those basins (*8*). Increases in river N drive the eutrophication of most estuaries, causing blooms of nuisance and even toxic algae, and threatening the sustainability of marine fisheries (*16*, *34*).

Other cycles. The cycles of carbon, water, and nitrogen are not alone in being altered by human activity. Humanity is also the largest source of oxidized sulfur gases in the atmosphere; these affect regional air quality, biogeochemistry, and climate. Moreover, mining and mobilization of phosphorus and of many metals exceed their natural fluxes; some of the metals that are concentrated and mobilized are highly toxic (including lead, cadmium, and mercury) (*35*). Beyond any doubt, humanity is a major biogeochemical force on Earth.

Synthetic organic chemicals. Synthetic organic chemicals have brought humanity many beneficial services. However, many are toxic to humans and other species, and some are hazardous in concentrations as low as 1 part per billion. Many chemicals persist in the environment for decades; some are both toxic and persistent. Long-lived organochlorine compounds provide the clearest examples of environmental consequences of persistent compounds. Insecticides such as DDT and its relatives, and industrial compounds like polychlorinated biphenyls (PCBs), were used widely in North America in the 1950s and 1960s. They were transported globally, accumulated in organisms, and magnified in concentration through food chains; they devastated populations of some predators (notably falcons and eagles) and entered parts of the human food supply in concentrations higher than was prudent. Domestic use of these compounds was phased out in the 1970s in the United States and Canada, and their concentrations declined thereafter. However, PCBs in particular remain readily detectable in many organisms, sometimes approaching thresholds of public health concern (*36*). They will continue to circulate through organisms for many decades.

Synthetic chemicals need not be toxic to cause environmental problems. The fact that the persistent and volatile chlorofluorocarbons (CFCs) are wholly nontoxic contributed to their widespread use as refrigerants and even aerosol propellants. The subsequent discovery that CFCs drive the breakdown of stratospheric ozone, and especially the later discovery of the Antarctic ozone hole and their role in it, represent great surprises in global environmental science (*37*). Moreover, the response of the international political system to those discoveries is the best extant illustration that global environmental change can be dealt with effectively (*38*).

Particular compounds that pose serious health and environmental threats can be and often have been phased out (although PCB production is growing in Asia). Nonetheless, each year the chemical industry produces more than 100 million tons of organic chemicals representing some 70,000 different compounds, with about 1000 new ones being added annually (*39*). Only a small fraction of the many chemicals produced and released into the environment are tested adequately for health hazards or environmental impact (*40*).

BIOTIC CHANGES

Human modification of Earth's biological resources—its species and genetically distinct populations—is substantial and growing. Extinction is a natural process, but the current rate of loss of genetic variability, of populations, and of species is far above background rates; it is ongoing; and it represents a wholly irreversible global change. At the same time, human transport of species around Earth is homogenizing Earth's biota, introducing many species into new areas where they can disrupt both natural and human systems.

Losses. Rates of extinction are difficult to determine globally, in part because the majority of species on Earth have not yet been identified. Nevertheless, recent calculations suggest that rates of species extinction are now on the order of 100 to 1000 times those before humanity's dominance of Earth (*41*). For particular well-known groups, rates of loss are even greater; as many as one-quarter of Earth's bird species have been driven to extinction by human activities over the past two millennia, particularly on oceanic islands (*42*) (Fig. 2). At present, 11% of the remaining birds, 18% of the mammals, 5% of fish, and 8% of plant species on Earth are threatened with extinction (*43*). There has been a disproportionate loss of large mammal species because of hunting; these species played a dominant role in many ecosystems, and their loss has resulted in a fundamental change in the dynamics of those systems (*44*), one that could lead to further extinctions. The largest organisms in marine systems have been affected similarly, by fishing and whaling. Land transformation is the single most important cause of extinction, and current rates of land transformation eventually will drive many more species to extinction, although with a time lag that masks the true dimensions of the crisis (*45*). Moreover, the effects of other components of global environmental change—of altered carbon and nitrogen cycles, and of anthropogenic climate change—are just beginning.

As high as they are, these losses of species understate the magnitude of loss of genetic variation. The loss to land transformation of locally adapted populations within species, and of genetic material within populations, is a human-caused change that reduces the resilience of species and ecosystems while precluding human use of the library of natural products and genetic material that they represent (*46*).

Although conservation efforts focused on individual endangered species have yielded some successes, they are expensive—and the protection or restoration of whole ecosystems often represents the most effective way to sustain genetic, population, and species diversity. Moreover, ecosystems themselves may play important roles in both natural and human-dominated landscapes. For example, mangrove ecosystems protect coastal areas from erosion and provide nurseries for offshore fisheries, but they are threatened by transformation in many areas.

Invasions. In addition to extinction, humanity has caused a rearrangement of Earth's biotic systems, through the mixing of floras and faunas that had long been isolated geographically. The magnitude of transport of species, termed "biological invasion," is enormous (*47*); invading species are present almost everywhere. On many islands, more than half of the plant species are nonindigenous, and in many continental areas the figure is 20% or more (*48*) (Fig. 2).

As with extinction, biological invasion occurs naturally—and as with extinction, human activity has accelerated its rate by orders of magnitude. Land transformation interacts strongly with biological invasion, in that human-altered ecosystems generally provide the primary foci for invasions, while in some cases land transformation itself is driven by biological invasions (*49*). International commerce is also a primary cause of the breakdown of biogeographic barriers; trade in live organisms is massive and global, and many other organisms are inadvertently taken along for the ride. In freshwater systems, the combination of upstream land transformation, altered hydrology, and numerous deliberate and accidental species introductions has led to particularly widespread invasion, in continental as well as island ecosystems (*50*).

In some regions, invasions are becoming more frequent. For example, in the San Francisco Bay of California, an average of one new species has been established every 36 weeks since 1850, every 24 weeks since 1970, and every 12 weeks for the last decade (*51*). Some introduced species quickly become invasive over large areas (for example, the Asian clam in the San Francisco Bay), whereas others become widespread only after a lag of decades, or even over a century (*52*).

Many biological invasions are effectively irreversible; once replicating biological material is released into the environment and becomes successful there, calling it back is difficult and expensive at best. Moreover, some species introductions have consequences. Some degrade human health and that of other species; after all, most infectious diseases are invaders over most of their range. Others have caused economic losses amounting to billions of dollars; the recent invasion of North America by the zebra mussel is a well-publicized example. Some disrupt ecosystem processes, altering the structure and functioning of whole ecosystems. Finally, some invasions drive losses in the biological diversity of native species and populations; after land transformation, they are the next most important cause of extinction (*53*).

CONCLUSIONS

The global consequences of human activity are not something to face in the future—as Fig. 2 illustrates, they are with us now. All of these changes are ongoing, and in many cases accelerating; many of them were entrained long before their importance was recognized. Moreover, all of these seemingly disparate phenomena trace to a single cause—the growing scale of the human enterprise. The rates, scales, kinds, and combinations of changes occurring now are fundamentally different from those at any other time in history; we are changing Earth more rapidly than we are understanding it. We live on a human-dominated planet—and the momentum of human population growth, together with the imperative for further economic development in most of the world, ensures that our dominance will increase.

The [information in these pages] summarize our knowledge of and provide specific policy recommendations concerning major human-dominated ecosystems. In addition, we suggest that the rate and extent of human alteration of Earth should affect how we think about Earth. It is clear that we control much of Earth, and that our activities affect the rest. In a very real sense, the world is in our hands—and how we handle it will determine its composition and dynamics, and our fate.

Recognition of the global consequences of the human enterprise suggests three complementary directions. First, we can work to reduce the rate at which we alter the Earth system. Humans and human-dominated systems may be able to adapt to slower change, and ecosystems and the species they support may cope more effectively with the changes we impose, if those changes are slow. Our footprint on the planet (*54*) might then be stabilized at a point where enough space and resources remain to sustain most of the other species on Earth, for their sake and our own. Reducing the rate of growth in human effects on Earth involves slowing human population growth and using resources as efficiently as is practical. Often it is the waste products and by-products of human activity that drive global environmental change.

Second, we can accelerate our efforts to understand Earth's ecosystems and how they interact with the numerous components of human-caused global change. Ecological research is inherently complex and demanding: It requires measurement and monitoring of populations and ecosystems; experimental studies to elucidate the regulation of ecological processes; the development, testing, and validation of regional and global models; and integration with a broad range of biological, earth, atmospheric, and marine sciences. The challenge of understanding a human-dominated planet further requires that the human dimensions of global change—the social, economic, cultural, and other drivers of human actions—be included within our analyses.

Finally, humanity's dominance of Earth means that we cannot escape responsibility for managing the planet. Our activities are causing rapid, novel, and substantial changes to Earth's ecosystems. Maintaining populations, species, and ecosystems in the face of those changes, and maintaining the flow of goods and services they provide humanity (*55*), will require active management for the foreseeable future. There is no clearer illustration of the extent of human dominance of Earth than the fact that maintaining the diversity of "wild" species and the functioning of "wild" ecosystems will require increasing human involvement.

REFERENCES

1. Intergovernmental Panel on Climate Change, *Climate Change 1995* (Cambridge Univ. Press, Cambridge, 1996), pp. 9–49.
2. United Nations Environment Program, *Global Biodiversity Assessment*, V. H. Heywood, Ed. (Cambridge Univ. Press, Cambridge, 1995).
3. D. Skole and C. J. Tucker, *Science* **260,** 1905 (1993).
4. J. S. Olson, J. A. Watts, L. J. Allison, *Carbon in Live Vegetation of Major World Ecosystems* (Office of Energy Research, U.S. Department of Energy, Washington, DC, 1983).
5. P. M. Vitousek, P. R. Ehrlich, A. H. Ehrlich, P. A. Matson, *Bioscience* **36,** 368 (1986); R. W. Kates, B. L. Turner, W. C. Clark, in (35), pp. 1–17; G. C. Daily, *Science* **269,** 350 (1995).
6. D. A. Saunders, R. J. Hobbs, C. R. Margules, *Conserv. Biol.* **5,** 18 (1991).
7. J. Shukla, C. Nobre, P. Sellers, *Science* **247,** 1322 (1990).
8. R. W. Howarth *et al., Biogeochemistry* **35,** 75 (1996).
9. W. B. Meyer and B. L. Turner II, *Changes in Land Use and Land Cover: A Global Perspective* (Cambridge Univ. Press, Cambridge, 1994).
10. S. R. Carpenter, S. G. Fisher, N. B. Grimm, J. F. Kitchell, *Annu. Rev. Ecol. Syst.* **23** 119 (1992); S. V. Smith and R. W. Buddemeier, *ibid.*, p. 89; J. M. Melillo, I. C. Prentice, G. D. Farquhar, E.-D. Schulze, O. E. Sala, in (1), pp. 449–481.
11. R. Leemans and G. Zuidema, *Trends Ecol. Evol.* **10,** 76 (1995).
12. World Resources Institute, *World Resources 1996–1997* (Oxford Univ. Press, New York, 1996).
13. D. Pauly and V. Christensen, *Nature* **374,** 257 (1995).
14. Food and Agricultural Organization (FAO), *FAO Fisheries Tech. Pap. 335* (1994).
15. D. L. Alverson, M. H. Freeberg, S. A. Murawski, J. G. Pope, *FAO Fisheries Tech. Pap. 339* (1994).
16. G. M. Hallegraeff, *Phycologia* **32,** 79(1993).
17. D. S. Schimel *et al.*, in *Climate Change 1994: Radiative Forcing of Climate Change*, J. T. Houghton *et al.*, Eds. (Cambridge Univ. Press, Cambridge, 1995), pp. 39–71.
18. R. J. Andres, G. Marland, I. Y. Fung, E. Matthews, *Global Biogeochem. Cycles* **10,** 419 (1996).
19. G. W. Koch and H. A. Mooney, *Carbon Dioxide and Terrestrial Ecosystems* (Academic Press, San Diego, CA, 1996); C. Körner and F. A. Bazzaz, *Carbon Dioxide, Populations, and Communities* (Academic Press, San Diego, CA, 1996).
20. S. L. Postel, G. C. Daily, P. R. Ehrlich, *Science* **271,** 785 (1996).
21. J. N. Abramovitz, *Imperiled Waters, Impoverished Future: The Decline of Freshwater Ecosystems* (Worldwatch Institute, Washington. DC, 1996).
22. M. I. L'vovich and G. F. White, in (35), pp. 235–252; M. Dynesius and C. Nilsson, *Science* **266,** 753 (1994).

23. P. Micklin, *Science* **241,** 1170 (1988); V. Kotlyakov, *Environment* **33,** 4 (1991).
24. C. Humborg, V. Ittekkot, A. Cociasu, B. Bodungen, *Nature* **386,** 385 (1997).
25. P. H. Gleick, Ed., *Water in Crisis* (Oxford Univ. Press, New York, 1993).
26. V. Gornitz, C. Rosenzweig, D. Hillel, *Global Planet Change* **14,** 147 (1997).
27. P. C. Milly and K. A. Dunne, *J. Clim.* **7,** 506 (1994).
28. J. N. Galloway, W. H. Schlesinger, H. Levy II, A. Michaels, J. L. Schnoor, *Global Biogeochem. Cycles* **9,** 235 (1995).
29. J. N. Galloway, H. Levy II, P. S. Kasibhatla, *Ambio* **23,** 120 (1994).
30. V. Smil, in (*35*), pp. 423–436.
31. P. M. Vitousek *et al., Ecol. Appl.,* in press.
32. J. D. Aber, J. M. Melillo, K. J. Nadelhoffer, J. Pastor, R.D. Boone, *ibid.* **1,** 303 (1991).
33. D. Tilman, *Ecol. Monogr.* **57,** 189 (1987).
34. S. W. Nixon *et al., Biogeochemistry* **35,** 141 (1996).
35. B. L. Turner II *et al.,* Eds., *The Earth As Transformed by Human Action* (Cambridge Univ. Press, Cambridge, 1990).
36. C. A. Stow, S. R. Carpenter, C. P. Madenjian, L. A. Eby, L. J. Jackson, *Bioscience* **45,** 752 (1995).
37. F. S. Rowland, *Am. Sci.* **77,** 36(1989): S. Solomon, *Nature* **347,** 347 (1990).
38. M. K. Tolba *et al.,* Eds., *The World Environment 1972–1992* (Chapman & Hall, London, 1992).
39. S. Postel, *Defusing the Toxics Threat: Controlling Pesticides and Industrial Waste* (Worldwatch Institute, Washington, DC, 1987).
40. United Nations Environment Program (UNEP), *Saving Our Planet—Challenges and Hopes* (UNEP, Nairobi, 1992)
41. J. H. Lawton and R. M. May, Eds., *Extinction Rates* (Oxford Univ. Press, Oxford, 1995); S. L. Pimm, G. J. Russell, J. L. Gittleman, T. Brooks, *Science* **269,** 347 (1995).
42. S. L. Olson, in *Conservation for the Twenty-First Century,* D. Western and M. C. Pearl, Eds. (Oxford Univ. Press, Oxford, 1989), p. 50; D. W. Steadman, *Science* **267,** 1123 (1995).
43. R. Barbault and S. Sastrapradja, in (2), pp. 193–274.
44. R. Dirzo and A. Miranda, in *Plant-Animal Interactions,* P. W. Price, T. M. Lewinsohn, W. Fernandes, W. W. Benson, Eds. (Wiley Interscience, New York, 1991), p. 273.
45. D. Tilman, R. M. May, C. Lehman, M. A. Nowak, *Nature* **371,** 65 (1994).
46. H. A. Mooney, J. Lubchenco, R. Dirzo, O. E. Sala, in (2), pp. 279–325.
47. C. Elton, *The Ecology of Invasions by Animals and Plants* (Methuen, London, 1958); J. A. Drake *et al.,* Eds., *Biological Invasions. A Global Perspective* (Wiley, Chichester, UK, 1989).
48. M. Rejmanek and J. Randall, *Madrono* **41,** 161 (1994).
49. C. M. D'Antonio and P. M. Vitousek, *Annu. Rev. Ecol. Syst.* **23,** 63 (1992).
50. D. M. Lodge, *Trends Ecol. Evol.* **8,** 133 (1993).
51. A. N. Cohen and J. T. Carlton, *Biological Study: Nonindigenous Aquatic Species in a United States Estuary: A Case Study of the Biological Invasions of the San Francisco Bay and Delta* (U.S. Fish and Wildlife Service, Washington, DC, 1995).
52. I. Kowarik, in *Plant Invasions—General Aspects and Special Problems,* P. Pysek, K. Prach, M. Rejmánek, M. Wade, Eds. (SPB Academic, Amsterdam, 1995), p. 15.
53. P. M. Vitousek, C. M. D'Antonio, L. L. Loope, R. Westbrooks, *Am. Sci.* **84,** 468 (1996).

54. W. E. Rees and M. Wackernagel, in *Investing in Natural Capital: The Ecological Economics Approach to Sustainability*, A. M. Jansson, M. Hammer, C. Folke, R. Costanza, Eds. (Island, Washington, DC, 1994).

55. G. C. Daily, Ed., *Nature's Services* (Island, Washington, DC, 1997).

56. J. Lubchenco *et al., Ecology* **72,** 371 (1991); P. M. Vitousek, *ibid.* **75,** 1861 (1994).

57. S. M. Garcia and R. Grainger, *FAO Fisheries Tech. Pap. 359* (1996).

58. We thank G. C. Daily, C. B. Field, S. Hobbie, D. Gordon, P. A. Matson, and R. L. Naylor for constructive comments on this paper, A. S. Denning and. S. M. Garcia for assistance with illustrations, and C. Nakashima and B. Lilley for preparing text and figures for publication.

CHAPTER 3 Ecology and Ecosystems

3.1 G. EVELYN HUTCHINSON

Homage to Santa Rosalia, or Why Are There So Many Kinds of Animals?

The distinguished American ecologist G. Evelyn Hutchinson (1903–1991) was a true renaissance man. His expertise extended beyond the sciences to literature and the arts, and he authored several literary works in addition to his scientific books and articles. Although he made many significant contributions to ecology—especially in the field of limnology, the study of freshwater lakes and ponds—perhaps his greatest achievement was the inspiration he instilled in many of his students at Yale University, who went on to launch their own successful careers in ecology. Hutchinson's broad approach to ecology, which emphasized the importance of mastering the fundamentals of biochemistry, geology, zoology and botany, became known as the "Hutchinson School." He was one of the first prominent scientists to warn that the human race was gaining, through modern technology, the potential to destroy its own environment.

Hutchinson's literary skills are evident in the following selection from "Homage to Santa Rosalia, or Why Are There So Many Kinds of Animals?" *The American Naturalist* (May–June 1959). He approaches the fundamental ecological question of why there exists a great diversity of species, with his characteristically broad, systematic mode of analysis. After considering the requirements of food chains and webs, the role of natural selection, size effects, and other factors that may either limit or promote diversity, he comes to what is still a controversial conclusion among ecologists. Hutchinson contended that there is species diversity partly because ecosystem complexity increases stability.

Key Concept: factors affecting species diversity

When you did me the honor of asking me to fill your presidential chair, I accepted perhaps without duly considering the duties of the president of a society, founded largely to further the study of evolution, at the close of the year that marks the centenary of Darwin and Wallace's initial presentation of the theory of natural selection. It seemed to me that most of the significant aspects of modern evolutionary theory have come either from geneticists, or from those heroic museum workers who suffering through years of neglect, were able to establish about 20 years ago what has come to be called the "new systematics." You had, however, chosen an ecologist as your president and one of that school at times supposed to study the environment without any relation to the organism.

A few months later I happened to be in Sicily. An early interest in zoogeography and in aquatic insects led me to attempt to collect near Palermo, certain species of water-bugs, of the genus Corixa, described a century ago by Fieber and supposed to occur in the region, but never fully reinvestigated. It is hard to find suitable localities in so highly cultivated a landscape as the Concha d'Oro. Fortunately, I was driven up Monte Pellegrino, the hill that rises to the west of the city, to admire the view. A little below the summit, a church with a simple baroque facade stands in front of a cave in the limestone of the hill. Here in the 16th century a stalactite encrusted skeleton associated with a cross and twelve beads was discovered. Of this skeleton nothing is certainly known save that it is that of Santa Rosalia, a saint of whom little is reliably reported save that she seems to have lived in the 12th century, that her skeleton was found in this cave, and that she has been the chief patroness of Palermo ever since. Other limestone caverns on Monte Pellegrino had yielded bones of extinct pleistocene Equus, and on the walls of one of the rock shelters at the bottom of the hill there are beautiful Gravettian engravings. Moreover, a small relic of the saint that I saw in the treasury of the Cathedral of Monreale has a venerable and petrified appearance, as might be expected. Nothing in her history being known to the contrary, perhaps for the moment we may take Santa Rosalia as the patroness of evolutionary studies, for just below the sanctuary, fed no doubt by the water that percolates through the limestone cracks of the mountain, and which formed the sacred cave, lies a small artificial pond, and when I could get to the pond a few weeks later, I got from it a hint of what I was looking for.

Vast numbers of Corixidae were living in the water. At first I was rather disappointed because every specimen of the larger of the two species present was a female, and so lacking in most critical diagnostic features, while both sexes of the second slightly smaller species were present in about equal number. Examination of the material at leisure, and of the relevant literature, has convinced me that the two species are the common European *C. punctata* and *C. affinis*, and that the peculiar Mediterranean species are illusionary. The larger *C. punctata* was clearly at the end of its breeding season, the smaller *C. affinis* was probably just beginning to breed. This is the sort of observation that any naturalist can and does make all the time. It was not until I asked myself why the larger species should breed first, and then the more general question as to why there should be two and not 20 or 200 species of the genus in the pond, that ideas suitable to present to you began to emerge. These ideas finally prompted the very general question as to why there are such an enormous number of animal species.

There are at the present time supposed to be (Muller and Campbell, 1954; Hyman, 1955) about one million described species of animals. Of these about three-quarters are insects, of which a quite disproportionately large number are members of a single order, the Coleoptera.[1] The marine fauna although it has at its disposal a much greater area than has the terrestrial, lacks this astonishing diversity (Thorson, 1958). If the insects are excluded, it would seem to be more diverse. The proper answer to my initial question would be to develop a theory at least predicting an order of magnitude for the number of species of 10^6 rather than 10^8 or 10^4. This I certainly cannot do. At most it is merely possible to point out some of the factors which would have to be considered if such a theory was ever to be constructed. . . .

In any study of evolutionary ecology, food relations appear as one of the most important aspects of the system of animate nature. There is quite obviously much more to living communities than the raw dictum "eat or be eaten," but in order to understand the higher intricacies of any ecological system, it is most easy to start from this crudely simple point of view.

FOOD CHAINS

Animal ecologists frequently think in terms of food chains, of the form *individuals of species S_1 are eaten by those of S_2, of S_2 by S_3, of S_3 by S_4*, etc. In such a food chain S_1 will ordinarily be some holophylic organism or material derived from such organisms. The simplest case is that in which we have a true *predator chain* in Odum's convenient terminology, in which the lowest link is a green plant, the next a herbivorous animal, the next a primary carnivore, the next a secondary carnivore, etc. A specially important type of predator chain may be designated Eltonian, because in recent years C. S. Elton has emphasized its widespread significance, in which the predator at each level is large and rarer than its prey. This phenomenon was recognized much earlier, notably by A. R. Wallace in his contribution to the 1858 communication to the Linnean Society of London.

In such a system we can make a theoretical guess of the order of magnitude of the diversity that a single food chain can introduce into a community. If we assume that in general 20 per cent of the energy passing through one link can enter the next link in the chain, which is overgenerous (Slobodkin in an unpublished study finds 13 per cent as a reasonable upper limit) and if we suppose that each predator has twice the mass, (or 1.26 the linear dimensions) of its prey, which is a very low estimate of the size difference between links, the fifth animal link will have a population of one ten thousandth (10−4 of the first, and the fiftieth animal link, if there was one, a population of 10−49 the size of the first. Five animal links are certainly possible, a few fairly clear cut cases having been in fact recorded. If, however, we wanted 50 links, starting with a protozoan or rotifer feeding on algae with a density of 10^6 cells per ml, we should need a volume of 10^{26} cubic kilometers to accommodate on an average one specimen of the ultimate predator, and this is vastly greater than the volume of the world ocean. Clearly the Eltonian food-chain of itself cannot give any great diversity, and the same is almost certainly true of the other types of food chain, based on detritus feeding or on parasitism.

Natural Selection

Before proceeding to a further consideration of diversity, it is, however, desirable to consider the kinds of selective force that may operate on a food chain, for this may limit the possible diversity.

It is reasonably certain that natural selection will tend to maintain the efficiency of transfer from one level to another at a maximum. Any increase in the predatory efficiency of the nth link of a simple food chain will however always increase the possibility of the extermination of the $(n-1)^{th}$ link. If this occurs either the species constituting the nth link must adapt itself to eating the $(n-2)^{th}$ link or itself become extinct. This process will in fact tend to shortening of food chains. A lengthening can presumably occur most simply by the development of a new terminal carnivore link, as its niche is by definition previously empty. In most cases this is not likely to be easy. The evolution of the whale-bone whales, which at least in the case of *Balaenoptera borealis,* can feed largely on copepods and so rank on occasions as primary carnivores, presumably constitutes the most dramatic example of the shortening of a food chain. Mechanical considerations would have prevented the evolution of a larger rarer predator, until man developed essentially non-Eltonian methods of hunting whales.

Effect of Size

A second important limitation of the length of a food chain is due to the fact that ordinarily animals change their size during free life. If the terminal member of a chain were a fish that grew from say one cm to 150 cms in the course of an ordinary life, this size change would set a limit by competition to the possible number of otherwise conceivable links in the 1–150 cm range. At least in fishes this type of process (metaphoetesis) may involve the smaller specimens belonging to links below the larger and the chain length is thus lengthened, though under strong limitations, by cannibalism.

We may next enquire into what determines the number of food chains in a community. In part the answer is clear, though if we cease to be zoologists and become biologists, the answer begs the question. Within certain limits, the number of kinds of primary producers is certainly involved, because many herbivorous animals are somewhat eclectic in their tastes and many more limited by their size or by such structural adaptations for feeding that they have been able to develop.

Effects of Terrestrial Plants

The extraordinary diversity of the terrestrial fauna, which is much greater than that of the marine fauna, is clearly due largely to the diversity provided by terrestrial plants. This diversity is actually two-fold. Firstly, since terrestrial plants compete for light, they have tended to evolve into structures growing into a gaseous medium of negligible buoyancy. This has led to the formation of specialized supporting, photosynthetic, and reproductive structures which inevitably differ in chemical and physical properties. The ancient Danes and Irish are supposed to have eaten elm-bark, and sometimes sawdust, in periods of stress, has been hydrolyzed to produce edible carbohydrate; but usually man, the most omnivorous of all animals, has avoided almost all parts of trees except fruits as sources of food, though various individual species of animals can deal with practically every tissue of many arboreal species. A major source of terrestrial diversity was thus introduced by the evolution of almost 200,000 species of flowering plants, and the three quarters of a million insects supposedly known today are in part a product of that diversity. But of itself merely providing five or ten kinds of food of different consistencies and compositions does not get us much further than the five or ten links of an Eltonian pyramid. On the whole the problem still remains, but in the new form: why are there so many kinds of plants? As a zoologist I do not want to attack that question directly, I want to stick with animals, but also to get the answer. Since, however, the plants are part of the general system of communities, any sufficiently abstract properties of such communities are likely to be relevant to plants as well as to herbivores and carnivores. It is, therefore, by being somewhat abstract, though with concrete zoological details as examples, that I intend to proceed.

INTERRELATIONS OF FOOD CHAINS

Biological communities do not consist of independent food chains, but of food webs, of such a kind that an individual at any level (corresponding to a link in a single chain) can use some but not all of the food provided by species in the levels below it. . . .

MacArthur concludes that in the evolution of a natural community two partly antagonistic processes are occurring. More efficient species will replace less efficient species, but more stable communities will outlast less stable communities. In the process of community formation, the entry of a new species may involve one of three possibilities. It may completely displace an old species. This of itself does not necessarily change the stability, though it may do so if the new species inherently has a more stable population than the old. Secondly, it may occupy an unfilled niche, which may, by providing new partially independent links, increase stability. Thirdly, it may partition a niche with a pre-existing species. Elton in a fascinating work largely devoted to the fate of species accidentally or purposefully introduced by man, concludes that in very diverse communities such introductions are difficult. Early in the history of a community we may suppose many niches will be empty and invasion will proceed easily; as the community becomes more diversified, the process will be progressively more difficult. Sometimes an extremely successful invader may oust a species but add little or nothing to stability, at other times the invader by some specialization will be able to compete successfully for the marginal parts of a niche. In all cases it is probable that invasion is most likely when one or more species happen to be fluctuating and are underrepresented at a given moment. As the communities build up, these opportunities will get progressively rarer. In this way a complex community containing some highly specialized species is constructed asymptotically.

Modern ecological theory therefore appears to answer our initial question at least partially by saying that there is a great diversity of organisms because communities of many diversified organisms are better able to persist than are communities of fewer less diversified organisms. Even though the entry of an invader which takes over part of a niche will lead to the reduction in the *average* population of the species originally present, it will also lead to an increase in stability reducing the risk of the original population being at times underrepresented to a dangerous degree. In this way loss of some niche space may be compensated by reduction in the amplitude of fluctuations in a way that can be advantageous to both species. The process however appears likely to be asymptotic and we have now to consider what sets the asymptote, or in simpler words why are there not more different kinds of animals?

LIMITATION OF DIVERSITY

It is first obvious that the processes of evolution of communities must be under various sorts of external control, and that in some cases such control limits the possible diversity. Several investigators, notably Odum and MacArthur, have pointed out that the more or less cyclical oscillations observed in arctic and boreal fauna may be due in part to the communities not being sufficiently complex to damp out oscillations. It is certain that the fauna of any such region is qualitatively poorer than that of warm temperate and tropical areas of comparable effective precipitation. It is probably considered to be intuitively obvious that this should be so, but on analysis the obviousness tends to disappear. If we can have one or two species of a large family adapted to the rigors of Arctic existence, why can we not have more? It is reasonable to suppose that the total biomass may be involved. If the fundamental productivity of an area is limited by a short growing season to such a degree that the total biomass is less than under more favorable conditions, then the rarer species in a community may be so rare that they do not exist. It is also probable that certain absolute limitations on growth-forms of plants, such as those that make the development of forest impossible above a certain latitude, may in so acting, severely limit the number of niches....

G. Evelyn Hutchinson

NICHE REQUIREMENTS

The various evolutionary tendencies, notably metaphoetesis, which operate on single food chains must operate equally on the food-web, but we also have a new, if comparable, problem as to how much difference between two species at the same level is needed to prevent them from occupying the same niche. Where metric characters are involved we can gain some insight into this extremely important problem by the study of what Brown and Wilson have called *character displacement* or the divergence shown when two partly allopatric species of comparable niche requirements become sympatric in part of their range.

... In the case of the aquatic insects with which I began my address, we have over most of Europe three very closely allied species of Corixa, the largest *punctata,* being about 116 per cent longer than the middle sized species *macrocephala,* and 146 per cent longer than the small species *affinis.* In northwestern Europe there is a fourth species, *C. dentipes,* as large as *C. punctata* and very similar in appearance. A single observation (Brown) suggests that this is what I have elsewhere termed a fugitive species, maintaining itself in the face of competition mainly on account of greater mobility. According to Macan while both *affinis* and *macrocephala* may occur with *punctata* they never are found with each other, so that all three species never occur together. In the eastern part of the range, *macrocephala* drops out, and *punctata* appears to have a discontinuous distribution, being recorded as far east as Simla, but not in southern Persia or Kashmir, where *affinis* occurs. In these eastern localities, where it occurs by itself, *affinis* is larger and darker than in the west, and superficially looks like *macrocephala.*

This case is very interesting because it looks as though character displacement is occurring, but that the size differences between the three species are just not great enough to allow them all to co-occur. Other characters than size are in fact clearly involved in the separation, *macrocephala* preferring deeper water than *affinis* and the latter being more tolerant of brackish conditions. It is also interesting because it calls attention to a marked difference that must occur between hemimetabolous insects with annual life cycles involving relatively long growth periods, and birds or mammals in which the period of growth in length is short and of a very special nature compared with the total life span. In the latter, niche separation may be possible merely through genetic size differences, while in a pair of animals like *C. punctata* and *C. affinis* we need not only a size difference but a seasonal one in reproduction; this is likely to be a rather complicated matter. For the larger of two species always to be larger, it must never breed later than the smaller one. I do not doubt that this is what was happening in the pond on Monte Pellegrino, but have no idea how the difference is achieved....

MOSAIC NATURE OF THE ENVIRONMENT

A final aspect of the limitation of possible diversity, and one that perhaps is of greatest importance, concerns what may be called the mosaic nature of the environment. Except perhaps in open water when only uniform quasi-horizontal surfaces are considered, every area colonized by organisms has some local diversity. The significance of such local diversity depends very largely on the size of the organisms under consideration. In another paper MacArthur and I (Hutchinson and MacArthur, 1959) have attempted a theoretical formulation of this property of living communities and have pointed out that even if we consider only the herbivorous level or only one of the carnivorous levels, there are likely, above a certain lower limit of size, to be more species of small or medium sized organisms than of large organisms. It is difficult to go much beyond crude qualitative impressions in testing this hypothesis, but we find that for mammal faunas, which contain such diverse organisms that they may well be regarded as models of whole faunas, there is a definite hint of the kind of theoretical distribution that we deduce. In qualitative terms the phenomenon can be exemplified by any of the larger species of ungulates which may require a number of different kinds of terrain within their home ranges, any one of which types of terrain might be the habitat of some small species. Most of the genera or even subfamilies of very large terrestrial animals contain only one or two sympatric species. In this connection I cannot refrain from pointing out the immense scientific importance of obtaining a really full insight into the ecology of the large mammals of Africa while they can still be studied under natural conditions. It is indeed quite possible that the results of studies on these wonderful animals would in long-range though purely practical terms pay for the establishment of greater reservations and National Parks than at present exist.

In the passerine birds the occurrence of five or six closely related sympatric species is a commonplace. In the mammal fauna of western Europe no genus appears to contain more than four strictly sympatric species. In Britain this number if not reached even by Mustela with three species, on the adjacent parts of the continent there may be three sympatric shrews of the genus Crocidura and in parts of Holland three of Microtus. In the same general region there are genera of insects containing hundreds of species, as in Athela in the Coleoptera and Dasyhelea in the Diptera Nematocera. The same phenomenon will be encountered whenever any well-studied fauna is considered. Irrespective of their position in a food chain, small size, by permitting animals to become specialized to the conditions offered by small diversified elements of the environmental mosaic, clearly makes possible a degree of diversity quite unknown among groups of larger organisms.

We may, therefore, conclude that the reason why there are so many species of animals is at least partly because a complex trophic organization of a community is more stable than a simple one, but that limits are set by the tendency of food chains to shorten or become blurred, by unfavorable physical factors, by space, by the fineness of possible subdivision of niches, and by those characters of the environmental mosaic which permit a greater diversity of small than of large allied species.

NOTES

G. Evelyn Hutchinson

1. There is a story, possibly apocryphal, of the distinguished British biologist, J. B. S. Haldane, who found himself in the company of a group of theologians. On being asked what one could conclude as to the nature of the Creator from a study of his creation, Haldane is said to have answered, "An inordinate fondness for beetles."

3.2 EUGENE P. ODUM

Great Ideas in Ecology for the 1990s

Eugene P. Odum was very influential in elevating ecology from a subtopic within biology—as it was generally taught in courses before the 1950s—to an important integrative biological discipline. His textbook *Fundamentals of Ecology*, published in 1953, became a widely used standard in introductory ecology courses. Odum's research on birds and wetland ecology at the University of Georgia and at the Woods Hole Marine Biological Laboratory in Falmouth, Massachusetts, gave him deep insight into the interdependence among microorganisms, plants, herbivores, and carnivores.

In the following selection from "Great Ideas in Ecology for the 1990s," *Bioscience* (July/August 1992), Odum considers the important question of designing a basic ecology course, with the goal of improving the environmental literacy of undergraduate students. he presents descriptions of 20 "great ideas" in ecology that he suggests should be included in such a course.

Key Concept: basic ecological concepts

In a commentary entitled "Science literacy" (Pool 1991), there is a table of "Science's top 20 greatest hits" as chosen by biologist Robert Hazen and physicist James Trefil. They suggested that these "great ideas" might be the basis for a course in general science, and comments from readers were invited.

In addition to the two laws of thermodynamics, Hazen and Trefil's list includes three other concepts that could be construed as ecological. These concepts are "everything on earth operates in cycles," "all forms of life evolved by natural selection," and "all life is connected."

For many years, I have contended that ecology is no longer a subdivision of biology but has emerged from its roots in biology to become a separate discipline that integrates organisms, the physical environment, and humans—in line with *oikos*, root of the word *ecology* (Odum 1977). From this view, the ecosystem level becomes the major focus. Populations are considered as ecosystem components and landscapes as associations of interacting ecosystems. This viewpoint is now generally accepted, as was indicated by a recent British Ecological Society survey in which members were asked to list what they considered the most important ecological concepts. The ecosystem was the concept most frequently listed (Cherrett 1989).

At the time that the science literacy article appeared, I was drawing up a list of basic concepts in ecology that might be included in courses designed to improve environmental literacy among undergraduates here at the University of Georgia. Here are 20 of my "great ideas" in ecology, as distinguished from "great ideas" in biology (e.g., DNA, genetic code, and general theory of natural selection). The last five items in my list relate to human ecology and the ecology-economics interface, which must be major foci in environmental literacy education in view of the increasingly serious global impacts resulting from human activities. The references I have selected for each concept may not be the best ones, and certainly they are not the only ones.

Concept 1. An ecosystem is a thermodynamically open, far from equilibrium, system. Input and output environments are an essential part of this concept. For example, in considering a forest tract, what is coming in and going out is as important as what is inside the tract. The same holds for a city. It is not a self-contained unit ecologically or economically; its future depends as much on the external life-support environment as on activities within city limits (Odum 1983, Patton 1972, Prigogine et al. 1972).

Concept. 2. The source-sink concept: one area or population (the source) exports to another area or population (the sink). This statement is a corollary to concept 1. It is applicable at ecosystem as well as population levels. At the ecosystem level, an area of high productivity (salt marsh, for example) may feed an area of low productivity (adjacent coastal waters). At the population level, a species in one area may have a higher reproduction rate than needed to sustain the population, and surplus individuals may provide recruitment for an adjacent area of low reproduction. Food chains may also involve sources and sinks (see concept 12; Lewin 1989, Pulliam 1988).

Concept 3. In hierarchical organization of ecosystems, species interactions that tend to be unstable, nonequilibrium, or even chaotic are constrained by the slower interactions that characterize large systems. Short-term interactions, such as interspecific competition—the evolutionary arms race between a parasite and its host, herbivore-plant interactions, and predator-prey activities—tend to be oscillatory or cyclic. Large, complex systems—such as oceans, the atmosphere, soils, and large forests—tend to go from randomness to order and will tend to have more steady-state characteristics, for example, the atmosphere's gaseous balances.

Accordingly, large ecosystems tend to be more homeostatic than their components. This principle may be the most important of all, because it warns that what is true at one level may or may not be true at another level of organization. Also, if we are serious about sustainability, we must raise our focus in management and planning to large landscapes and beyond (Allen & Starr 1982, Kauffman 1990, O'Neill et al. 1986, Prigogine and Stengers 1984, Ulanowicz 1986).

Concept 4. The first signs of environmental stress usually occur at the population level, affecting especially sensitive species. If there is sufficient redundancy, other species may fill the functional niche occupied by the sensitive species. Even so, this early warning should not be ignored, because the backup components may not be as efficient. When the stress produces detectable ecosystem-level effects, the health and survival of the whole system is in jeopardy. This idea is a corollary of item 3: parts are less stable than wholes (Odum 1985, 1990, Schindler 1990).

Concept 5. Feedback in an ecosystem is internal and has no fixed goal. There are no thermostats, chemostats, or other set-point controls in the biosphere. Cybernetics at the ecosystem level thus differs from that at the organism level (body temperature control, for example) or that of human-made mechanical systems (temperature control of a building, for example) where the control is external with a set point. Ecosystem control, where manifested, is the result of a network of internal feedback processes as yet little understood—another corollary of concept 3 (Patten and Odum 1981).

Concept 6. Natural selection may occur at more than one level. This idea is another corollary to concept 3. Accordingly, coevolution, group selection, and traditional Darwinism are all part of the hierarchical theory of evolution. Not only is the evolution of a species affected by the evolution of interacting species, but a species that benefits its community has survival value greater than a species that does not (Axelrod 1984, 1980, Axelrod and Hamilton 1981, Gould 1982, Wilson 1976, 1980).

Concept 7. There are two kinds of natural selection, or two aspects of the struggle for existence: organism versus organism, which leads to competition, and organism versus environment, which leads to mutualism. To survive, an organism does not compete with its environment as it might with another organism, but it must adapt to or modify its environment and its community in a cooperative manner. This concept was first suggested by Peter Kropotkin soon after Darwin (Gould 1988, Kropotkin 1902).

Concept 8. Competition may lead to diversity rather than to extinction. Although competition plays a major role in shaping the species composition of biotic communities, competition exclusion (in which one species eliminates another, as in a flour beetle microcosm) is probably the exception rather than the rule in the open systems of nature. There, species are often able to shift their functional niches to avoid the deleterious effects of competition (den Boer 1986).

Concept 9. Evolution of mutualism increases when resources become scarce. Cooperation between species for mutual benefit has special survival value when resources become tied up in the biomass, as in mature forests, or when the soil or water is nutrient poor, as in some coral reefs or rainforests (Boucher et al. 1982, Odum and Biever 1984). The recent shift from confrontation to cooperation among the world's superpower nations may be a parallel in societal evolution (Kolodziej 1991).

Concept 10. Indirect effects may be as important as direct interactions in a food web and may contribute to network mutualism. When food chains function in food web networks, organisms at each end of a trophic series (for example, plankton and bass in a pond) do not interact directly but indirectly benefit each other.

Bass benefit by eating planktiverous fish supported by the plankton, whereas plankton benefit when bass reduce the population of its predators. Accordingly, there are both negative (predator-prey) and positive (mutualistic) interactions in a food web network (Patton 1991, Wilson 1986).

Concept 11. Since the beginning of life on Earth, organisms have not only adapted to physical conditions but have modified the environment in ways that have proven to be beneficial to life in general (e.g., increase O_2 and reduce CO_2). This modified Gaia hypothesis is now accepted by many scientists. Especially important is the theory that microorganisms play major roles in vital nutrient cycles (especially the nitrogen cycle) and in atmospheric and oceanic homeostasis (Cloud 1988, Lovelock 1979, 1988, Kerr 1988, Margulis and Olendzenski 1991).

Concept 12. Heterotrophs may control energy flow in food webs. For example, in warm waters, bacteria may function as a sink in that they short-circuit energy flow so that less energy reaches the ocean bottom to support demersal fisheries. In cooler waters, bacteria are less active, allowing more of the fruits of primary production to reach the bottom (Pomeroy 1974, Pomeroy and Deibel 1986, Pomeroy and Wiebe 1988). Small hetertrophs may play similar controlling roles in terrestrial ecosystems such as grasslands (Dyer et al. 1982, 1986, Seastadt and Crossley 1984). This concept is a corollary of concept 11.

Concept 13. An expanded approach to biodiversity should include genetic and landscape diversity, not just species diversity. The focus on preserving biodiversity must be at the landscape level, because the variety of species in any region depends on the size, variety, and dynamics of patches (ecosystems) and corridors (Odum 1982, Turner 1988, Wilson 1988).

Concept 14. Ecosystem development or autogenic ecological succession is a two-phase process. Early or pioneer stages tend to be stochastic as opportunistic species colonize, but later stages tend to be more self-organized (perhaps another corollary of concept 3; Odum 1989a).

Concept 15. Carrying capacity is a two-dimensional concept involving number of users and intensity of per capita use. These characteristics track in a reciprocal manner —as the intensity of per capita impact goes up, the number of individuals that can be supported by a given resource base goes down (Catton 1987). Recognition of this principle is important in estimating human carrying capacity at different quality-of-life levels and in determining how much buffer natural environment to set aside in land-use planning.

Concept 16. Input management is the only way to deal with nonpoint pollution. Reducing waste in developed countries by source reduction of the pollutants will not only reduce global-scale pollution but will spare resources needed to improve quality of life in undeveloped countries (Odum 1987, 1989b).

Concept 17. An expenditure of energy is always required to produce or maintain an energy flow or a material cycle. According to this net-energy concept, communities and systems, whether natural or human-made, as they become larger and more complex, require more of the available energy for maintenance (the so-called complexity theory). For example, when a city doubles in size, more than double the energy (and taxes) is required to maintain order (Odum and Odum 1981, Pippenger 1978).

Concept 18. There is an urgent need to bridge the gaps between human-made and natural life-support goods and services and between nonsustainable short-term and sustainable long-term management. Agroecosystems, tropical forests, and cities are of special concern. H. T. Odum's "emergy" concept and Daly and Cobb's index of sustainable economic welfare are examples of recent attempts to bridge these gaps (Daly and Cobb 1989, Folke and Kaberger 1991, Holden 1990, Odum 1988).

Concept 19. Transition costs are always associated with major changes in nature and in human affairs. Society has to decide who pays, for example, the cost of new equipment, procedures, and education in changing from high-input to low-input farming or in converting from air polluting to clean power plants (Renner 1991, Spencer et al. 1986).

Concept 20. A parasite-host model for man and the biosphere is a basis for turning from exploiting the earth to taking care of it (going from dominionship to stewardship, to use a biblical metaphor). Despite, or perhaps because of, technological achievements, humans remain parasitic on the biosphere for life support. Survival of a parasite depends on reducing virulence and establishing reward feedback that benefits the host (Alexander 1981, Anderson and May 1981, 1982, Levin and Pimentel 1981, Pimentel 1968, Pimentel and Stone 1968, Washburn et. al. 1991). Similar relationships hold for herbivory and predation (Dyer et al. 1986, Lewin 1989, Owen and Wiegert 1976). In terms of human affairs, this concept involves reducing wastes and destruction of resources to reduce human virulence, promote the sustainability of renewable resources, and invest more in Earth care.

CONCEPT COMPARISONS

In my list, I have covered the ecological items in Hazen and Trefil's "great ideas in science." Thermodynamics is represented in concept 1, natural selection in concepts 6 and 7 (and indirectly in most others), and cyclic behavior and connectiveness in concept 3.

The British Ecological Society survey listed approximately 50 wide-ranging items. The editor (Cherrett 1989) divided the 600 or so responding ecologists into two groups: practical holists and theoretical reductionists. I do not believe these dichotomies between holism and reductionism and that between theoretical and practical are very helpful. The most exciting of the concepts listed apply to all levels or to the interaction of levels, and any and all may have practical as well as theoretical aspects. I believe that my more comprehensive 20 concepts cover most of what is in the 50-item list.

I am sure that there are other concepts that might be added to my list, and I suspect that some of my choices may be considered by some ecologists as too hypothetical to have been included. I invite comments to further explore what concepts are most important to public knowledge of ecology (i.e., environmental literacy).

Alexander, M. 1981. Why microbial parasites and predators do not eliminate their prey and hosts. *Annu. Rev. Microbiol.* 35: 113–133.

Allen, T. F. H., and T. B. Starr, 1982. *Hierarchy: Perspectives for Ecological Complexity.* University of Chicago Press, Chicago.

Anderson, R. M., and R. M. May. 1981. The population dynamics of microparasites and their invertebrate hosts. *Philos. Trans. R. Soc. Lond. Biol. Sci. B* 291: 451–524.

_____. 1982. Coevolution of hosts and parasites, *Parasitology* 85: 411–426.

Axelrod, R. 1984. *Evolution of Cooperation.* Basic Books, New York.

Axelrod, R., and W. D. Hamilton. 1981. The evolution of cooperation. *Science* 211: 1390–1396.

Boucher, D. S., S. James, and K. H. Keeler. 1982. The ecology of mutualism. *Annu. Rev. Ecol. Syst.* 13: 315–347.

Catton, W. R. 1987. The world's most polymorphic species: carrying capacity transgressed two ways. *BioScience* 37: 413–419.

Cherrett, J. M., ed. 1989. *Ecological Concepts: The Contribution of Ecology to an Understanding of the Natural World.* Blackwell Scientific Publ., London.

Cloud, P. E. 1988. Gaia modified. *Science* 240: 1716.

Daly, H. E., and B. J. Cobb. 1989. *For the Common Good: Redirecting the Economy Towards Community, the Environment, and a Sustainable Future.* Beacon Press, Boston.

den Boer, P. J. 1986. The present status of the competition exclusion principle. *Trends Ecol. Evol.* 1: 25–28.

Dyer, M. I., D. L. DeAngelis, W. M. Post. 1986. A model of herbivore feedback in plant productivity. *Math Biosci.* 79: 171–184.

Dyer, M. I., J. K. Detling, J. K. Coleman, and D. W. Hilbert. 1982. The role of herbivores in grasslands. Pages 255–295 in J. R. Estes, R. J. Tyri, and J. N. Brunken, eds. *Grasses and Grasslands,* University of Oklahoma Press, Norman.

Folke, C., and T. Kaberger. 1991. *Linking the Natural Environment and the Economy: Essays from the Eco-Eco Group.* Kluvier Academic Publ., Boston.

Gould, S. J. 1982, Darwinism and the expansion of evolutionary theory. *Science* 216: 380–387.

_____. 1988. Kropotkin was no crackpot. *Nat. Hist.* 97(7): 12–21.

Holden, C. 1990. Multidisciplinary look at a finite world. *Science* 249: 18–19.

Kauffman, S. 1990. Spontaneous order, evolution and life. *Science* 247: 1543–1544.

Kerr, R. A. 1988. No longer willful, Gaia becomes respectable. *Science* 240: 393–395.

Kolodziej, E. A. 1991. The cold war as cooperation. *Bulletin of the American Academy of Arts and Sciences* 44(7): 9–39.

Kropotkin, P. 1902. *Mutual Aid: A Factor of Evolution.* William Heinmann, London. Reprinted 1935, Extending Horizon Books, Boston.

Levin, S., and D. Pimentel. 1981. Selection of intermediate rates of increase in parasite-host systems. *Am. Nat.* 117: 308–315.

Lewin, R. 1989. Sources and sinks complicate ecology. *Science* 243: 477–478.

Lovelock, J. E. 1979. *Gaia: A New Look at Life on Earth.* Oxford University Press, New York.

_____. 1988. *The Ages of Gaia.* W. W. Norton, New York.

Margulis, L., and L. Olendzenski. 1991. *Environmental Evolution.* MIT Press, Cambridge, MA.

Odum, E. P. 1977. The emergence of ecology as a new integrative science. *Science* 195: 1289–1293.

———. 1982. Diversity and the forest ecosystem. Pages 35–41 in J. L. and J. H. Cooley, eds. *Proceedings of a Workshop in Natural Diversity in Forest Ecosystems.* US Forest Service, SE Forest Experiment Station, Athens, GA.

———. 1983. *Basic Ecology.* Saunders Publ., Philadelphia.

———. 1985. Trends expected in stressed ecosystem. *BioScience* 35: 419–422.

———. 1987. Reduced-input agriculture reduces nonpoint pollution. *J. Soil Water Conserv.* 42: 412–414.

———. 1989a. *Ecology and Our Endangered Life-support Systems.* Sinauer Assoc., Sunderland, MA.

———. 1989b. Input management of production systems. *Science* 243: 177–182.

———. 1990. Field experimental tests of ecosystem-level hypotheses. *Trends Ecol. Evol.* 5: 204–205.

Odum, E. P., and L. J. Biever. 1984. Resource quality, mutualism and energy partitioning in food chains. *Am. Nat.* 124: 360–376.

Odum, H. T. 1988. Self-organization, transformity and information. *Science* 242: 1132–1139.

Odum, H. T., and E. C. Odum. 1981. *Energy Basis for Man and Nature.* McGraw-Hill, New York.

O'Neill, R. V., D. L. DeAngelis, J. B. Waide, and T. F. H. Allen. 1986. *A Hierarchical Concept of Ecosystems.* Princeton University Press, Princeton, NJ.

Owen, D. F., and R. G. Weigert. 1976. Do consumers maximize plant fitness? *Oikos* 27: 489–492.

Patten, B. C. 1978. Systems approach to the concept of environment. *Ohio J. Sci.* 78: 206–222.

———. 1991. Network ecology: indirect determination of the life-environment relationship in ecosystems. Pages 288–351 in M. Higashi and T. P. Burns, eds. *Theoretical Studies of Ecosystems: The Network Perspective.* Cambridge University Press, New York.

Patten, B. C., and E. P. Odum. 1981. The cybernetic nature of ecosystems. *Am. Nat.* 118: 886–895.

Pimentel, D. 1968. Population regulation and genetic feedback. *Science* 159: 1432–1437.

Pimentel, D., and F. A. Stone. 1968. Evolution and population ecology of parasite-host systems. *Can. Entomol.* 100: 655–662.

Pippenger, N. 1978. Complexity theory. *Sci. Am.* 238(6): 114–124.

Pomeroy, L. R. 1974. The ocean's food web: a changing paradigm. *BioScience* 24: 499–504.

Pomeroy, L. R., and D. Deibel. 1986. Temperature regulation of bacterial activity during the spring bloom in Newfoundland coastal waters. *Science* 233: 359–361.

Pomeroy, L. R., and W. J. Wiebe. 1988. Energetics of microbial food webs. *Hydrobiologia* 159: 7–18.

Pool, R. 1991. Science literacy: the enemy is us. *Science* 251: 266–267.

Prigogine, I., F. Nicoles, and A. Babloyantz. 1972. Thermodynamics and evolution. *Physics Today* 25(11): 23–38; 25(12): 138–142.

Prigogine, I., and I. Stengers. 1984. *Order Out of Chaos: Man's New Dialogue with Nature,* Bantam, New York.

Pulliam, H. R. 1988. Sources, sinks and population regulation. *Am. Nat.* 132: 652–661.

Renner, M. 1991. *Jobs in a Sustainable Economy.* Worldwatch paper 104, Worldwatch Institute, Washington, DC.

Schindler, D. W. 1990. Experimental perturbations of whole lakes as tests of hypotheses concerning ecosystem structure and function. *Oikos* 57: 25–41.

Seastadt, T. R., and D. A. Crossley. 1984. The influence of anthropods on ecosystems. *BioScience* 34: 157–161.

Spencer, D. F., S. B. Alpert, and H. H. Gilman. 1986. Cool water: demonstration of a clean and efficient new coal technology. *Science* 232: 609–612.

Turner, M. G. 1988. Landscape ecology: the effect of pattern on process. *Annu. Rev. Ecol. Syst.* 20: 171–197.

Ulanowicz, R. E. 1986. *Growth and Development: Ecosystem Phenomenology.* Springer-Verlag, New York.

Washburn, J. O., D. R. Mercer, and J. R. Anderson. 1991. Regulatory role of parasites: impact on host population shifts with resource availability. *Science* 253: 185–188.

Wilson, D. S. 1976. Evolution on the level of communities. *Science* 192: 1358–1360.

____. 1980. *The Natural Selection of Populations and Communities.* Benjamin Cummings, Menlo Park, CA.

____. 1986. Adaptive indirect effects. Pages 437–444 in J. Diemaon and T. J. Case. eds. *Community Ecology.* Harper and Row, New York.

Wilson, E. O., ed. 1988. *Biodiversity.* National Academy Press, Washington, DC.

CHAPTER 4 The Hydrosphere and the Geosphere

4.1 JOHN TEAL AND MILDRED TEAL

Life and Death of the Salt Marsh

John and Mildred Teal began their intensive study of North American coastal wetlands at the Sapelo Island Marine Institute of the University of Georgia in 1955. They credit Dalhousie University in Halifax, Nova Scotia, Canada, with providing them the opportunity to explore the northern marshes of the Maritimes—the Canadian provinces of New Brunswick, Nova Scotia, and Prince Edward Island. The observations they made during these investigations were subsequently analyzed and systematized with the help of colleagues at the Woods Hole Oceanographic Institute in Falmouth, Massachusetts. This research led to a deep understanding of the flora and fauna that are supported by the salt marshes, their overall ecological importance, and their sensitivity to human intrusion.

The essence of the Teals' findings is described in their classic book *Life and Death of the Salt Marsh* (Ballantine Books, 1969), from which the following selection is taken. It is one of the most frequently referenced works in environmental and ecological courses that include discussions of wetland ecosystems. Through their work, the Teals provided much of the foundation for the present efforts of environmental activists and grassroots organizations to convince elected officials of the urgent need for wetland protection.

Key Concept: the ecological significance of salt marshes

INTRODUCTION

Along the eastern coast of North America, from the north where ice packs grate upon the shore to the tropical mangrove swamps tenaciously holding the land together with a tangle of roots, lies a green ribbon of soft, salty, wet, low-lying land, the salt marshes.

The ribbon of green marshes, part solid land, part mobile water, has a definite but elusive border, now hidden, now exposed, as the tides of the Atlantic fluctuate. At one place and tide there is a line at which you can say, "Here begins the marsh." At another tide, the line, the "beginning of the marsh," is completely inundated and looks as though it had become part of the sea. The marsh reaches as far inland as the tides can creep and as far into the sea as marsh plants can find a roothold and live in saline waters.

The undisturbed salt marshes offer the inland visitor a series of unusual perceptions. At low tide, the wind blowing across *Spartina* grass sounds like wind on the prairie. When the tide is in, the gentle music of moving water is added to the prairie rustle. There are sounds of birds living in the marshes. The marsh wren advertises his presence with a reedy call, even at night, when most birds are still. The marsh hen, or clapper rail, calls in a loud, carrying cackle. You can hear the tiny, high-pitched rustling thunder of the herds of crabs moving through the grass as they flee before advancing feet or the more leisurely sound of movement they make on their daily migrations in search of food. At night, when the air is still and other sounds are quieted, an attentive listener can hear the bubbling of air from the sandy soil as a high tide floods the marsh.

The wetlands are filled with smells. They smell of the sea and salt water and of the edge of the sea, the sea with a little iodine and trace of dead life. The marshes smell of *Spartina*, a fairly strong odor mixed from the elements of sea and the smells of grasses. These are clean, fresh smells, smells that are pleasing to one who lives by the sea but strange and not altogether pleasant to one who has always lived inland.

Unfortunately, in marshes which have been disturbed, dug up, suffocated with loads of trash and fill, poisoned and eroded with the wastes from large cities, there is another smell. Sick marshes smell of hydrogen sulfide, a rotten egg odor. This odor is very faint in a healthy marsh.

As the sound and smell of the salt marsh are its own, so is its feel. Some of the marshes can be walked on, especially the landward parts. In the north, the *Spartina patens* marsh is covered with dense grass that may be cut for salt hay. Its roots bind the wet mud into a firm surface. But the footing is spongy on an unused hay marsh as the mat of other years' grass, hidden under the green growth, resists the walker's weight and springs back as he moves along.

In the southern marshes, only one grass covers the entire marsh area, *Spartina alterniflora*. On the higher parts of the marsh, near the land, the roots have developed into a mass that provides firm footing although the plants are much more separated than in the northern hay marshes and you squish gently on mud rather than grass. It is like walking on a huge trampoline. The ground is stiff. It is squishy and wet, to be sure, but still solid as you walk about. However, jump and you can feel the ground give under the impact and waves spread out in all directions. The ground is a mat of plant roots and mud on top of a more liquid layer underneath which gives slightly by flowing to all sides when you jump down on it.

As you walk toward the edge of the marsh, the seaward edge, each step closer to open water brings a change in footing. The mud has less root material in it and is less firmly bound together. It begins to ooze around your shoes. On the edges of the creeks, especially the larger ones, there may be natural levees where the ground is higher. Here the rising tide meets its first real resistance as it spills over the creek banks and has to flow between the close-set plants. Here it is slowed and drops the mud it may be carrying. Here too, especially after a series of tides, lower than usual, the ground is firm and even dry and hard.

Down toward the creek, where the mud is watered at each tide, the soil is as muddy as you can find anywhere. When you try to walk across to the water at low tide, across the exposed mud where the marsh grass does not yet grow, hip boots are not high enough to keep you from getting muddy. The boots are pulled off on the first or second step when they have sunk deep into the clutching zone. There are no roots to give solidarity, nothing but the mud and water fighting a shifting battle to hold the area.

At low tide the salt marsh is a vast field of grasses with slightly higher grasses sticking up along the creeks and uniformly tall grass elsewhere. The effect is like that of a great flat meadow. At high tide, the look is the same, a wide flat sea of grass but with a great deal of water showing. The marsh is still marsh, but spears of grass are sticking up through water, a world of water where land was before, each blade of grass a little island, each island a refuge for the marsh animals which do not like or cannot stand submersion in salt water....

SOLUTIONS AND SUGGESTIONS

The dangers to salt marshes stem from human activities, not natural processes. We destroy wetlands and shallow water bottoms directly by dredging, filling, and building. Indirectly we destroy them by pollution. Much of this destruction is simply foolish. The marsh would often have been much more valuable as a marsh than it is in its subsequent desecrated form.

The increase in population pressure along the coast will inevitably destroy more and more of the frail marsh estuarine system. We do not propose the preservation of the marshes simply for the sake of their preservation. Instead, we regard them in light of their benefit to the growing population. The benefit of marshes will accrue to everyone, not only those who venture onto the surface of marshes but to fishermen along the coast and to consumers of fishery products who may live far inland.

Some destruction is inevitable. Even for those marshes preserved as wildlife areas, an access must be constructed so that people who want to enjoy these pieces of nature can do so. Roads must be built to the marshes, along the edges of marshes, and to impoundments that are designed for mosquito control and waterfowl hunting. Also, building roads to boat-launching ramps so that the network of creeks and rivers in the wetlands can be enjoyed is not only a convenience but a preservative: damage to the marshes will be less if adequate access is provided from the waterside.

But after having conceded that we cannot avoid destroying some marshes, how are we to decide which should be destroyed and which preserved? And by what means shall we preserve them? It is obvious that overall planning is necessary. The very minimum of planning could be approached on the state level, but a more rational approach demands planning on the national level, as it is the whole marsh system with its high productivity, rather than individual marshes, that needs preservation. Overall planning demands that we have a classification of the value and importance of every area of marsh along the coast. . . .

Whatever method is used to preserve marshes, it must include safeguards against the increased pressures to develop because of the ever increasing population. There have been too many cases in which the last land in the town, land reserved for park and playground, has been diverted to industrial use. The diversion occurred because the industry threatened to move to another town or even another state if it were not allowed to secure the land. This sort of corporate blackmail is hard to withstand and will inevitably bring pressures on organizations controlling the marshes.

Pressure even comes from the state officials who are trying to encourage industries to come to their area by offering filled marsh for building. The battle between the forces of development and conservation need be won only once by the developers but must be fought and won every year for conservation to triumph.

In the last hundred years, our nation has been called on to preserve certain unique natural resources such as those contained in many of our national parks: Grand Canyon, Yosemite Valley, the Yellowstone hot springs, Mammoth Cave, the Everglades, and the Cape Cod seashore. Now we are confronted with the problem of preserving a different sort of natural resource. This resource is much more extensive—the ribbon of green marshes along the eastern coast of North America, which must be preserved almost in its entirety if its preservation is to have any real meaning.

Geologists, Engineers, and a Rising Sea Level

Orrin H. Pilkey (b. 1934) is the James B. Duke Professor of Geology and director of the Program for the Study of Developed Shorelines at Duke University. He also serves on the Committee on Beach Nourishment and Protection of the National Academy of Sciences' National Research Council. A winner of both the Society for Sedimentary Geology's Francis Shepard Medal for excellence in marine geology (1987) and the American Geological Institute's award for outstanding contributions to public understanding of geology (1993), Pilkey has earned a reputation as one of the most outspoken critics of efforts to manage erosion along America's coastline.

In the following selection taken from "Geologists, Engineers, and a Rising Sea Level," *Northeastern Geology,* (vol. 3, nos. 3/4, 1981), Pilkey focuses on attempts to stabilize rapidly eroding barrier beaches, although he maintains that his analysis applies to any sandy shoreline. Pilkey's sardonic, but serious review of the methods employed by the U.S. Army Corps of Engineers leads him to conclude that shoreline engineering is not only ineffective in the long run, but that in the short run it is a contributing factor to beach destruction. In his view, much of the Corps' work involves the use of taxpayers' money to offer temporary protection to coastal property owners at a price that is far in excess of the worth of the property, and at the additional environmental cost of promoting the destruction of public beachfront. He suggests that if the American public were properly educated they would keep engineers away from ocean shoreline.

Pilkey, along with coastal policy reform activist Katharine L. Dixon, has expanded on these arguments and analyses in their recently published book *The Corps and the Shore* (Island Press, 1996).

Key Concept: most engineering efforts to stabilize coastal beaches are ecologically counterproductive

INTRODUCTION

If an engineer is asked for his opinion on an open ocean shoreline erosion problem, he will suggest a solution with alacrity. With equal alacrity, a geologist will point out that there is no solution. Both are right. The engineer is right in a 20 to 30 year sense, and the geologist in a 50 to 150 year sense. No wonder that most pragmatic politicians' responses to a shoreline geologist's suggestion, is, "Who needs geologists?"

Most stretches of the American shoreline are eroding. Erosion is particularly widespread along our barrier island coasts. Many natural factors are involved in causing erosion, among which are sand supply, storms, changes in wave regime, and changes in bottom topography. Along developed shores man often becomes the principal cause of erosion. Sea walls, jetties, and groins all reduce shoreline flexibility and ultimately cause problems.

Underlying and reinforcing all causes of shoreline erosion is the sea level rise. Both the present rapid rates of shoreline retreat and the wide occurrence of such retreat can be laid at the doorstep of the sea level rise. Unfortunately none of the various shoreline stabilization schemes used by engineers responds to the sea level rise, the major cause of the problem they seek to alleviate. The long range result of treating the symptoms instead of the cause is destruction of the beach.

Understanding the disagreement between geologists and engineers is important because the future of the American recreational shoreline is at stake.

In the following discussion emphasis is placed on the effect of engineering on open ocean barrier island beaches. This is because most of the U.S. Gulf and Atlantic open ocean shoreline is on barrier islands and also because these systems, compared to rocky mainland shores, are more uniform and perhaps less complex. Nonetheless, most of the principles developed apply to any type of sandy shoreline.

THE SEA LEVEL RISE AND BARRIER ISLANDS AND BEACHES

Evidence from the continental shelf developed by Swift (1975) and Duane and Field (1976) indicates that barrier islands fronting all of the American coastal plain coasts probably formed at the shelf edge. After formation the islands migrated to their present position in response to the Holocene sea level rise. During their period of cross shelf migration the barrier islands must have been skinny (Table 1), judging from the fact that barrier islands today that are migrating rapidly are very skinny (e.g., Capes Island, South Carolina). It does not follow, however, that all skinny islands are migrating rapidly.

As the sea level rise slowed down, 4 to 5,000 years ago, many barrier islands began to fatten or grow seaward, e.g., Bogue Banks, North Carolina; Galveston Island, Texas. Fattening occurred on islands with large sand supplies. With few exceptions, the fattening of American barriers halted a few

TABLE 1

A Sketch History of Barrier Islands

Origin–15,000 years ago–
  Islands formed at shelf edge.
Skinny island phase–15,000–4,000 years ago–
  Islands migrated or hopped across the shelf in response to the Holocene sea level rise.
Fat island phase–4,000–300 years ago–
  Many islands grew seaward in response to the slowdown in the sea level rise.
Slimming down phase–300 years ago to present–
  Islands are slimming down in response to a sea level rise. The rise has accelerated within the last 50 years. Slimming prepares islands to migrate once more.

hundred years ago to be replaced by frontside retreat, probably in response to an increased rate of sea level rise once again.

Tide gauge studies by Hicks (1972) indicate that 50 or so years ago sea level rise rate took a major leap. It is now eustatically rising at a rate of close to one foot per century. Whether this sea level rise is a momentary "blip" on the sea level curve or the start of a major long-continued rise is unknown. Even the cause of the 50-year jump in rise rate remains a mystery; is the ice melting, is the melting due to the greenhouse effect or is a mid ocean ridge bulging?

The net result of the recent sea level rise is to cause erosion on both sides of many American islands. From a broad viewpoint this erosion can be seen as a "slimming down" of the islands, preparatory to continued landward migration, assuming that the sea level rise continues unabated. Rapidly migrating islands must widen themselves on their backsides over a broad front (Leatherman, 1979). The only way to do this is to allow complete cross-island overwash along a broad front; hence the need to slim down.

The evidence (other than tide gauge records) of a eustatic sea level rise is best seen on barrier islands. The most pertinent evidence is the fact that almost every "old" community on the front side of a barrier island is in trouble. If an "old" town has no frontside problems, it inevitably means that no one built on the frontside or more likely that a block or two of houses has been allowed to peacefully fall in when their time came. In other words, the community kept engineers off their beaches.

Other lines of evidence pointing to a eustatic sea level rise are the widespread direct (aerial photos, charts) and indirect (stumps, saltmarsh peat on beaches) indicators of frontside erosion. Recently on Whale Beach, New Jersey, marsh mud containing cow hooves and colonial implements appeared after a northeaster (Norbert Psuty, pers. comm.). This indicated a migration of

FIGURE 1

The Effect of Sea Level Rise on Shoreline Retreat

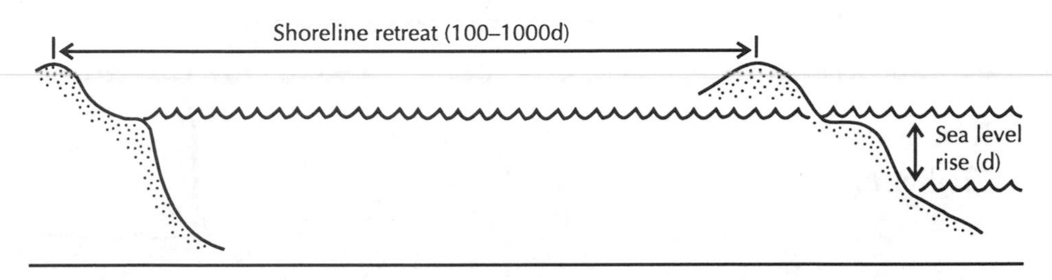

at least one island width since a colonist dumped his garbage in the salt marsh. In North Carolina widespread killing of certain tree species with low salt tolerance along the fringes of Pamlico and Albemarle Sounds reflects flooding by a rising sea.

Because of all the factors involved, it is difficult to directly relate sea level rise with shoreline erosion rate over a time span of a few years. However, Rosen (1978) states that the sea level rise accounts for all shore retreat in the Virginia Chesapeake Bay shoreline. Morton (1979) finds a very general correspondence between Texas shoreline movement and sea level change.

Two additional important points must be made regarding the relationship between the sea level rise and shoreline erosion. First of all, the rate of retreat should be roughly a function of the slope of the land surface over which the sea is flooding. Steep slopes produce slow migration rates; gentle coastal plain surfaces should produce a shoreline retreat 100 to 1000 times the sea level rise (Fig. 1). A one foot rise should cause a retreat of 100 to 1000 feet. The second point is that the retreat of a beach is not just a surf zone retreat. Instead the entire shoreface probably moves landward as suggested in Swift's model of continental shelf evolution. Bruun (1962) proposed that a beach moves up and back and retains its same surf zone profile as the sea level rises. This is not in conflict with Swift's ideas. Swift has simply looked at shoreline movement on a larger scale (Fig. 2).

WHAT ENGINEERS DO

To begin with, engineers respond to a perceived public need. They respond to a public outcry to "save the beach" or "save our cottages." Frequently the former phrase is used when the latter is really meant. In the long run abundant evidence indicates that engineering destroys open ocean beaches. However,

FIGURE 2

Comparison of the Bruun Rule and the Swift Model of Shoreface Response Elements to the Sea Level Rise

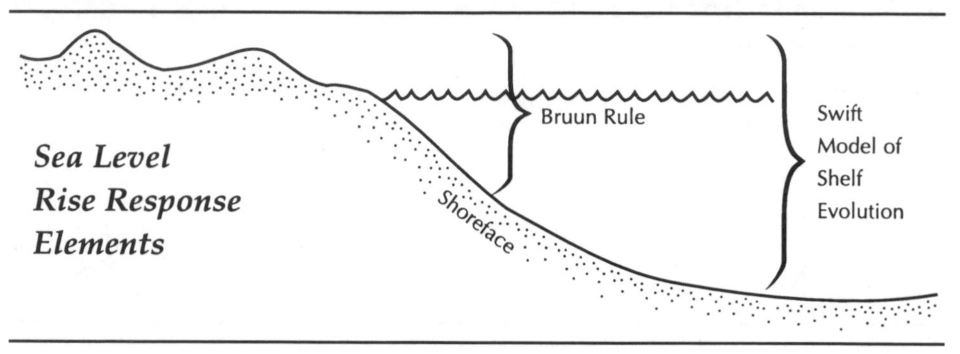

TABLE 2

The Truths of the Shoreline

I. There is no shoreline erosion problem until someone builds something on the beach to measure it by.
II. Construction on the beach reduces flexibility and in itself causes erosion.
III. The interests of beach property owners should not be confused with the national interest.
IV. Once you start stabilization, you can't stop.
V. The cost of saving beach property is, in the long run, greater than the value of the property to be saved.
VI. In order to save the beach, you destroy it.

Source: Pilkey *et al.*, 1980.

if short run preservation of beach front buildings is deemed the purpose of shoreline engineering, then engineers are much more successful.

It is important to point out that not all shoreline engineering structures that one sees on beaches were placed there by trained engineers. However, on the open ocean, high energy shoreline, few structures have been built without some expert design help. It is also important to note that sometimes engineers are forced for economic and political reasons to carry out projects that they know full well will cause serious environmental damage.

Table 2 is a summary of the truths of the shoreline; a set of carefully researched, incontrovertible, unassailable, precise generalizations. What engineers basically do to the shoreline in order to stabilize it is shown in Table 3. None of these approaches in any way addresses the sea level rise.

The gentlest thing that can be done by an engineer to a beach is replenishment. Replenished sands disappear faster than natural beaches as a rule for

TABLE 3

What Engineers Do to Beaches

I	Replace sand	replenishment
II	Trap sand	groins, jetties breakwaters
III	Block wave energy	sea walls, bulkheads, revetments

Note: I is a nonstructural "solution," II and III are structural "solutions."

two reasons. The most obvious problem is that, for good economic reasons, sand is only pumped onto the upper shoreface. Thus the beach is steepened and in order to come to a natural equilibrium the sand is soon scattered across the shoreface, most of which is below the tide line. Perhaps a more fundamental problem is that because of the sea level rise the "equilibrium island" or equilibrium profile should continue to be displaced landward of the stabilized shoreline and each successive replenished beach should be more unstable (and disappear more quickly). The replenished beach at Wrightsville Beach, North Carolina, disappeared 10 times faster than the natural erosion rate. It remains to be seen how rapidly the second replenishment project (1980) disappears.

Since replenishment is a relatively gentle thing to do to a beach, it is not unreasonable to ask, "Why replenish all of our problem areas?" The first and most obvious reason is cost. *Fifteen* miles of Miami Beach were replenished at a cost of 65 million dollars. A second factor is the effect of replenishment on development. The initial replenishment often takes place when development is relatively light. The new beach, however, adds time which is used by developers to increase density of development. The increased density and building size provides an increasing political power base which fosters additional replenishment projects but ultimately (in 50 years?) eliminates all alternatives to shoreline management except the inevitable sea wall.

Groins and jetties and other structures emplaced perpendicular to the beach hold sand in place by reducing longshore currents. Breakwaters work in the same way. The problem here is that the long range and long distance effect of sand trapping is impossible to predict. Morton (1979) estimates that sand trapping structures on the relatively undeveloped Texas shoreline have already tied up 50 percent of the longshore drift sand supply from eroding headlands. One would suspect that every grain of sand has been trapped on the much more heavily developed and heavily engineered New Jersey shore. Seasonal effects are often poorly understood. Engineers have proposed to place groins off Ocean City, Maryland. Statements have been made to the effect that no damage should accrue to the Delaware shoreline since it is "upstream" and net sand movement is to the south, away from Delaware. However, during the winter large amounts of sand travel north from Maryland to Delaware and hence the groins are likely to cause damage to Delaware beaches by cutting off their winter supply.

Designing groins and evaluating their potential effects on the basis of known net transportation rates is a common engineering error. Consider, for example, the part played in the shoreline longshore drift equilibrium picture by the 5 year, 10 year, 20 year, etc. storm. Storms that move sand in strange directions and quantities can't be plugged into formulas and hence are considered to be natural disasters by engineers. Geologists consider storms to be part of the system to be worked with even if we can't predict their role. Paul Komar (pers. comm.) points out that jetties caused severe beach erosion and loss of the town of Bay Ocean, Oregon, even though there is no net transport of sand on Tillamook Spit. The jetties caused the inner shelf to change shape, the changed shape refracted waves, refracted waves caused erosion, and buildings fell in.

The worst thing that can be done to a natural shoreline system is construction of a sea wall. As pointed out in the Corps' Shoreline Protection Manual (Corps of Engineers, 1975), wave reflection and other effects eventually remove beaches from the front of sea walls.

Discussing the effect of sea walls is a good time to bring in the concept of New Jerseyization, the end point of the walled open ocean shoreline in a rising sea level. New Jerseyization, a 50 to 150 year phenomena, can be viewed in any number of "old" New Jersey islands or shore communities such as Sea Bright, Monmouth Beach, Long Branch, Cape May, and many others. New Jerseyization is a phenomenon not restricted to New Jersey, however. Other American beach communities such as Ocean City, Maryland; Wrightsville Beach, North Carolina; Folly Beach, South Carolina; much of south Florida; Holly Beach, Louisiana; and South Padre Island, Texas (to name a few) are well on their way.

The classical New Jerseyized beach is no longer a beach except perhaps at low tide. The upper foreshore is scattered with debris from destroyed sea walls and groins of an earlier generation. Swimming is dangerous, but fishing is probably improved because of the multitude of habitats. The present sea wall is a massive and impressive structure. However, winds of 25 mph are capable of producing sea wall topping waves and during genuine storms, the volume of water pouring over and through the wall is impressive. The problem is that the shoreface has steepened considerably and all waves, big or small, break directly on the wall. By the time a developed shoreline has matured to the point of New Jerseyization, the equilibrium profile (where the beach wants to be) is 10's or hundreds of meters landward, the shoreface has steepened to the point where replenishment is no longer an economically viable alternative, and the island or beach has become a fortress. All the community can do hereafter is build bigger and better seawalls.

In spite of the massive evidence that New Jerseyization is the long range inevitable endpoint of open beach stabilization, sea walls are sprouting like dandelions. In 1979 David fathered new ones on at least three islands in Georgia. The attitude of the cottage owner is: I can't just let my house fall in, can I? But it is shortsighted in the extreme to respond to the few individuals whose homes are endangered when one considers that the long range effect will be to destroy a resource that is used by a much larger public.

The greatest sea wall in America's Atlantic or Gulf Coasts is the massive Galveston wall. It was built in response to the Hurricane of 1900 that killed 6,000 people in Galveston. Since the flooding that killed most of the people probably came from the backside, the structure was built on the front side. Over the years what was once a wide beach in front of the wall has largely disappeared. At the base of one end of the wall, a row of boulder rip rap was placed to protect the wall from undermining. A while later a second row of rip rap was emplaced to protect the first row of rip rap. Now a third row of rip rap is planned to protect the second row of rip rap which was emplaced to protect the first row of rip rap which protects the base of the sea wall. The sequence of riprapping is a vivid demonstration of the shoreface steepening effect of a stabilized shoreline in a rising sea level....

The Final Word: Open ocean shoreline stabilization in the long range results in loss of the taxpayer's beach in order to protect the property of a very few at a cost in taxpayers' dollars greater than the property is worth and, of course, the very few whose property is being saved cause the problem to begin with just by being there!

DO ENGINEERS EVER DO ANYTHING USEFUL?

Thousands of houses along miles and miles of stabilized American shorelines owe their very existence to engineering structures. The rising sea level and the migrating barrier island would have long since caught up with them had it not been for the sea walls and revetments fronting the beach. On the other hand many of these communities have been New Jerseyized; all that's left of their original *raison d'être* is the sea breeze.

What would have happened if the shorelines had not been stabilized or engineered? The answer is that the front row or two of houses would no longer exist; third avenue would now be front street, and a thriving beach and beautiful sea view would be there for all to enjoy. No private or public funds would have been wasted on engineering schemes and the future would not require a heavy tax burden to build bigger and better seawalls. Big storms striking non-engineered beach communities will destroy houses, but the same storm on an engineered beach will also destroy houses.

What are we to do with the beaches of New York City such as Coney Island where a hundred thousand or more people may frolic on a sunny day? Should we allow them to respond to the sea level rise and watch as ferris wheels, roller coasters, and apartment buildings fall in? Of course not. Costly beach preservation procedures are fully justifiable to preserve and even enhance these critical natural resources. The question is, where does one draw the tax money line between 100,000 swimmers a day on Coney Island, New York, and 100 beach cottage owners on Wrightsville Beach, North Carolina?

IS THERE A BETTER WAY?

The Corps of Engineers carries out demonstration projects along "quiet" water shorelines to experiment with new ideas on shoreline stabilization. Although conceived by some of the best brains in shoreline engineering and in spite of the fact that a great deal of ingenuity is exhibited in increasing effectiveness at lower costs, the structures still block wave energy, replace sand, or trap sand (Table 3). No one seems to sit back and ask the perfectly logical question: "How can we respond to the sea level rise?"

One unique approach that *would* take into account the rising sea was conceived by some National Park Service bureaucrats from Fire Island National Seashore. They have suggested that instead of pumping sand on the front side of the island, in the future it should be pumped on the backside. The sand would come from maintenance dredging of the intracoastal waterway. *This would migrate the island artificially.* There are, of course, some political problems with this. Home owners on front sides will be required to be good sports as their houses fall in, but their sportsmanship will be greatly aided by tax-supported federal flood insurance which will pay for their lost houses. Engineers have proposed alternative schemes for Fire Island management involving, you guessed it, sea walls, groins, jetties, and beach replenishment at a great cost to the taxpayers.

What other approaches can be taken as alternatives to stabilization? How about moving houses back as the shoreline retreats? The wrecking ball may have to be used on big structures, an approach actually suggested by some New Jersey officials. How about declaring front row conservation easements after the next hurricane; i.e., don't allow reconstruction next to the beach? How about establishing minimum construction standards for seawalls and groins (an idea given to me by engineer Jay Langfelder)? This would allow people to stabilize their beach with structures that would withstand the 5-year storm but not the 10-year storm. The homeowner thus will gamble on when the 10-year storm will occur while hopefully such temporary stabilization will not wipe out the beach. Care must be taken not to allow the temporary structures simply to be replaced. Stabilization schemes billed as temporary or stop gap have a consistent history of not only becoming permanent but also "bigger and better."

Best of all, why not do nothing? Is this really an unreasonable or irrational or irresponsible approach in view of the long range costs and effect of non-do-nothing engineering? The National Park Service came to the do-nothing alternative after long study. On Cape Cod National Seashore parking lots are being designed to be constructed with beach gravel so that when (not if) they fall in, no one except a harried bureaucrat will know.

Table 4 summarizes the good and bad points of the major approaches to shoreline management including doing nothing.

TABLE 4

Summary of the Long Range Economic and Environmental Effects of the Various Alternatives Open to Barrier Island Managers

	Preserves the beach?	Preserves houses?	Responds to Sea Level Rise?	Cost to Taxpayers?
Beach replenishment	Yes	Temporarily	No	Yes
Groins, jetties	Temporarily	Temporarily	No	Yes
Seawalls	No	Yes	No	Yes
Do-Nothing	Yes	No	Yes	No
Artificial Migration	Yes	No	Yes	Yes
Minimum construction standards	Probably	Temporarily	Yes	No

GEOLOGISTS AND ENGINEERS: DIFFERENT WAYS OF LIFE

A major problem in communication exists between engineers and geologists at the shoreline. Engineers faced with design criteria are forced to quantify the natural environment. Geologists tend to do a lot of arm waving instead, but nonetheless, have a better intuition for the long range view of shoreline processes. Table 5 is another carefully researched, incontrovertible summary of the differences between the engineering and the geological way of life.

The need to quantify is understandable. The mistake comes when the engineers actually believe their numbers. The complexities of a sandy shoreface are not about to be pinned down in equations that are meaningful in any long range sense.

Accepting numbers as "solid facts" tends to cause engineers to minimize potential damage caused by their structures. In fact, it is crucial to remember that anything and everything done to stabilize a beach is a long range (and perhaps short range) "environmental mistake." Perhaps the "environmental mistake" is acceptable in the equation relating human needs, politics, and economics. However, if the potential environmental damage of beach stabilization is simply brushed aside by confident engineers with impressive computer models, poor decisions will be made by beach community decision makers. . . .

The following is a quote from the author of a prominent textbook on coastal engineering:

> "The subject of sea level hardly warrants a comment. [Problems caused by rising sea level] will be solved by future generations."

The first part of the quote probably represents a minority view among coastal engineers. Most believe that the sea level rise is an important cause of erosion. It is my opinion, however, that a good portion of the "old guard" of coastal engineering still does not accept a widespread sea level rise or even

TABLE 5

The Differences Between Coastal Engineers and Geologists

	Engineers	Geologists
Reason for shoreline study	Stabilization	Stabilization; understanding of depositional-erosional processes; ancient environmental interpretation
People versus environment	On the side of people	On the side of the environment
How quantitative?	Highly quantitative	Arm wavers
Belief in beach "numbers"	Highly confident	Totally skeptical
View of the beach	A beach is a beach	A beach is a component of the continental shelf and of a barrier island chain, etc.
View of nature	Something to be defeated	Something to be lived with
Time viewpoint	Very short	Very long
Confidence in ultimate success of shoreline stabilization	Over confident	Totally skeptical

the evidence of ubiquitous erosion. Solving erosion on a crisis by crisis basis without recognition of the overall problem can only result in a disaster for the American shoreline. The second part of the quote regarding future generations is the very essence of engineering mentality! Future technology will come to the rescue so damn the torpedoes! However, for miles and miles and miles of New Jerseyized shoreline there is no economically feasible solution. The future is here! We *are* the future generation and now what?

There is no question that given enough money an engineer would be able to solve *any* shoreline problem. If money were no object, even a geologist could solve the problems. With replenishment costs running at 1 to 5 million dollars per mile and climbing at a rate similar to the rate of shoreline development, the point of no return has been reached.

Short sighted approaches to the open ocean shoreline continue to abound. Recently the Corps announced a *permanent* solution to Ocean City's severe erosion problem in the form of a massive steel bulkhead accompanied by replenishment. If a decade or two is "permanent," then the statement is not misleading. But citizens of Ocean City should have been told that the solution would mean they will be New Jerseyized at great cost to themselves. Engineers on Folly Beach, South Carolina, have proposed a small beach replenishment project by taking sand from the lower beach and putting it on the upper beach. This will very likely increase the erosion rate!

Another quotation from an engineer shown below illustrates another point of difference between geologists and engineers:

> "This project (the new Miami Beach) should last indefinitely providing a major storm doesn't come by."
>
> —Miami District Corps of Engineers Official

To a geologist the storm is a perfectly predictable and essential part of the natural system. Engineers know about storms too. But when they come, all bets are off on design life because of this tragic act of nature that's gone and messed things up.

Geologists also feel that most engineers have a narrow view of the shoreline systems. A beach is a beach, an inlet is an inlet, and they are not perceived to be part of a large and complex barrier island chain-continental shelf system. This results in a practice best described as "bandaid engineering." That is, the immediate problem is "solved" with minimum regard or understanding of the impact on other beaches, islands, lagoons, inlets, etc.

THE BOTTOM LINE

The time has come to bite the bullet and either do nothing at the shoreline or at least respond to what's causing the problem—the sea level rise. Suggestions to do nothing usually elicit comments such as this:

> "This do-nothing philosophy may have suited the times of King Canute in the 11th century when population was limited and few people had time or inclination to visit the coast, but it is unacceptable to the general public in 20th century America where the federal government has taken the responsibility to provide for the general public welfare."
>
> —Corps of Engineers (CERC) geologist

The do-nothing suggestion made to a beach cottage owner elicits far more colorful comments with even longer sentences.

Nonetheless, a strong case can be made that shoreline stabilization does not provide for the general public welfare. In fact, it has been a massive long range failure on American open ocean shorelines. The American public must be told what the long range costs and effects of stabilization will be. The net result of such education almost surely will be to keep the engineer off and away from the open ocean shoreline.

REFERENCES

BRUUN, P. , 1962, Sea level rise as a cause of shore erosion: Jour. Waterways and Harbor Div. A.S.C.E. Proc. 88, p. 117–130.

CORPS OF ENGINEERS, 1975, Shore Protection Manual. Vols. I, II, III.

FIELD, M.E., and DUANE, D. B., 1976, Post Pleistocene history of the U.S. inner continental shelf: Significance to origin of barrier islands: Geol. Soc. America Bull., v. 88, p. 734–736.

HICKS, S. D., 1972, On the classification and trends of long period sea level series: Shore and Beach, v. 40, p. 20–23.

LEATHERMAN, S. P., 1979, Barrier Island Handbook: Privately published, 101 p.

MORTON, R. A., 1979, Temporal and spatial variations in shoreline changes and their implications, examples from the Texas Gulf coast: Jour. Sed. Petrology, v. 49, p. 1101–1112.

PILKEY, O. H., NEAL, W. J., and RIGGS, S., 1980, From Currituck to Calabash: Living with North Carolina's Barrier Islands (2nd ed.): N. C. Sci. and Tech. Res. Center, 370 p.

ROSEN, P. S., 1978, A regional test of the Bruun rule on shoreline erosion: Marine Geology, v. 20, p. M7–M16.

SWIFT, D. J. P., 1975, Barrier island genesis: Sed. Geology, v. 14, p. 1–45.

______, 1976, Continental shelf sedimentation, *in* Stanley, D. J., and Swift, D. J. P. , eds., Marine Sediment Transport and Environmental Management: New York, John Wiley, p. 311–350.

PART TWO

On the Internet . . .

Sites appropriate to Part Two

This is the home page for the United States Department of Energy's Energy Efficiency and Renewable Energy Network. The Web site is a comprehensive resource for conservation and renewable energy information.

`http://www.eren.doe.gov`

The National Renewable Energy Laboratory is America's leading laboratory for renewable energy and energy efficiency research. This is the laboratory's home page.

`http://nrelinfo.nrel.gov`

The International Network for Sustainable Energy (INFORSE) includes more than 170 Non-Governmental Organizations (NGOs) worldwide, which work to promote sustainable energy and social development. The INFORSE Web site provides information on global developments in sustainable energy.

`http://www.inforse.dk`

This is the home page of "Energy Quest" and provides energy education resources from the California Energy Commission.

`http://www.energy.ca.gov/education/index.html`

CHAPTER 5 Energy and Ecosystems

5.1 CHANCEY JUDAY

The Annual Energy Budget of an Inland Lake

The study of the role played by energy in an ecosystem and the energy interchanges that occur among the biotic (living) and nonbiotic components of the system is a key ecological organizing principle used in understanding the dynamics of nature's biogeochemical cycles. Chancey Juday was a pioneer in the study of freshwater biotic communities. He was responsible for developing a program that made the University of Wisconsin the center for such studies in the United States.

Juday's study "The Annual Energy Budget of an Inland Lake," *Ecology* (October 1940), from which the following selection has been taken, was a groundbreaking piece of work. It was the first time a scientist attempted to examine the energy input and subsequent energy uses and transformations that occur within a complex biotic community. Juday's work represented an important step toward transforming ecology into a more quantitative discipline, one in which sophisticated analyses of entire ecosystems were possible.

Key Concept: the roles of energy in ecosystem dynamics

INTRODUCTION

The variation in the quantity of solar radiation delivered to the surface of an inland lake during the course of the year is the principal factor in determining the physical, chemical and biological cycle of changes that take place within the water. This is true especially of lakes which are situated in temperate latitudes where there are considerable differences between summer and winter temperatures of the air and of the water.

In the deeper lakes which become covered with ice for a few to several weeks during the winter, the annual cycle consists of four phases which correspond roughly to the four seasons of the year. (1) There is a winter stagnation period in which the water is inversely stratified while the lake is covered with ice; that is, the warmest water is at the bottom and the coldest at the surface. (2) There is an overturning and circulation of the entire body of water following the disappearance of the ice in spring.

(3) As the temperature rises above the point of maximum density (4° C.) in the spring, the free circulation of the water is hindered as a result of the difference in density between the warm upper layer and the cold lower stratum; this is due to the fact that most of the warming takes place in a comparatively thin upper stratum. From 65 to 90 per cent or more of the sun and sky radiation is cut off by the upper meter of water, depending upon the color and the transparency. This large reduction is due chiefly to reflection at the surface, to the rapid absorption of the radiation by the water itself and by the stains and suspensoids that may be present in the water.

With the further rise in the temperature of the upper water as the season advances, a summer stratification is established. This is a direct stratification in which there is a warm upper stratum, a cold lower stratum and a transition zone between them in which the temperature changes rapidly from that of the warm upper water to that of the cold stratum below; naming them in order from surface to bottom, they are known as the epilimnion, the thermocline or mesolimnion, and the hypolimnion. In the deeper lakes these three strata persist throughout the summer.

(4) The fourth stage is represented by the autumnal cooling and overturning, which is followed by a complete circulation of the water until the lake becomes covered with ice.

The circulation of the water in spring and autumn distributes the dissolved substances uniformly from surface to bottom, but a more or less marked difference in the chemical character of the upper and lower strata develops during the two stratification periods, especially in summer.

The seasonal changes in the physical and chemical characteristics of the water have an important effect upon aquatic life. The spring rise in temperature speeds up life processes and the increase in the amount of solar radiation at this time makes conditions more favorable for photosynthesis. Likewise the circulation of the water in spring and autumn brings dissolved substances into the upper stratum, which is the zone of photosynthesis, where they can be readily

TABLE 1

Quantity of Solar and Sky Radiation Used by Lake Mendota in Various Physical and Biological Processes

Melting of ice in spring	3,500
Annual heat budget of water	24,200
Annual heat budget of bottom	2,000
Energy lost by evaporation	29,300
Annual surface loss	28,500
Loss by conduction, convection and radiation	30,324
Biological energy budget (maximum)	1,048

The results are indicated in gram calories per square centimeter of surface.

obtained by the aquatic plants. The phytoplankton responds promptly to these favorable growing conditions and, as a result, the standing crop of plankton is usually much larger during the two circulation periods than it is at the time of summer and winter stratification.

The annual cycle of changes that is induced by the march of the seasons naturally raises the question of the amount of energy actually involved in the phenomenon as a whole and also in the various phases of the cycle. This problem is a very complex one and a large amount of data is necessary to evaluate each of the several items.

Limnological studies have been in progress on Lake Mendota at Madison, Wisconsin, for many years and the data accumulated in these investigations make it possible to give approximations of the quantity of energy involved in a number of the items. Quantitative studies of some of the factors have not been made up to the present time and a more complete assessment of the energy budget of the lake will have to await such investigations.

The annual energy budget of a lake may be regarded as comprising the energy received from sun and sky each year and the expenditures or uses which the lake makes of this annual income of radiation. In general the annual income and outgo substantially balance each other. This is true more particularly of the physical energy budget. Considerable biological material produced in one energy year lives over into the next, but this overlapping crop of organisms is much the same in quantity from year to year so that it plays approximately the same annual rôle. For this reason it does not require any special consideration. There is a certain amount of organic material contributed to the bottom deposits in the deeper water and to peat formation in the shallow water which lasts for long periods of time, but the annual energy value of these materials is so small in most cases that they may be neglected. . . .

The respective amounts of energy included in the four items of the physical energy budget . . . are indicated in table 1. The sum of these four items is 71,000 calories, which is approximately 60 per cent of the mean quantity of energy delivered to the surface of Lake Mendota annually by sun and sky. . . .

BIOLOGICAL ENERGY BUDGET

A certain amount of the solar radiation that passes into the water of Lake Mendota is utilized by aquatic plants in the process of photosynthesis. The products of this assimilation, namely, proteins, fats and carbohydrates, thus constitute the primary accumulation or storage of the energy derived from the sub-surface illumination. Since this organic material manufactured by the plants serves, either directly or indirectly, as a source of food for all of the non-chlorophyllaceous organisms that inhabit the lake, these latter forms, therefore, constitute a secondary stage in the storage of the energy accumulated by the aquatic plants. The original amount of energy represented by these secondary organisms varies with the different forms, depending upon the number of links in their respective food chains; in general they represent a comparatively small proportion of the primary organic material manufactured by the plants.

Chemical analyses of the various aquatic organisms have now progressed far enough to enable one to compute their energy values from the standards that have been established by food chemists. The standard values are 5,650 calories per gram of protein, 9,450 calories per gram of fat and 4,100 calories per gram of carbohydrate, on a dry weight basis. These values do not represent the total quantity of energy utilized by the aquatic organisms, however, because a part of the synthesized material is oxidized in the metabolic processes of the living organisms. These metabolic oxidations result in the production of heat which is transmitted to the water, but the quantity of heat derived from this source is extremely small in comparison with that which comes from direct insolation.

The amount of organic matter consumed in the metabolism of plants is much smaller than that in animals because several grams of plant material may be consumed in the production of one gram of animal tissue even in animals that feed directly on plants; the predaceous animals represent a still larger quantity of the original photosynthesized material.

... [I]t may be estimated that the average turnover in the organic matter of the mean standing crop of plankton takes place about every two weeks throughout the year. It would be more frequent than this in spring and summer, and less frequent in winter. A turnover of 26 times per year would give an annual yield of 6,240 kilograms of dry organic matter per hectare of surface as indicated in table 2. This material would consist of 2,704 kilograms of protein, 431 kilograms of fat and 3,105 kilograms of carbohydrate. Approximately 94 per cent of the organic matter comes from the phytoplankton and 6 per cent from the zooplankton....

Schuette and some of his students (1922–29) made chemical analyses of certain species of ... plants and several others have been analyzed in more recent years, so that the energy values of all of the more common forms can now be computed from the data in hand. The results of such computations are given in table 2. The 512 kilograms of dry organic matter per hectare consisted of 64 kilograms of protein, 6 kilograms of ether extract or fat and 442 kilograms of carbohydrate.

The bottom deposits, especially in the deeper water, contain a rather large population of bacteria.... While these organisms are present in considerable numbers, they are so small in size that they add very little to the crop of organic

TABLE 2

Annual Production of Plankton, Bottom Flora, Bottom Fauna and Fish, as Well as Crude Protein, Ether Extract (Fat), and Carbohydrate Constituents of the Organic Matter

	Dry Organic Matter	*Crude Protein*	*Ether Extract*	*Carbohydrate*
Total plankton	6,240	2,704	431	3,105
Phytoplankton	5,850	2,501	383	2,966
Zooplankton	390	203	48	139
Bottom flora	512	64	6	442
Bottom fauna	45	33	4	8
Fish	5	3.4	1	0.6
Dissolved organic matter	1,523	334	68	1,121
Total organic matter	8,325	3,138.4	510	4,676.6

The results are stated in kilograms per hectare on a dry, ash-free basis. The plankton yield is based on a turnover every two weeks during the year. The average quantity of dissolved organic matter is included also.

matter in the lake; so they have been disregarded. Likewise fungi are fairly abundant in the bottom deposits, but no quantitative study of them has yet been made; it seems probable that their contribution to the organic content of the lake is negligible from an energy standpoint....

The macroscopic bottom fauna yielded 45 kilograms of dry organic matter per hectare; of this amount 33 kilograms consisted of protein, 4 kilograms of ether extract or fat and 8 kilograms of carbohydrate. While some of these organisms live more than one year, others pass through two or three generations in a year; the two groups of organisms are generally considered as balancing each other, so that the above quantities may be taken as the annual crop of this material as shown in table 2.

Considerable numbers of protozoa and other microscopic animals have been found in the bottom deposits, but no quantitative study of them has been made. It seems probable, however, that these minute forms would not add an appreciable amount of organic matter to the total weight of the bottom fauna.

FISH. No accurate census of the fish caught by anglers in Lake Mendota each year has ever been made so that the assessment of this part of the biological crop can be estimated only roughly....

The average yield of carp between 1933 and 1936, inclusive, was 16 kilograms per hectare; adding the carp crop to that of the game and pan fish gives an annual fish yield of 22 kilograms per hectare, live weight. On a dry, ash-free basis, the total yield amounts to a little more than 5 kilograms per hectare as indicated in table 2. By far the greater part of this material consists of protein and fat.

TABLE 3

Energy Values of the Organic Matter in the Organisms, Together With the Estimated Amounts of Energy Represented in Their Metabolism, and in the Dissolved Organic Matter

Phytoplankton	299
Metabolism	100
Zooplankton	22
Metabolism	110
Bottom flora	22
Metabolism	7
Bottom fauna and fish	3
Metabolism	15
Dissolved organic matter	71
Total	649

The values are stated in gram calories per square centimeter of lake surface. The results for phytoplankton and zooplankton are based on a turnover every two weeks during the year.

ENERGY VALUE OF ANNUAL CROP. Table 2 shows that the total quantity of stored and accumulated energy in the form of dry organic matter in the annual crop of plants and animals amounts to 6,802 kilograms per hectare; of this quantity protein constitutes a little more than 2,804 kilograms, ether extract or fat 442 kilograms and carbohydrates 3,556 kilograms. On the basis of the energy equivalents of these three classes of organic matter,... the total energy value of the annual crop amounts to 346 gram calories per square centimeter of lake surface (table 3).

In addition to the organic material in the plants and animals, the water contains a certain amount of organic matter which cannot be recovered with a high speed centrifuge. It is either in true solution or is in such a finely divided state that it cannot be obtained with a centrifuge; for lack of a better term it has been called "dissolved organic matter" as compared with the "particulate organic matter" which can be recovered from the water with a centrifuge. The water of Lake Mendota contains 10 to 14 milligrams per liter, dry weight, of this dissolved organic matter; the mean of some 60 determinations is 12 milligrams per liter. When computed to an area basis, the average weight of this material is 1,523 kilograms per hectare, dry weight, of which 334 kilograms are protein, 68 kilograms fat and 1,121 kilograms carbohydrate. The energy value of this dissolved organic matter is about 71 gram calories per square centimeter.

UTILIZATION OF SOLAR ENERGY. The chlorophyll-bearing aquatic plants are responsible for the utilization of the sub-surface radiation; that is, the sun furnishes the power and the chlorophyll and associated pigments of the plants serve as the machines for the manufacture of the fundamental organic matter of the lake. Table 2 shows that the phytoplankton and the large aquatic plants constitute the major item in the annual yield of biological material. Together they contribute 6,362 kilograms of dry organic matter per hectare as compared with 440 kilograms of zooplankton, bottom fauna and fish; that is, the plant contribution is 93 per cent and the animal part is 7 per cent of the total organic matter.

Table 3 gives the energy value of the various constituents of the annual biological crop. The two groups of plants, namely phytoplankton and large aquatics, have an energy value of 321 gram calories as compared with 25 gram calories per square centimeter in the animals. The 321 gram calories represented in the organic matter of the plants is only 0.27 of one per cent of the mean annual radiation delivered to the surface of the lake, namely 118,872 calories. Two corrections need to be made in this result, however. (1) As already indicated some 28,500 calories of solar energy are lost at the surface of the water and thus do not reach the aquatic vegetation. Deducting this amount leaves 90,372 calories which pass into the water and thus become available for the plants. On this basis the percentage of utilization is increased to a little more than 0.35 of one per cent of the available radiation. (2) A certain amount of the organic matter synthesized by the plants is used in their metabolism and this does not appear in the percentage of utilization given above. Experiments show that some of the algae utilize in their metabolic processes about one-third of the organic matter that they synthesize. No data are available for the large aquatics, but assuming that they also utilize a similar proportion in their metabolism, the two groups of plants would represent a utilization of 428 calories which is equivalent to 0.47 of one per cent of the annual quantity of solar energy that actually enters the water.

This percentage is based on an average turnover in the phytoplankton every two weeks throughout the year, but there is some evidence that the turnover takes place more frequently, especially from April to October. With an average turnover once a week in the organic matter of the phytoplanton during the year, the energy value of this crop would be 798 calories, including metabolism; adding to this amount the 29 calories in the annual crop of bottom flora gives a total of 827 calories which is utilized by the plants. This is 0.91 of one per cent of the 90,372 calories of energy that penetrate the water and become available to the plants; in round numbers this may be regarded as a utilization of one per cent.

This small percentage of utilization of solar energy by aquatic plants shows that Lake Mendota is not a very efficient manufacturer of biological products in so far as utilizing the annual supply of solar and sky radiation is concerned; on the other hand it belongs to the group of highly productive lakes.

While the aquatic plant crop appears to be inefficient in its utilization of solar energy, it compares very favorably with some of the more important land crops in this respect. Transeau ('26) states that only 1.6 per cent of the total available energy is used by the corn plant in photosynthesis during a growing period of 100 days, or from June 1 to September 8. Spoehr ('26) gives a table or Pütter's calculations for various crop plants in which the general average of the utilization of solar energy is about 3.0 per cent; summer wheat is given as 3.2 per cent, potatoes 3.0 per cent and beets 2.1 per cent. These computations for cultivated crops, however, take into account only the quantity of solar radiation available during comparatively brief growing periods and thus do not cover the entire year as indicated for the aquatic plants.

ENERGY VALUE OF ANIMALS. The organic content of the animal population of the lake represents a conversion and further storage of the material manufactured by the plants, but no direct utilization of solar energy is involved in the transformation. It may be regarded as an expensive method of prolonging the existence of a certain portion of the original plant material. As previously indicated, it may take five grams of plant food to produce one gram of animal tissue, so that the plant equivalent of the animal crop may be reckoned as five times as large as the organic content of the animals; in the predatory animals, however, it would be much larger.

Table 3 shows that the energy value of the bottom and fish population is 25 gram calories per square centimeter; on the five fold basis, this would represent the conversion of at least 125 gram calories of original plant organic matter. This utilization is approximately 40 per cent of the potential energy stored in the annual plant crop of 321 gram calories which is based on a turnover in the phytoplankton every two weeks during the year. A turnover in the phytoplankton every week would give a plant crop of 620 calories and an animal utilization of a little more than 20 per cent.

DISSOLVED ORGANIC MATTER. The energy value of the dissolved organic matter is indicated as 71 gram calories per square centimeter in table 3. This material is constantly being supplied to the water by the various organisms and the standing crop of it remains fairly uniform in quantity during the different seasons of the year as well as in different years. While there is a regular turnover in this organic matter, it needs to be taken into account only once in computing the organic crop of the lake because it has its source in the plants and animals for which an annual yield has already been computed.

SUMMARY

1. The mean annual quantity of sun and sky radiation delivered to the surface of Lake Mendota over a period of 28 years was 118,872 gram calories per square centimeter of surface.
2. In the physical energy budget, the melting of the ice utilized 3,500 calories; the annual heat budget of the water was 24,200 calories and of the bottom 2,000 calories; the loss of energy by evaporation amounted to 29,300 calories; the surface loss by reflection, upward scattering and absorption was 28,500 calories and about 31,000 calories were lost by conduction, convection and radiation.
3. On the basis of a turnover in the organic content of the plankton every two weeks during the year, the energy value of the annual crop of plants and animals was 346 gram calories per square centimeter; of this amount, 321 calories were contributed by the plants and 25 calories by the animals. Adding to this the organic matter utilized by the plants and animals in their metabolic processes (232 calories) gives the annual crop an energy value of 578 gram calories. In addition the dissolved organic matter had a value of 71 gram calories.

4. Assuming an average turnover of once a week in the organic matter of the phytoplankton, instead of every two weeks, would raise the energy budget of the annual crop of plants and animals to 977 gram calories, including the metabolized material. Adding the 71 calories in the dissolved organic matter gives a total of 1,048 calories in the biological energy budget.
5. An average turnover of once a week in the phytoplankton would give an annual utilization by these organisms and by the large aquatic plants of about one per cent of the sub-surface solar energy.

Energy and the Environment

Physicist John M. Fowler has devoted most of his professional career to the promotion of effective science education for both scientists and the general public. He served for several years as the executive officer of the American Physical Society's Commission on College Physics and more recently as director of special projects for the National Science Teachers Association. He was also an active participant in the Scientist's Institute for Public Information, an organization devoted to providing the public with the scientific knowledge and understanding required to make intelligent decisions on issues of science and public policy. One of Fowler's specific educational preoccupations has been with promoting the public's understanding of the key role played by the production and use of energy in our increasingly technological society.

Most of the world's environmental problems, including air, water, and land pollution, are directly affected by the particular energy sources that have been developed and the efficiency with which they are used. Fowler wrote *Energy and the Environment* (McGraw-Hill, 1975), from which the following selection has been taken, when the United States was experiencing its first "energy crisis" as a result of the 1973–1974 Arab oil embargo. There was suddenly serious debate about the future of America's growing dependence on petroleum. At the same time, a movement was gaining momentum and swelling the ranks of grassroots organizations concerned about the growing signs of environmental deterioration. Through his book, Fowler was one of the first to successfully explain the links between these two issues in a manner that was accessible to the nonscientific, literate public.

Key Concept: limits on the efficiency of energy conversion

ENERGY FOR LIFE

Let's... take up the trail of energy through the eons. We left it 5 billion years ago with the formation of the sun. We will skip over the 4.5 billion years during which the earth was formed, mountains were raised and washed away by the great seas, and life began. We will take up the trail, again, 300 million years ago, in the middle part of the Late Paleozoic Era. Plants have emerged from the sea and the tropical marshlands are covered by ferns, horsetails, and mosses grown to enormous size. Man is still far in the future, his genes are beginning their evolution in some of the creatures of these jungles.

A small fraction of the sun's energy which beat down on these strange jungles was used by the huge fern trees and other vegetation in their life processes. They converted the radiant kinetic energy of sunlight into *chemical potential energy,* the third great form of potential energy.

The energy-converting process of photosynthesis can be summarized in chemical shorthand as

$$CO_2 + H_2O + \text{energy} \rightarrow C_x(H_2O)_y + O_2$$

which, in words, states that in a plant, carbon dioxide (CO_2) and water (H_2O) are combined, with the addition of energy from sunlight, to form the carbohydrate group $C_x(H_2O)_y$ and oxygen (O_2). $C_x(H_2O)_y$ is a general formula for the carbohydrate group which is an important part of the molecule of sugars and starches. Ordinary sugar, for instance, is $C_{12}(H_2O)_{11}$. This is an oversimplification of a step-wise process which proceeds more correctly as

$$CO_2 + H_2O + \text{energy} \rightarrow \text{intermediate products}$$
$$\text{intermediate products} + \text{energy} \rightarrow C_x(H_2O)_y + O_2$$

The radiant energy from the sun, kinetic energy (energy on the move), is caught by plants and used to break up the molecules of CO_2 and H_2O and rearrange their atoms. To form carbohydrates from CO_2 and H_2O, forces must be operating so that molecules are pulled apart and atoms moved around. Since the forces which hold atoms and molecules together are electrical forces, the chemical potential energy which we have just introduced is a form of *electrical potential energy.*

It is certainly potential energy. The carbohydrate formed with the help of the sun's energy is either food or fuel, depending on the use we make of it. Eating or burning turns the photosynthesis reactions around; carbohydrates in the plant sugars or starches combine with oxygen to re-form CO_2 and H_2O, releasing energy in the process.

$$C_x(H_2O)_y + O_2 \rightarrow CO_2 + H_2O + \text{energy}$$

The fern and moss jungle of 300 million years ago grew in a rich environment. Much of the world at that time had an almost tropical climate, wet and warm. In addition, the atmosphere was very rich in CO_2, released from the depths of the earth by many active volcanoes and hot springs.

Let us now focus down on one spot on Paleozoic earth, a swampy river delta in southern Illinois only a few feet above the level of the ocean which covered Missouri to the west and parts of Kentucky to the south. This swampy delta was formed of sediment washed down from the mountain ranges to the east and north. It was covered by a gloomy jungle where, for thousands of years, fern trees and huge mosses grew, died, and fell into the shallow water from which they emerged. Some of them decayed. Bacteria consumed their carbohydrates and recombined the carbon C with O_2 to form CO_2 which was returned to the atmosphere to feed future plants in the familiar cycle of life and death. Some of the plants, however, were covered with water, and the oxygen-requiring bacteria could not work on them. Over tens of thousands of years, huge amounts of this undecayed plant material built up into the spongy mass we call *peat.* (A modern peat bog, 6 feet deep, can be found in regions of the Dismal Swamp of Virginia and North Carolina.)

In peat, the *anaerobic* bacteria (those which do not require oxygen), plus the pressure and accompanying heat caused by the overbearing sediment, drove off some of the water, oxygen, nitrogen, and miscellaneous plant products so that the percentage of energy-rich carbon was increased. Peat, when burned as fuel, as it is in parts of the world, releases about 6,000 Btu per pound.

In our Illinois river delta the thick layer of peat was finally buried under tons of sea-born sediment and further changes took place. The pressure increased greatly and, along with it, the temperature. The peat was greatly compacted, perhaps by as much as a factor of 16 (in other words, a thickness of 16 feet became compressed to a thickness of one foot). More and more of the water, nitrogen, oxygen, and other materials were driven out so that more and more pure carbon was released from its earlier molecular combination. The percentage of carbon increased from an original 50 percent in the living material to 75 or 80 percent. The peat was now compacted into a hard, black mineral; the southern Illinois coal basin was formed.

What had happened to the energy? The radiant energy from the sun was converted to chemical potential energy in the plants. Through the compaction and reactions under pressure and heat, a great concentration of this energy had taken place. Thus, in the high-grade coals, there are 11,000 to 12,000 Btu of energy per pound compared with, for instance, the 5,000 or so Btu per pound of the original wood.

Man Enters the Scene

The Illinois coal deposits, with their tremendous energy locked in the carbon atoms, lay buried for millions of years. As the earliest settlers came to Illinois, they found scattered about black outcroppings of this coal thrust to the surface by mammoth foldings of the earth's crust. Those early settlers knew what it was; coal had been used in Europe as early as the twelfth and thirteenth centuries. But they were not very interested in it as fuel, for all about them was the abundance of wood—wood which needed to be cleared from the land.

Not until the settlers had chopped their way across the land and wood no longer lay at their doorstep did they turn to coal. As the country became more and more industrialized, coal consumption grew, from perhaps 8 million tons in 1850 to 56 million tons in 1875 and 270 million tons in 1900. It reached a peak of 633 million tons in 1945 but production has since decreased, amounting to about 450 million tons in 1970.

Let us go back to our saga of energy and follow a ton of coal mined from that Illinois deposit in the 1970s. Had it been pulled from the ground thirty years earlier, odds are it would either have been made into coke for the smelting of iron ore or would have ended up in the firebox of a steam-driven locomotive. In 1970, however, its likely destination was an electric power plant, perhaps one on the outskirts of Chicago; the one whose plume of smoke you might have noticed while flying over that city.

In that ton of coal were some 24 million Btu of energy. Dumped into the huge furnace of the power plant the coal burned efficiently; most of the heat energy released (85 to 90 percent) went into the boiler which turned water into steam. The other 10 to 15 percent of the released heat, along with the sulphur impurities in the coal, went up the stack. (This wasted heat which warmed the air above the power plant, is of interest in itself. . . . Inside the boiler, the superheated steam with a temperature of 1000°F, was sent against the giant turbines to turn them, thus turning the electric generators which they drove to produce electricity.

Our story is almost over. We have followed this infinitesimal bit of the energy from that "Cosmic Egg" down through the ages from sun to plants to coal and now to electricity. We must keep careful track of what is left. Ten percent of the 24 million Btu (that is, 2.4 million Btu) were lost as heat to the atmosphere. We lost a lot more in the turbine; the steam went in at 1000°F, but it left still hot; it was considerably above the 212°F boiling point of water when it was exhausted from the turbine. Some 60 percent of the heat energy was not used to turn the turbine but instead was carried to a nearby river by the power plant's cooling system. Thirteen million Btu were lost that way.

The generator converted the kinetic energy of the turbine into that most important form of kinetic energy, electricity. This was done very efficiently, so that practically all of the remaining 8.6 million Btu went out into the high voltage transmission lines as electrical energy to be distributed throughout Chicago.

Who can say where it ended? Electricity has many varied uses. We can estimate with some assurance that 10 percent of it (about 1 million Btu of energy) was lost in the transmission and distributing lines; it was converted to heat and warmed the surrounding atmosphere. The remaining 7.7 million Btu was used throughout the city, in air conditioners, cooking stoves, water heaters, motors, industrial plants, and the like. In the end, all the kinetic energy of this electricity ended as heat energy, directly converted in ranges and heaters or converted by friction in motors.

Let's look back now with the help of Table 1. We began with 24 million Btu of chemical potential energy in our ton of coal, lost about 2.5 million Btu up the stack as heat when we burned it, lost 13 million Btu as heat to the river due to the inefficiency of the turbine, approximately another million Btu went as heat into the atmosphere from losses in the transmission lines, and finally the remaining nearly 8 million Btu did its job and ended up again as heat in the air above Chicago.

We need but one more night to finish the story, a clear night with no clouds between Chicago and deep space. The heated air and the warm water of the river now finally give back to the universe the loan made so long ago. The heat is radiated away from the earth and travels with the speed of light out into those seemingly empty infinite depths; but space is not empty. Here and there in those depths are atoms and molecules of gas and dust, the raw material of the stars,

TABLE 1

Where the Coal's Heat Went

Distribution	*Amount (M Btu)**	
Original total heat	24	
Lost with stack gases (10%)		2.4
Lost to cooling water (60% of remainder)		13.0
Electrical output	8.6	
Lost in transmission (10%)		0.9
Total losses		16.3
Useful work	7.7	

*M Btu = million Btu

thinner now in the aging galaxy. These widely scattered particles absorb the heat energy traveling out from earth and increase their motion by the amount of that energy. The universe becomes a bit warmer. . . .

Energy cannot be created or destroyed; that was the statement of the First Law of Thermodynamics. None of the energy we have followed was destroyed, or created; it was only converted from one form to another. But at each conversion there was a *heat tax* imposed; some of the energy went from a "useful" form (either kinetic or potential) into heat energy, in which its usefulness has been decreased. The conversion to heat energy is a one-way street. While it is possible to convert any form of energy to heat with 100 percent efficiency, to reverse the process is difficult. As we saw in the case of the steam turbine, in the conversion to mechanical energy of the heat released by burning coal, we lost 60 percent of it.

The one-way nature of energy conversion is also a law of nature. For our study of the "energy crisis," more important than the First Law we have just reviewed, is the *Second Law of Thermodynamics*, which states that:

> No device can be constructed which, operating in a cycle (like an engine), accomplishes *only* the extraction of heat energy from a reservoir and its complete conversion to mechanical energy (work).

There are many other ways to state this important law. . . . Its consequence, however, should already be clear. Heat energy cannot be completely converted to mechanical energy. In any conversion some of it is irrevocably lost; it remains in the form of heat and cannot be reclaimed for useful purposes.

CHAPTER 6 Renewable and Nonrenewable Energy

6.1 AMORY B. LOVINS

Soft Energy Paths: Toward a Durable Peace

In 1976 the prestigious journal *Foreign Affairs* published an energy policy article by a little-known physcist trained at Harvard University and Oxford University named Amory B. Lovins. The article was vigorously denounced by the energy establishment. In the article, Lovins, who was the British representative of the environmental organization Friends of the Earth, made a radical proposal: he argued that the United States should make a transition from dependence on fossil fuels and nuclear power, which he referred to as the "hard path," to an energy supply system based on increased efficiency coupled with renewable sources powered by the sun, which he labeled the "soft path." Although his ideas were ridiculed by scientists and executives of electric utilities and the hard path energy industry, the wisdom and creativity of his arguments soon began to be recognized by independent energy planners in government and the private sector who became early clients of the private consulting firm that Lovins established.

In *Soft Energy Paths: Toward a Durable Peace* (Ballinger, 1977), from which the following selection has been taken, Lovins expands on his proposal. He explains why the soft path would be far better for the environment, would result in a more secure and more economical energy system, and would reduce the threat of international conflict over energy resources. He also explains why a choice must be made between the hard and the soft paths.

After the publication of *Soft Energy Paths,* Lovins quickly became the energy guru of the environmental movement. Over time the accuracy of his forecasts and his stream of innovative proposals won him the grudging respect and even the qualified acceptance of the energy establishment that initially treated him as a pariah. His recent clients have included the Pacific Gas and Electric Company, General Motors and the utility industry's Electric Power Research Institute.

Key Concept: the advantages of energy efficiency and renewable energy sources

TECHNOLOGY IS THE ANSWER! (BUT WHAT WAS THE QUESTION?)

The energy problem, according to conventional wisdom, is how to increase energy supplies (especially domestic supplies) to meet projected demands. The solution to this problem is familiar: ever more remote and fragile places are to be ransacked, at ever greater risk and cost, for increasingly elusive fuels, which are then to be converted to premium forms—electricity and fluids—in ever more costly, complex, centralized, and gigantic plants. The side effects of these efforts become increasingly intolerable even as their output allegedly becomes ever more essential to our way of life and our very survival. As population in most industrial countries rises by less than a fifth over the next few decades, we are told that our use of energy must double and our use of electricity treble. Not fulfilling such prophecies, it is claimed, would mean massive unemployment, economic depression, and freezing in the dark.

But where do these projected "energy needs" come from? Herman Daly provides a pungent but broadly accurate summary:

> Recent growth rates of population and per capita energy use [or of population, per capita GNP, and energy use per unit of GNP] are projected up to some arbitrary, round-numbered date. Whatever technologies are required to produce the projected amount are automatically accepted, along with their social implications, and no thought is given to how long the system can last once the projected levels are attained. Trend is, in effect, elevated to destiny, and history either stops or starts afresh on the bi-millenial year, or the year 2050 or whatever.
>
> This approach is unworthy of any organism with a central nervous system, much less a cerebral cortex. To those of us who also have souls it is almost incomprehensible in its inversion of ends and means.
>
> . . . [It says] that there is no such things as enough[;] that growth in population and per capita energy use are either desirable or inevitable[;] that it is useless to worry about the future for more than 20 years, since all reasonably discounted costs and benefits become nil over that period[;] and that the increasing scale of technology is simply time's arrow of progress, and refusal to follow it represents a failure of nerve.

Most thoughtful analysts now see that this approach is rapidly grinding to a halt. It is looking *politically* unworkable: most people, for example, who are on the receiving end of offshore and Arctic oil operations, coal stripping, and the plutonium economy have greeted these enterprises with a comprehensive lack of enthusiasm, because they directly perceive the prohibitive social and environmental costs. Extrapolative policy seems *technically* unworkable: there is mounting evidence that even the richest and most sophisticated countries lack the skills, industrial capacity, and managerial ability to sustain such rapid expansion of untried and unforgiving technologies. And it seems *economically* unworkable: for excellent reasons, such free market mechanisms as still operate have persistently shown themselves unwilling to allocate to the extremely capital-intensive, high risk supply technologies the money needed to build them. The inexorable disintegration of current policy thus makes us reexamine its premises.

The basic tenet of high-energy projections is that the more energy we use, the better off we are. But how much energy we use to accomplish our social goals could instead be considered a measure less of our success than of our failure—just as the amount of traffic we must endure to get where we want to go is a measure not of well-being but rather of our failure to establish a rational settlement pattern. As the U.S. National Research Council CONAES study states: "The first, and dominant, 'facet of the solution' [to the energy problem] relates to the issue of *how fast*, and indeed *whether, our use of energy may need to grow and*, ultimately, *how much* energy our society will require to sustain the way of life that it chooses. Energy is but a means to social ends; it is not an end in itself."

Thus before we conclude that technology—any technology—is the answer, we need to remind ourselves what the question was. In order not to talk nonsense about future energy requirements, we must, as Daly points out, first ask:

1. *Who* is going to require the energy?
2. *How much* energy?
3. *What kind* of energy?
4. For *what purpose*?
5. For *how long*?

... Underlying much of the energy debate is a tacit, implicit divergence about what the energy problem "really" is. Public discourse suffers because our society has mechanisms only for resolving conflicting interests, not conflicting views of reality, so we seldom notice that those perceptions differ markedly. I see no basis for deciding (in the absence of another century's experience mellowed by 20/20 hindsight) which of the several prevalent world views, if any, is most useful, let alone which is "right" or "wrong." As a basis for mutual understanding, therefore, instead of leaving my world view to be guessed at (as most energy writers do), I shall make explicit a few of my underlying opinions—not on every aspect of the whole universe of perceptions that must support any coherent view of our energy future, but at least on a few basic values. Attempting this is unusual and difficult but important. Briefly, then, I think that:

1. we are more endangered by too much energy too soon than by too little too late, for we understand too little the wise use of power;
2. we know next to nothing about the carefully designed natural systems and cycles on which we depend; we must therefore take care to preserve resilience and flexibility, and to design for large safety margins (whose importance we do not yet understand), recognizing the existence of human fallibility, malice, and irrationality (including our own) and of present trends that erode the earth's carrying capacity;
3. people are more important than goods; hence energy, technology, and economic activity are means, not ends, and their quantity is not a measure of welfare; hence economic rationality is a narrow and often defective test of the wisdom of broad social choices, and economic costs and prices, which depend largely on philosophical conventions, are neither revealed truth nor a meaningful test of rational or desirable behavior;
4. though the potential for growth in the social, cultural, and spiritual spheres is unlimited, resource-crunching material growth is inherently limited (a consequence of the round-earth theory) and, in countries as affluent as the U.S., should be not merely stabilized but returned to sustainable levels at which the net marginal utility of economic activity (to borrow for a moment the economist's abstractions) is clearly positive;
5. since sustainability is more important than the momentary advantage of any generation or group, long-term discount rates should be zero or even slightly negative, reinforcing a frugal, though not penurious, ethic of husbanding;
6. the energy problem should be not how to expand supplies to meet the postulated extrapolative needs of a dynamic economy, but rather how to accomplish social goals elegantly with a minimum of energy and effort, meanwhile taking care to preserve a social fabric that not only tolerates but encourages diverse values and lifestyles;
7. the technical, economic, and social problems of fission technology are so intractable, and technical efforts to palliate those problems are politically so dangerous, that we should abandon the technology with due deliberate speed;
8. many other energy technologies are exceedingly unattractive and should be developed and deployed sparingly or not at all (such as nuclear fusion, large coal-fired power stations and conversion plants, many current coal-mining technologies, urban-sited terminals for liquefied natural gas, much Arctic and offshore petroleum extraction, most "unconventional" hydrocarbons, and many "exotic" large-scale solar technologies such as solar satellites and monocultural biomass plantations);
9. ordinary people are qualified and responsible to make these and other energy choices through the democratic political process, and on the social and ethical issues central to such choices the opinion of any technical expert is entitled to no special weight; for although hu-

manity and human institutions are not perfectable, legitimacy and the nearest we can get to wisdom both flow, as Jefferson believed, from the people, whereas pragmatic Hamiltonian concepts of central governance by a cynical elite are unworthy of the people, increase the likelihood and consequences of major errors, and are ultimately tyrannical;

10. issues of material growth are inseparable from the more important issues of distributional equity, both within and among nations; indeed, high growth in overdeveloped countries is inimical to development in poor countries;
11. for poor countries, the self-reliant ecodevelopment concepts inherent in the New Economic Order approach are commendable and practicable while the patterns of industrial development that served the OECD countries in the different circumstances of the past two centuries are not: indeed, so much have conditions changed that ecodevelopment concepts are now the most appropriate for the rich countries too;
12. national interests lie less in traditional geopolitical balancing acts than in striving to attain a just and equitable, therefore peaceful, world order, even at the expense of temporary commercial advantage....

ASYMMETRIES IN POLICY

The future is no more certain today than it has ever been, and we hear much about making decisions under uncertainty. But to do this properly we must be wary of the danger of not being imaginative enough to see how undetermined the future is and how far we can shape it. As Kenneth Boulding remarks, deciding under uncertainty is bad enough, but deciding under an illusion of certainty is catastrophic.

It is therefore important not to reject out of hand futures that are clearly possible on the basis of Boulding's First Law ("Anything that exists is possible"). As Daly reminds us,

> ... the "low quad scenario" (one-half current U.S. per capita [primary] energy usage) *exists* today in Western Europe. It also existed in the United States as recently as 1960. Therefore the common notion that the low demand scenario is "far out"[,] or merely a hypothetical polar case, is due to inability to recognize the obvious. It is the high demand, hard technology scenarios that have never before been experienced and are completely hypothetical. Yet our "crackpot realists" all treat the hypothetical high energy projection as if it were empirically verified, and the empirically verified low energy scenario as if it were the flimsiest conjecture!
>
> ... Many people do not like to face up to this basic choice because it is not a question of rationality of means, but of sanity of ends. Taking a position requires moral self-definition, imposes responsibility, and may involve one in conflict. At this stage of the discussion refusal to take a position, accompanied by the usual

call for "more studies," serves only to increase the already excessive output of unconscionable mush....

HOW CAN WE GET THERE FROM HERE?

Who needs how much of what kind of energy for what purpose for how long?... [A]ddressing this basic question suggests a different path along which our energy system can evolve from now on: a way of redirecting our efforts, thus disproportionately freeing resources for other tasks that can use them more effectively. To do this requires us to take three initial steps that will enable ordinary market and social processes to complete the job.

The first of these steps is correcting the institutional barriers... that now impede conservation and rational supply technologies. The second step is removing the subsidies now given to conventional fuel and power industries—now estimated at well over $10 billion per year in the United States alone—and vigorously enforcing antitrust laws. The third step is gradually making energy prices consistent with what it will cost in the long run to replace our dwindling stocks of cheap fuels....

Changing any policy, even one that is plainly unworkable, is never easy. It entails doubt, conflict, trial, error, and hard work. Taking the initial steps toward a soft energy path, and following up to be sure they work, will not be easy—only easier than not taking them. But if wisely handled they can have enormous political appeal. Instead of trading off one constituency against another—unemployment versus inflation, economic growth versus environmental quality, inconvenience versus vulnerability—a soft path offers advantages for every constituency.... [A] soft path simultaneously offers jobs for the unemployed, capital for businesspeople, environmental protection for conservationists, enhanced national security for the military, opportunities for small-business to innovate and for big business to recycle itself, exciting technologies for the secular, a rebirth of spiritual values for the religious, traditional virtues for the old, radical reforms for the young, world order and equity for globalists, energy independence for isolationists, civil rights for liberals, states' rights for conservatives.

Thus, though present policy is consistent with the perceived short-term interests of a few powerful institutions, a soft path is consistent with far more strands of convergent social change at the grass roots. It goes with, not against, our political grain. And it is compatible with innovations in a great many other areas of public policy that we ought to be making anyhow for other reasons, and that plainly have the country behind them. If we free some log jams of outmoded perceptions that stifle our present approach to the energy problem, we can release a flood of change for the better.

Perhaps our salvation will yet be that the basic issues in energy strategy, far from being too complex and technical for ordinary people to understand, are on the contrary too simple and political for experts to understand. We must concentrate on these simple yet powerful ideas, not only if we are to gain a fuller understanding of the consequences of choice, but also if we are to appreciate the very wide range of choices that is available. And here too we must be symmetrical: if we do not like some aspects of the soft energy path, we must consider whether we prefer the hard path and all its consequences. Robert Frost, in the [following] poem . . . , had an unmentioned third choice—bushwhacking through the shrubbery—but if we have that choice too, nobody has yet discovered it. The soft and the hard energy paths, and a myriad variations on their themes, appear to be the only choices there are, and we must decide which we prefer.

THE ROAD NOT TAKEN[1]

Two roads diverged in a yellow wood,
And sorry I could not travel both
And be one traveller, long I stood
And looked down one as far as I could
To where it bent in the undergrowth;

Then took the other, as just as fair,
And having perhaps the better claim,
Because it was grassy and wanted wear;
Though as for that the passing there
Had worn them really about the same,

And both that morning equally lay
In leaves no step had trodden black.
Oh, I kept the first for another day!
Yet knowing how way leads on to way,
I doubted if I should ever come back.

I shall be telling this with a sigh
Somewhere ages and ages hence:
Two roads diverged in a wood, and I—
I took the one less travelled by,
And that has made all the difference.

—ROBERT FROST

NOTES

1. From THE POETRY OF ROBERT FROST edited by Edward Connery Lathem. Copyright 1916, © 1969 by Holt, Rinehart and Winston. Copyright 1944 by Robert Frost. Reprinted by permission of Holt, Rinehart and Winston, Publishers, and Jonathan Cape Ltd, Publishers.

6.2 CHRISTOPHER FLAVIN AND SETH DUNN

Reinventing the Energy System

The Arab oil embargo of 1973–1974 and the disruption in petroleum supplies due to the overthrow of the Shah of Iran in 1978 both resulted in gasoline shortages and long lines at the gas pumps in the United States. Headline news stories about an energy crisis increased dramatically. Recognition of future vulnerability to perhaps more drastic crises resulted in numerous governmental and private studies of energy supply and demand. These studies focused on energy usage and resources in the rest of the world as well as in the United States. Several of the best-publicized of these studies, such as the Energy Project initiated at the Harvard Business School in 1972, concluded that the days of cheap petroleum were over. Although this conclusion appears to have been premature, there is little doubt that limits on recoverable petroleum resources will reduce the world's reliance on this convenient energy source in the future.

Christopher Flavin is the senior vice president and director of research programs at the Worldwatch Institute and Seth Dunn is a research associate. Both are members of the Institute's energy and climate change research team. In "Reinventing the Energy System," in Lester B. Brown et al., *State of the World 1999: A Worldwatch Institute Report on Progress Toward a Sustainable Society* (W. W. Norton, 1999), from which the following selection is taken, Flavin and Dunn argue that the combined effects of changing societal needs, the development of new technologies, and serious global environmental problems are likely to speed the transition to a new world energy system in the early part of the twenty-first century. They predict an accelerating shift to the more efficient use of energy and the replacement of fossil fuels by renewable sources of the sort advocated by Amory B. Lovins in *Soft Energy Paths: Toward a Durable Peace* (Friends of the Earth, 1977). Flavin and Dunn state that this energy transition may "reestablish the positive but too often neglected connections between energy, human well-being, and the environment."

Key Concept: forging an energy system to meet world needs in the twenty-first century

Christopher Flavin and Seth Dunn

When the American Press Association gathered the country's "best minds" on the eve of the 1893 Chicago World's Fair and asked them to peer a century into the future, the nation's streets were filled with horse-drawn carriages and illuminated at night by gas lights that were still considered a high-tech novelty. And coal—whose share of commercial energy use had risen from 9 percent in 1850 to more than 60 percent in 1890—was expected to remain dominant for a long time to come.

The commentators who turned their crystal balls toward the nation's energy system foresaw some major changes—but missed others. They anticipated, for example, that "Electrical power will be universal.... Steam and all other sorts of power will be displaced." But while some wrote of trains traveling 100 miles an hour and moving sidewalks, none predicted the ascent of oil, the proliferation of the automobile, or the spread of suburbs and shopping malls made possible by cars. Their predictions also missed the many ways in which inexpensive energy would affect lives and livelihoods through the advent of air-conditioning, television, and continent-bridging jet aircraft. Nor did they foresee that oil and other fossil fuels would one day be used on such a scale as to raise sea levels, disrupt ecosystems, or increase the intensity of heat waves, droughts, and floods.

To most of today's energy futurists, the current system might seem even more solid and immutable than the nineteenth-century system appeared 100 years ago. The internal combustion engine has dominated personal transportation in industrial countries for more than eight decades, and electricity is now so taken for granted that any interruption in its supply is considered an emergency. Today the price of energy is nearly as low—in terms of consumer purchasing power—as it has ever been, and finding new energy sources that are more convenient, reliable, and affordable than fossil fuels is beyond the imagination of many experts. Former Eastern Bloc countries seek economic salvation in oil booms, while China and other developing nations are rushing to join the oil era—pouring hundreds of billions of dollars into the construction of coal mines, oil refineries, power plants, automobile factories, and roads.

Fossil fuels—coal, oil, and natural gas—that are dug or pumped from the ground, then burned in engines or furnaces, provide 90 percent or more of the energy in most industrial countries and 75 percent of energy worldwide. (See Table 1.) They are led by petroleum, the most convenient and ubiquitous among them—an energy source that has shaped the twentieth century, and that now seems irreplaceable. But as the Chicago World's Fair writings remind us, energy forecasts can overlook what later seems obvious. A close examination of technological, economic, social, and environmental trends suggests that we may already be in the early stages of a major global energy transition—one that is likely to accelerate early in the next century.

To understand energy in world history is to expect the unexpected. And as we live in a particularly dynamic period, the least likely scenario may be that the energy picture 100 years from now will closely resemble that of today. Although the future remains, as always, far from crystal clear, the broad outlines of a new energy system may now be emerging, thanks in part to a series of revolutionary new technologies and approaches. These developments suggest

TABLE 1

World Energy Use, 1900 and 1997

Energy Source	1900 Total	1900 Share	1997 Total	1997 Share
	(million tons of oil equivalent)	(percent)	(million tons of oil equivalent)	(percent)
Coal	501	55	2,122	22
Oil	18	2	2,940	30
Natural gas	9	1	2,173	23
Nuclear	0	0	579	6
Renewables[1]	383	42	1,833	19
Total	911	100	9,647	100

[1] Includes biomass, hydro, wind, geothermal, and solar energy.

that our future energy economy may be highly efficient and decentralized, using a range of sophisticated electronics. The primary energy resources for this system may be the most abundant ones on Earth: the sun, the wind, and other "renewable" sources of energy. And the main fuel for this twenty-first-century economy could be hydrogen, the lightest and most abundant element in the universe.

This transition would in some sense be a return to our roots. *Homo sapiens* has relied for most of its existence on a virtually limitless flow of renewable energy resources—muscles, plants, sun, wind, and water—to meet its basic needs for shelter, heat, cooking, lighting, and movement. The relatively recent transition to coal that began in Europe in the seventeenth century marked a major shift to dependence on a finite stock of fossilized fuels whose remaining energy is now equivalent to less than 11 days of sunshine. From a millennial perspective, today's hydrocarbon-based civilization is but a brief interlude in human history.

The next century may be as profoundly shaped by the move away from fossil fuels as this century was marked by the move toward them. Although it may take several decades for another system to fully develop, the underlying markets could shift abruptly in the next few years, drying up sales of conventional power plants and cars in a matter of years and affecting the share prices of scores of companies. The economic health—and political power—of nations could be sharply boosted or diminished. And our industries, homes, and cities could be transformed in ways we can only begin to anticipate.

Through the ages, the evolution of human societies has both influenced and been influenced by changes in patterns of energy use. But the timing of this next transition will be especially crucial. Today's energy system completely bypasses roughly 2 billion people who lack modern fuels or electricity, and underserves another 2 billion who cannot afford most energy amenities, such as refrigeration or hot water. Moreover, by relying on the rapid depletion of non-renewable resources and releasing billions of tons of combustion gases into the atmosphere, we have built the economy on trends that cannot possibly be sustained for another century. The efforts made today to lay the foundations for a new energy system will affect the lives of billions of people in the twenty-first century and beyond.

PRIME MOVERS

Energy transitions do not occur in a vacuum. Past shifts have been propelled by technological change and a range of social, economic, and environmental forces. Understanding these developments is essential for mapping out the path that humanity may follow in the next 100 years. The emergence of an oil-based economy at the beginning of this century, for example, was influenced by rapid scientific advances, the growing needs of an industrial economy, mounting urban environmental problems in the form of smoke and manure, and the aspirations of millions for higher living standards and greater mobility.

Resource limits are one force that could help push the world away from fossil fuels in the coming decades.... Natural gas and coal are both available in sufficient amounts to last until the end of the twenty-first century or beyond—but oil is not. Just as seventeenth-century Britain ran out of cheap wood, today we face the danger of running out of inexpensive petroleum.

Although oil markets have been relatively stable for more than a decade, and real prices approached historical lows in 1998, estimates of the underlying resource base have increased very little. Most of the calm in the oil markets of the 1990s has been due to slower demand growth, not an increase in supply. Despite prodigious exploration efforts, known oil resources have expanded only marginally in the last quarter-century, though some nations have raised their official reserve figures in order to obtain larger OPEC production quotas. Approximately 80 percent of the oil produced today comes from fields discovered before 1973, most of which are in decline. Total world production has increased less than 10 percent in two decades.

In a recent analysis of data on world oil resources, geologists Colin Campbell and Jean Laherrere estimate that roughly 1 trillion barrels of oil remain to be extracted. Since 800 billion barrels have already been used up, this suggests that the original exploitable resource base is nearly half gone. As extraction of a nonrenewable resource tends to follow a bell-shaped curve, these figures can be extrapolated to project that world production will peak by 2010, and then begin to decline. Applying the more optimistic resource estimates of other oil experts would push back this production pinnacle by just a decade.

A peak in world oil production early in the new century would reverberate through the energy system. The problem is not just the large amount of oil currently used—67 million barrels daily—but the intent of many developing countries, most lacking much oil of their own, to increase their use of automobiles and trucks. Meeting the growing needs of China, India, and the rest of the developing world in the way industrial countries' demands are met today would require a tripling of world oil production, even assuming no increases in industrial-country use. Yet production capacity in 2020 is unlikely to be much above current levels—and may well be declining.

Long before we completely run out of fossil fuels, however, the environmental and health burdens of using them may force us toward a cleaner energy system. Fossil fuel burning is the main source of air pollution and a leading cause of water and land degradation. Combustion of coal and oil produces carbon monoxide and tiny particulates that have been implicated in lung cancer and other respiratory problems; nitrogen and sulfur oxides create urban smog, and bring acid rain that has damaged forests extensively. Oil spills, refinery operations, and coal mining release toxic materials that impair water quality. Increasingly, oil exploration disrupts fragile ecosystems and coal mining removes entire mountains. Although modern pollution controls have improved air quality in most industrial countries in recent decades, the deadly experiences of London and Pittsburgh are now being repeated in Mexico City, São Paulo, New Delhi, Bangkok, and many other cities in the developing world. Each year, coal burning is estimated to kill 178,000 people prematurely in China alone.

Beyond these localized problems, it is the cumulative, global environmental effects that now are calling the fossil fuel economy into question. More than 200 years have passed since we began burning the sequestered sunlight of fossilized plants that took millions of years to accumulate, but only recently has it become evident that the carbon those fuels produce is disrupting the Earth's radiation balance, causing the planet to warm. Fossil fuel combustion has increased atmospheric concentrations of the heat-trapping gas carbon dioxide (CO_2) by 30 percent since preindustrial times. CO_2 levels are now at their highest point in 160,000 years, and global temperatures at their highest since the Middle Ages. Experts believe human activities could be ending the period of relative climatic stability that has endured over the last 10,000 years, and that permitted the rise of agricultural and industrial society....

Stabilizing atmospheric CO_2 concentrations at safe levels will require a 60–80 percent cut in carbon emissions from current levels, according to the best estimates of scientists. The Kyoto Protocol to the U.N. Framework Convention on Climate Change, agreed to in December 1997, is intended to be a small step on this long journey—which would eventually end the fossil-fuel-based economy as we know it today.

Energy transitions are also shaped by the changing needs of societies. Historians argue that coal won out over wood and other renewable resources during the eighteenth and nineteenth centuries in part due to the requirements of the shift from a rural, agrarian society to an urban, industrial one. Abundant and concentrated forms of energy were required for the new industries and booming cities of the period. In this view, coal did not bring about the transition but adapted to it more quickly. Ironically, the success of watermills and windmills in promoting early industrialization led to expanding energy demands that could only be met by the coal-fired steam engine.

Today's fast-growth economic sectors are not the production of food or automobiles, but software, telecommunications, and a broad array of services—

from finance and news to education and entertainment. The Information Revolution will, like the Industrial Revolution, have its own energy needs—and will place a premium on reliability. Computer systems freeze up if power is cut off for a fraction of a second; heavy industries, such as chemical and steel production, now depend on semiconductor chips to operate. Yet the mechanical machines and networks of above-ground wires and pipelines that power current energy systems are vulnerable. Today's systems are also centralized, while much of the service economy can be conducted from far-flung locations that are connected through the Internet, and may require more localized, autonomous energy supplies than power grids or gas lines can provide. As with the water wheel, so with oil: the growing demands of the new economy might not be met by the energy system that helped launch it.

In the twenty-first century, the requirements of the developing world—where 80 percent or more of the new energy investment is expected to take place—are likely to be the leading driver of energy markets. Eighteenth-century Great Britain shifted to coal, and the twentieth-century United States to oil, in part to meet the demands of growing populations; similar changes might be expected as more than 5 billion people seek more convenient transportation, refrigeration, air-conditioning, and other amenities in the years ahead. Technologies that can meet the demands of developing nations at minimal cost may therefore assume prominent roles in the overall transition.

SYSTEMIC CHANGE

... Today a new energy system is gestating in the late-twentieth-century fields of electronics, synthetic materials, biotechnology, and software. The silicon semiconductor chip, promising increased processing power and miniaturization of electronic devices, allows energy use to be matched more closely to need. Wider use of these chips offers efficiency gains in appliances, buildings, industry, and transport, making it possible to control precisely nearly all energy-using devices. Electronic controls also enable a range of small-scale, modular technologies to challenge the large-scale energy devices of the twentieth century.

Breakthroughs in chemistry and materials science are also playing key roles in energy, providing sophisticated, lightweight materials that operate without the wear and tear of moving parts. Modern wind turbines use the same carbon-fiber synthetic materials found in bullet-proof vests, "gore-tex" synthetic membranes line the latest fuel cells, and new "super-insulation" that reduces the energy needs of buildings relies on the same aluminum foil vacuum process that keeps coffee fresh. The latest electrochemical window coatings can be adjusted to reflect or absorb heat and light in response to weather conditions and the time of day.

A particularly fertile area of advance is in lighting, where the search is on for successors to Thomas Edison's incandescent bulb. Improvements in small-scale electronic ballasts have given rise to the compact fluorescent lamp (CFL), which requires one quarter the electricity of incandescent bulbs and lasts 10 times as long. Manufacturers are now working on even more advanced models with tiny ballasts that work with any light socket, and that cost half as much as today's models. Yet the new light-emitting diode (LED), a solid-state semiconductor device that emits a very bright light when charged, is twice as efficient as CFLs and lasts 10 times as long. Today's LEDs produce red and yellow light, which limits their market to applications such as traffic signals and automobile taillights, but scientists believe that white-light versions will soon become practical.

Late-twentieth-century technology has also revived an ancient source of energy: the wind. The first windmills for grinding grain appeared in Persia just over 1,000 years ago, and eventually spread to China, throughout the Mediterranean, and to northern Europe, where the Dutch developed the massive machines for which the country is still known. Wind power emerged as a serious option for generating electricity when Danish engineers began to apply advanced engineering and materials in the 1970s. The latest versions, which are also manufactured by companies based in Germany, India, Spain, and the United States, have variable-pitch fiberglass blades that are as long as 40 meters, electronic variable speed drives, and sophisticated microprocessor controls. Wind power is now economically competitive with fossil fuel generated electricity, and the market, valued at roughly $2 billion in 1998, is growing more than 25 percent annually.

Use of the sun as an energy source is also being renewed by modern technology. The solar photovoltaic cell, a semiconductor device that turns the sun's radiation directly into electric current, is widely used in off-grid applications as a power source for satellites and remote communications systems, as well as in consumer electronic devices such as pocket calculators and watches. Improvements in cell efficiency and materials have lowered costs by 80 percent in the past two decades, and the cells are now being built into shingles, tiles, and window glass—allowing buildings to generate their own electricity. Markets are booming. The cost of solar cells will need to fall by another 50–75 percent in order to be fully competitive with coal-fired electricity, but automated manufacturing, larger factories, and more-efficient cells promise further cost reductions in the near future. Semiconductor research is also nurturing the development of a close cousin of the solar cell, the "thermophotovoltaic" cell, which can produce electricity from industrial waste heat.

The technology that could most transform the energy system, the fuel cell, was first discovered in 1829, five decades before the internal combustion engine. The fuel cell attracted considerable interest at the turn of the century but required efficiency improvements before its first modern application in the U.S. space program in the 1960s. Fuel cells use an electrochemical process that combines hydrogen and oxygen, producing water and electricity. Avoiding the inherent inefficiency of combustion, today's top fuel cells are roughly twice as efficient as conventional engines, have no moving parts, require little maintenance, are nearly silent, and emit only water vapor. Unlike today's power plants, they are nearly as economical on a small scale as on a large one. Indeed, they could turn the very notion of a power plant into something more closely resembling a home appliance.

Although the first fuel cells now run on natural gas—which can be separated into hydrogen and carbon dioxide—in the long term they may be fueled by pure hydrogen that is separated from water by using electricity, a process known as electrolysis. Researchers are also testing various catalysts that, when placed in water that is illuminated by sunlight, may one day produce inexpensive hydrogen. Chemists have recently developed a solar-powered "water splitter" that nearly doubles the efficiency of converting solar energy to hydrogen. Some scientists note that finding a cheap and efficient way to electrolyze water could make hydrogen as dominant an energy carrier in the twenty-first century as oil was in the twentieth. . . .

GREAT POWERS, GEOPOLITICAL PRIZES

. . . A solar-hydrogen economy would be based on resources that are more abundant and more evenly distributed. Some countries are better endowed than others: Mexico, India, and South Africa are particularly well positioned to deploy solar energy, while Canada, China, and Russia have especially large wind resources. But although some countries could export renewably generated electricity or hydrogen, few are likely to depend mainly on imports. The international energy balance might be more like the world food economy today, where some countries are net exporters and others importers, but the majority produce most of their own food. In other words, energy would become a more "normal" commodity, one not constantly on the verge of international crisis.

Since renewable energy resources are relatively evenly spread, leadership in the new industries is less likely to go to countries with the most resources than to those with the know-how, skilled labor force, openness to innovation, efficient financial structures, and strategic foresight to position themselves for the new era. Today, it is the world's three leading technological powers—Germany, Japan, and the United States—that are ahead in the development of many of the key devices. But nations need not be large or powerful to find a strategic niche, as demonstrated by Denmark's preeminence in wind power today. More than half the global wind power market is now supplied by Danish firms or licensees —an achievement made possible by a two-decade-long strategic partnership between government and industry.

The conditions for an energy transition are particularly ripe in developing countries, most of which are far better endowed with renewable energy sources than with fossil fuels. Most of these countries have embryonic energy systems and massively underserved populations, and therefore represent a potentially far larger market for innovative technologies. Developing nations are in position to bypass or "leapfrog" the twentieth-century systems that are quickly becoming outdated—and several of them, including Costa Rica, the Dominican Republic, and South Africa, have already plunged ahead with some of the new technologies. Given their large populations and surging energy demands, China and India are especially well positioned to become leading centers of the next energy system. This could mean a reversal in the flow of initiative and innovation between East and West—and could perhaps precipitate a broader shift in the world's economic and political center of gravity back to where it was a millennium ago: Asia. In the New World, Brazil, with its vast supplies of renewable resources, could also become a major player.

The relatively diffuse nature of renewable energy sources, and the need to accelerate their use worldwide, might help diminish international conflict and stimulate cooperation. The evolution of the energy system may be determined less by OPEC cartels and struggles over oil leases than by the ongoing international negotiations to protect the climate, as "de-carbonizing" the world economy becomes a greater "geopolitical imperative," yielding its own prizes. One small country that has already made such a strategic move is Iceland. In 1997 the small nation's Prime Minister announced a plan to convert Iceland to a "hydrogen economy" within 15 to 20 years; the government is working with Daimler-Benz and Ballard Power Systems to shift its fishing fleet to hydrogen, and its motor vehicle fleet to methanol and hydrogen. Icelandic officials are also exploring the prospects for exporting hydrogen to other countries.

ENERGY AND SOCIETY

... In addition to concentrating wealth and power, today's fossil-fuel-based system has engendered large imbalances in energy use and social well-being. Its benefits have not been extended to roughly 2 billion of the world's poor—a third of global population—who still rely on biomass for cooking and lack access to electricity. Today, the richest fifth of humanity consumes 58 percent of the world's energy, while the poorest fifth uses less than 4 percent. The United States, with 5 percent of the world's population, uses nearly one quarter of global energy supplies; on a per capita basis, it consumes twice as much energy as Japan and 12 times as much as China.

A more decentralized, renewable-resource-based energy system may have a better chance of spreading energy services more broadly. In fact, meeting the needs of the 2 billion people who do not have modern fuels or electricity and of another 2 billion who are badly underserved might become a new social imperative—akin to the push to electrify rural areas of the United States in the 1930s. Providing clean, advanced energy services would stimulate development in the poorer regions of the world, provide rural employment, and lessen the burden of daily wood gathering that now falls on hundreds of millions of women and children. The World Bank, which has devoted tens of billions of dollars to electrifying cities using central power plants over the past several decades, has recently undertaken a range of initiatives intended to provide decentralized, renewable power supplies to hundreds of millions of rural people.

Even with a shift to more energy-efficient technologies that rely on renewable resources, societies will have to confront basic consumption patterns in order to make the energy economy sustainable. In the United States, the energy efficiency gains of the past quarter-century have been overwhelmed by escalating consumer demand for energy services. U.S. per capita energy use neared its previous 1973 peak in the late 1990s, with gasoline use per person already at record levels. Increased driving, sports utility vehicles, larger homes, and "killer kitchens" with all the latest energy-hungry appliances have created an insatiable appetite for fuel.

The mass consumer culture of twentieth-century North America—and to a slightly lesser extent, Europe and Japan—has been predicated on a "high-energy society" that has viewed inexpensive, abundant energy as something of a constitutional right. But Americans' energy-intensive lifestyles, and the U.S.-led global energy consumption trend of the past century—a 10-fold increase, with a quadrupling since 1950—cannot possibly be a sustainable model for a population of more than 9 billion in the twenty-first century.

It will be far easier to meet the energy needs of the world in coming years if sufficiency replaces profligacy as the ethic of the next energy paradigm. This will require a breakthrough not so much in science or technology as in values and lifestyles. Modest changes, such as owning smaller cars and homes, or driving less and cycling more, would still leave us with lifestyles that are luxurious by historical standards but that are far more compatible with an energy system that can be sustained. Several studies show that societies that focus less on absolute consumption and more on improving human welfare can meet development goals with much lower energy requirements. Russia, for example, has higher per capita energy use but far lower living standards than Japan, whose economic success of the 1970s and 1980s was greatly assisted by its "delinking" of energy use and development.

The energetic challenge facing humanity is not unlike that confronting Russians a decade ago: creating a decentralized, demand-oriented system when a centrally planned, consumption-oriented economy has been the industrial norm for three generations. Like the Soviet system, the fossil-fuel-based model is losing authority as people become more aware of its negative social and environmental effects and the constrained choices that it offers. And like the reform movements that swept Central Europe in 1989, the new energy system must be built from the bottom up, by the actions of millions, through democratization of the energy decisionmaking process. Only through the efforts of a diverse cast of characters—activists protesting air pollution, consumers seeking lower energy bills, villagers demanding power, and industry captains pursuing profits—are societies likely to build a sustainable energy system.

Designing a new energy system suitable for the twenty-first century may help reestablish the positive but too often neglected connections between energy, human well-being, and the environment. Rather than treat energy as a commodity to be consumed without regard for its consequences, we might instead recover a much older notion of energy as something to be valued, saved, and used to meet our needs in ways that respect the realities of the natural world—thereby avoiding the kind of ecological catastrophe that has befallen civilizations that overdrew their environmental endowments. The sooner we can bring the fleeting hydrocarbon era to a close and accomplish the historic shift to a civilization based on the efficient use of renewable energy and hydrogen, the sooner we can stop drawing down the natural inheritance of future generations and begin investing in a livable planet.

PART THREE

Environmental Degradation

On the Internet . . .

Sites appropriate to Part Three

The National Wildlife Federation is dedicated to wilderness preservation and the protection of endangered species. The organization's home page contains links to information about these issues.

`http://www.nwf.org/nwf/`

This Web site contains text of the Clean Water Act and information concerning water pollution.

`http://www.usbr.gov/laws/cleanwat.html`

The Love Canal issue is explored on this comprehensive Web site.

`http://ublib.buffalo.edu/libraries/units/sel/exhibits/lovecanal.html`

Ozone Action is based in Washington, DC, and is a nonprofit public interest organization focused exclusively on two atmospheric threats: global warming and stratospheric ozone depletion. This is the organization's home page.

`http://www.ozone.org/index.html`

CHAPTER 7 Forests, Wilderness, and Wildlife

7.1 WILLIAM O. DOUGLAS

Sierra Club v. Morton

The courts have played a very significant role in the cause of environmental protection. In cases where conservation or antipollution legislation exists but is being ignored or violated, individuals or grassroots organizations have frequently been successful in bringing suit against the offending party or against the agency that is failing to enforce the law. More difficult is the task of preventing an act of environmental desecration that may do great ecological harm but that violates no specific state or federal law. In such cases it is generally necessary for the individuals or organizations seeking judicial restraint to demonstrate that they will suffer unreasonable personal injury or loss as a result of the project or activity in question.

This important problem is probably best illustrated by the 1972 U.S. Supreme Court case *Sierra Club v. Morton,* specifically, the dissenting opinion of Justice William O. Douglas (1898–1980), from which the following selection has been taken. The issue in question was the attempt by the Sierra Club to prevent Walt Disney Enterprises, Inc., from degrading the Mineral King Valley, which is adjacent to Sequoia National Park, by building a ski resort there. The Court denied the Sierra Club's petition because the club had not proven that it would be harmed by the Disney project and therefore did not have legal standing to sue.

Douglas's landmark dissent on this ruling was based in large part on arguments made by law professor Christopher Stone in his treatise entitled

"Should Trees Have Standing?" Douglas also gives proper credit to U.S. Forest Service officer Aldo Leopold, who had earlier argued for a land ethic that extends the boundaries of the community to include animate and inanimate components of the environment as well as human beings. That Justice Douglas took this position is no surprise in view of his reputation as an ardent conservationist and advocate of environmental protection. Indeed, Douglas had foreshadowed his position in this case by arguments he had made in several of the numerous books he authored, including *A Wilderness Bill of Rights,* published in 1964.

Key Concept: the legal defense of the environment

Mr. Justice DOUGLAS, dissenting.

I share the views of my Brother BLACKMUN and would reverse the judgment below.

The critical question of "standing" would be simplified and also put neatly in focus if we fashioned a federal rule that allowed environmental issues to be litigated before federal agencies or federal courts in the name of the inanimate object about to be despoiled, defaced, or invaded by roads and bulldozers and where injury is the subject of public outrage. Contemporary public concern for protecting nature's ecological equilibrium should lead to the conferral of standing upon environmental objects to sue for their own preservation. See Stone, Should Trees Have Standing?—Toward Legal Rights for Natural Objects, 45 S.Cal.L.Rev. 450 (1972). This suit would therefore be more properly labeled as Mineral King v. Morton.

Inanimate objects are sometimes parties in litigation. A ship has a legal personality, a fiction found useful for maritime purposes. The corporation sole —a creature of ecclessiastical law—is an acceptable adversary and large fortunes ride on its cases. The ordinary corporation is a "person" for purposes of the adjudicatory processes, whether it represents proprietary, spiritual, aesthetic, or charitable causes.

So it should be as respects valleys, alpine meadows, rivers, lakes, estuaries, beaches, ridges, groves of trees, swampland, or even air that feels the destructive pressures of modern technology and modern life. The river, for example, is the living symbol of all the life it sustains or nourishes—fish, aquatic insects, water ouzels, otter, fisher, deer, elk, bear, and all other animals, including man, who are dependent on it or who enjoy it for its sight, its sound, or its life. The river as plaintiff speaks for the ecological unit of life that is part of it. Those people who have a meaningful relation to that body of water—whether it be a fisherman, a canoeist, a zoologist, or a logger—must be able to speak for the values which the river represents and which are threatened with destruction.

I do not know Mineral King. I have never seen it nor traveled it, though I have seen articles describing its proposed "development" notably Hano, Protectionists vs. recreationists—The Battle of Mineral King, N.Y. Times Mag., Aug. 17, 1969, P. 25; and Browning, Mickey Mouse in the Mountains, Harper's, March 1972, p. 65. The Sierra Club in its complaint alleges that "[o]ne of the principal purposes of the Sierra Club is to protect and conserve the national resources of the Sierra Nevada Mountains." The District Court held that this uncontested allegation made the Sierra Club "sufficiently aggreived" to have "standing" to sue on behalf of Mineral King.

Mineral King is doubtless like other wonders of the Sierra Nevada such as Tuolumne Meadows and the John Muir Trail. Those who hike it, fish it, hunt it, camp in it, frequent it, or visit it merely to sit in solitude and wonderment are legitimate spokesmen for it, whether they may be few or many. Those who have that intimate relation with the inanimate object about to be injured, polluted, or otherwise despoiled are its legitimate spokesmen.

The Solicitor General... takes a wholly different approach. He considers the problem in terms of "government by the Judiciary." With all respect, the problem is to make certain that the inanimate objects, which are the very core of America's beauty, have spokesmen before they are destroyed. It is, of course, true that most of them are under the control of a federal or state agency. The standards given those agencies are usually expressed in terms of the "public interest." Yet "public interest" has so many differing shades of meaning as to be quite meaningless on the environmental front. Congress accordingly has adopted ecological standards in the National Environmental Policy Act of 1969, and guidelines for agency action have been provided by the Council on Environmental Quality of which Russell E. Train is Chairman.

Yet the pressures on agencies for favorable action one way or the other are enormous. The suggestion that Congress can stop action which is undesirable is true in theory; yet even Congress is too remote to give meaningful direction and its machinery is too ponderous to use very often. The federal agencies of which I speak are not venal or corrupt. But they are notoriously under the control of powerful interests who manipulate them through advisory committees, or friendly working relations, or who have that natural affinity with the agency which in time develops between the regulator and the regulated.

As early as 1894, Attorney General Olney predicted that regulatory agencies might become "industry-minded," as illustrated by his forecast concerning the Interstate Commerce Commission:

> "The Commission... is, or can be made, of great use to the railroads. It satisfies the popular clamor for a government supervision of railroads, at the same time that that supervision is almost entirely nominal. Further, the older such a commission gets to be, the more inclined it will be found to take the business and railroad view of things." M. Josephson, The Politicos 526 (1938).

Years later a court of appeals observed, "the recurring question which has plagued public regulation of industry [is] whether the regulatory agency is unduly oriented toward the interests of the industry it is designed to regulate, rather than the public interest it is designed to protect." Moss v. CAB, 139, U.S.App.D.C. 150, 152, 430 F.2d 891, 893....

The Forest Service—one of the federal agencies behind the scheme to despoil Mineral King—has been notorious for its alignment with lumber companies, although its mandate from Congress directs it to consider the various aspects of multiple use in its supervision of the national forests....

The voice of the inanimate object, therefore, should not be stilled. That does not mean that the judiciary takes over the managerial functions from the federal agency. It merely means that before these priceless bits of Americana (such as a valley, an alpine meadow, a river, or a lake) are forever lost or are so transformed as to be reduced to the eventual rubble of our urban environment, the voice of the existing beneficiaries of these environmental wonders should be heard.

Perhaps they will not win. Perhaps the bulldozers of "progress" will plow under all the aesthetic wonders of this beautiful land. That is not the present question. The sole question is, who has standing to be heard?

Those who hike the Appalachian Trail into Sunfish Pond, New Jersey, and camp or sleep there, or run the Allagash in Maine, or climb the Guadalupes in West Texas, or who canoe and portage the Quetico Superior in Minnesota, certainly should have standing to defend those natural wonders before courts or agencies, though they live 3,000 miles away. Those who merely are caught up in environmental news or propaganda and flock to defend these waters or areas may be treated differently. That is why these environmental issues should be tendered by the inanimate object itself. Then there will be assurances that all of the forms of life which it represents will stand before the court—the pileated woodpecker as well as the coyote and bear, the lemmings as well as the trout in the streams. Those inarticulate members of the ecological group cannot speak. But those people who have so frequented the place as to know its values and wonders will be able to speak for the entire ecological community.

Ecology reflects the land ethic; and Aldo Leopold wrote in A Sand County Almanac 204 (1949), "The land ethic simply enlarges the boundaries of the community to include soils, waters, plants, and animals, or collectively: the land."

That, as I see it, is the issue of "standing" in the present case and controversy.

7.2 WILLIAM CRONON

The Trouble With Wilderness; or, Getting Back to the Wrong Nature

Most environmentalists support the preservation of wilderness as an unquestioned ecological necessity. Thus, when respected University of Wisconsin environmental historian, William Cronon, wrote a short article, "The Trouble With Wilderness; or, Getting Back to the Wrong Nature," *The New York Times Magazine* (August 13, 1995), sharply attacking that perspective, he immediately became the focus of a heated controversy. Cronon further stoked the fires of that controversy by editing a book entitled *Uncommon Ground: Toward Reinventing Nature* (W.W. Norton, 1995), which contained essays by a select group of scholars from a variety of academic disciplines. These essays examine various aspects of the interface between human beings and nature. Most contain provocative, unconventional critiques of commonly held attitudes toward nature. Included in that book is a greatly expanded version of Cronon's *New York Times Magazine* article. It is from this version of the article that the following selection is taken.

Cronon attacks the very concept of wilderness. He views wilderness as being a human creation, reflecting particular cultural perspectives, rather than a pristine "natural" ecological state unaffected by human influence. He argues that the movement to preserve large tracts of land that are designated as wilderness is an elitist enterprise that is not only unnecessary, but ecologically counterproductive. Yet, Cronon considers himself to be an environmentalist and is even on the governing board of the Wilderness Society. In the face of his uncompromising denigration of the commonly held concept of wilderness, Cronon has confounded both his supporters and detractors by declaring that "wilderness is my religion." Such assertions have done little to reduce the animosity that Cronon's writings have engendered among many environmentalists who have accused him of misinterpreting the conservation movement, ignoring such serious issues as the need to preserve biodiversity, and paying too little attention to the political consequences of his work. With regard to this latter concern, it is a fact that whether wittingly or not, Cronon has become something of a hero to the antienvironmental lobby.

Key Concept: the concept of wilderness as a human construct

The time has come to rethink wilderness.

This will seem a heretical claim to many environmentalists, since the idea of wilderness has for decades been a fundamental tenet—indeed, a passion—of the environmental movement, especially in the United States. For many Americans wilderness stands as the last remaining place where civilization, that all too human disease, has not fully infected the earth. It is an island in the polluted sea of urban-industrial modernity, the one place we can turn for escape from our own too-muchness. Seen in this way, wilderness presents itself as the best antidote to our human selves, a refuge we must somehow recover if we hope to save the planet. As Henry David Thoreau, [a nineteenth-century writer and naturalist,] once famously declared, "In Wildness is the preservation of the World."

But is it? The more one knows of its peculiar history, the more one realizes that wilderness is not quite what it seems. Far from being the one place on earth that stands apart from humanity, it is quite profoundly a human creation—indeed, the creation of very particular human cultures at very particular moments in human history. It is not a pristine sanctuary where the last remnant of an untouched, endangered, but still transcendent nature can for at least a little while longer be encountered without the contaminating taint of civilization. Instead, it is a product of that civilization, and could hardly be contaminated by the very stuff of which it is made. Wilderness hides its unnaturalness behind a mask that is all the more beguiling because it seems so natural. As we gaze into the mirror it holds up for us, we too easily imagine that what we behold is Nature when in fact we see the reflection of our own unexamined longings and desires. For this reason, we mistake ourselves when we suppose that wilderness can be the solution to our culture's problematic relationships with the nonhuman world, for wilderness is itself no small part of the problem.

To assert the unnaturalness of so natural a place will no doubt seem absurd or even perverse to many readers, so let me hasten to add that the nonhuman world we encounter in wilderness is far from being merely our own invention. I celebrate with others who love wilderness the beauty and power of the things it contains. Each of us who has spent time there can conjure images and sensations that seem all the more hauntingly real for having engraved themselves so indelibly on our memories. Such memories may be uniquely our own, but they are also familiar enough to be instantly recognizable to others. Remember this? The torrents of mist shoot out from the base of a great waterfall in the depths of a Sierra canyon, the tiny droplets cooling your face as you listen to the roar of the water and gaze up toward the sky through a rainbow that hovers just out of reach. Remember this too: looking out across a desert canyon in the evening air, the only sound a lone raven calling in the distance, the rock walls dropping away into a chasm so deep that its bottom all but vanishes as you squint into the amber light of the setting sun. And this: the moment beside the trail as you sit on a sandstone ledge, your boots damp with the morning dew while you take in the rich smell of the pines, and the small red fox—or maybe for you it was a raccoon or a coyote or a deer—that suddenly ambles across your path, stopping for a long moment to gaze in your direction with cautious indifference before continuing on its way. Remember the feelings of such moments, and you will know as well as I do that you were in the presence of something irreducibly nonhuman, something profoundly Other than yourself. Wilderness is made of that too.

And yet: what brought each of us to the places where such memories became possible is entirely a cultural invention....

Wilderness had once been the antithesis of all that was orderly and good—it had been the darkness, one might say, on the far side of the garden wall—and yet now it was frequently likened to Eden itself. When John Muir, [a nineteenth-century writer and explorer,] arrived in the Sierra Nevada in 1869, he would declare, "No description of Heaven that I have ever heard or read of seems half so fine." He was hardly alone in expressing such emotions. One by one, various corners of the American map came to be designated as sites whose wild beauty was so spectacular that a growing number of citizens had to visit and see them for themselves. Niagara Falls was the first to undergo this transformation, but it was soon followed by the Catskills, the Adirondacks, Yosemite, Yellowstone, and others. Yosemite was deeded by the U.S. government to the state of California in 1864 as the nation's first wildland park, and Yellowstone became the first true national park in 1872.

By the first decade of the twentieth century, in the single most famous episode in American conservation history, a national debate had exploded over whether the city of San Francisco should be permitted to augment its water supply by damming the Tuolumne River in Hetch Hetchy valley, well within the boundaries of Yosemite National Park. The dam was eventually built, but what today seems no less significant is that so many people fought to prevent its completion. Even as the fight was being lost, Hetch Hetchy became the battle cry of an emerging movement to preserve wilderness. Fifty years earlier, such opposition would have been unthinkable. Few would have questioned the merits of "reclaiming" a wasteland like this in order to put it to human use. Now the defenders of Hetch Hetchy attracted widespread national attention by portraying such an act not as improvement or progress but as desecration and vandalism. Lest one doubt that the old biblical metaphors had been turned completely on their heads, listen to John Muir attack the dam's defenders. "Their arguments," he wrote, "are curiously like those of the devil, devised for the destruction of the first garden—so much of the very best Eden fruit going to waste; so much of the best Tuolumne water and Tuolumne scenery going to waste." For Muir and the growing number of Americans who shared his views, Satan's home had become God's own temple.

The sources of this rather astonishing transformation were many, but for the purposes of this [selection] they can be gathered under two broad headings: the sublime and the frontier. Of the two, the sublime is the older and more pervasive cultural construct, being one of the most important expressions of that broad transatlantic movement we today label as romanticism; the frontier is more peculiarly American, though it too had its European antecedents and parallels. The two converged to remake wilderness in their own image, freighting it with moral values and cultural symbols that it carries to this day. Indeed, it is not too much to say that the modern environmental movement is itself a grandchild of romanticism and post-frontier ideology, which is why it is no accident that so much environmentalist discourse takes its bearings from the wilderness these intellectual movements helped create. Although wilderness may today seem to be just one environmental concern among many, it in fact serves as the foundation for a long list of other such concerns that on their face seem quite remote from it. That is why its influence is so pervasive and, potentially, so insidious. . . .

The emotions [John] Muir describes in [his writings about] Yosemite could hardly be more different from [Henry David] Thoreau's on Katahdin or [William] Wordsworth's, [a nineteenth-century poet,] on the Simplon Pass. Yet all three men are participating in the same cultural tradition and contributing to the same myth: the mountain as cathedral. The three may differ in the way they choose to express their piety—Wordsworth favoring an awe-filled bewilderment, Thoreau a stern loneliness, Muir a welcome ecstasy—but they agree completely about the church in which they prefer to worship. Muir's closing words on North Dome diverge from his older contemporaries only in mood, not in their ultimate content:

> Perched like a fly on this Yosemite dome, I gaze and sketch and bask, oftentimes settling down into dumb admiration without definite hope of ever learning much, yet with the longing, unresting effort that lies at the door of hope, humbly prostrate before the vast display of God's power, and eager to offer self-denial and renunciation with eternal toil to learn any lesson in the divine manuscript.

. . . But the romantic sublime was not the only cultural movement that helped transform wilderness into a sacred American icon during the nineteenth century. No less important was the powerful romantic attraction of primitivism, dating back at least to [Jean Jacques] Rousseau, [a French philosopher (1712–1778),]—the belief that the best antidote to the ills of an overly refined and civilized modern world was a return to simpler, more primitive living. In the United States, this was embodied most strikingly in the national myth of the frontier. The historian Frederick Jackson Turner wrote in 1893 the classic academic statement of this myth, but it had been part of American cultural traditions for well over a century. As Turner described the process, easterners and European immigrants, in moving to the wild unsettled lands of the frontier, shed the trappings of civilization, rediscovered their primitive racial energies, reinvented direct democratic institutions, and thereby reinfused themselves with a vigor, an independence, and a creativity that were the source of American democracy and national character. Seen in this way, wild country became a place not just of religious redemption but of national renewal, the quintessential location for experiencing what it meant to be an American.

One of Turner's most provocative claims was that by the 1890s the frontier was passing away. Never again would "such gifts of free land offer themselves" to the American people. "The frontier has gone," he declared, "and with its going has closed the first period of American history." Built into the frontier myth from its very beginning was the notion that this crucible of American identity was temporary and would pass away....

This nostalgia for a passing frontier way of life inevitably implied ambivalence, if not downright hostility, toward modernity and all that it represented. If one saw the wild lands of the frontier as freer, truer, and more natural than other, more modern places, then one was also inclined to see the cities and factories of urban-industrial civilization as confining, false, and artificial. Owen Wister, [an American writer of cowboy fiction (1860–1938),] looked at the post-frontier "transition" that had followed "the horseman of the plains," and did not like what he saw: "a shapeless state, a condition of men and manners as unlovely as is that moment in the year when winter is gone and spring not come, and the face of Nature is ugly." In the eyes of writers who shared Wister's distaste for modernity, civilization contaminated its inhabitants and absorbed them into the faceless, collective, contemptible life of the crowd. For all of its troubles and dangers, and despite the fact that it must pass away, the frontier had been a better place. If civilization was to be redeemed, it would be by men like the Virginian who could retain their frontier virtues even as they made the transition to post-frontier life.

The mythic frontier individualist was almost always masculine in gender: here, in the wilderness, a man could be a real man, the rugged individual he was meant to be before civilization sapped his energy and threatened his masculinity. Wister's contemptuous remarks about Wall Street and Newport suggest what he and many others of his generation believed—that the comforts and seductions of civilized life were especially insidious for men, who all too easily became emasculated by the feminizing tendencies of civilization. More often than not, men who felt this way came, like Wister and [Theodore] Roosevelt, from elite class backgrounds. The curious result was that frontier nostalgia became an important vehicle for expressing a peculiarly bourgeois form of antimodernism. The very men who most benefited from urban-industrial capitalism were among those who believed they must escape its debilitating effects. If the frontier was passing, then men who had the means to do so should preserve for themselves some remnant of its wild landscape so that they might enjoy the regeneration and renewal that came from sleeping under the stars, participating in blood sports, and living off the land. The frontier might be gone, but the frontier experience could still be had if only wilderness were preserved.

Thus the decades following the Civil War saw more and more of the nation's wealthiest citizens seeking out wilderness for themselves. The elite passion for wild land took many forms: enormous estates in the Adirondacks and elsewhere (disingenuously called "camps" despite their many servants and amenities), cattle ranches for would-be rough riders on the Great Plains, guided big-game hunting trips in the Rockies, and luxurious resort hotels wherever railroads pushed their way into sublime landscapes. Wilderness suddenly emerged as the landscape of choice for elite tourists, who brought with them strikingly urban ideas of the countryside through which they traveled. For them, wild land was not a site for productive labor and not a permanent home; rather, it was a place of recreation. One went to the wilderness not as a producer but as a consumer, hiring guides and other backcountry residents who could serve as romantic surrogates for the rough riders and hunters of the frontier if one was willing to overlook their new status as employees and servants of the rich.

In just this way, wilderness came to embody the national frontier myth, standing for the wild freedom of America's past and seeming to represent a highly attractive natural alternative to the ugly artificiality of modern civilization. The irony, of course, was that in the process wilderness came to reflect the very civilization its devotees sought to escape. Ever since the nineteenth century, celebrating wilderness has been an activity mainly for well-to-do city folks. Country people generally know far too much about working the land to regard *un*worked land as their ideal. In contrast, elite urban tourists and wealthy sportsmen projected their leisure-time frontier fantasies onto the American landscape and so created wilderness in their own image.

There were other ironies as well. The movement to set aside national parks and wilderness areas followed hard on the heels of the final Indian wars, in which the prior human inhabitants of these areas were rounded up and moved onto reservations. The myth of the wilderness as "virgin," uninhabited land had always been especially cruel when seen from the perspective of the Indians who had once called that land home. Now they were forced to move elsewhere, with the result that tourists could safely enjoy the illusion that they were seeing their nation in its pristine, original state, in the new morning of God's own creation. Among the things that most marked the new national parks as reflecting a post-frontier consciousness was the relative absence of human violence within their boundaries. The actual frontier had often been a place of conflict, in which invaders and invaded fought for control of land and resources. Once set aside within the fixed and carefully policed boundaries of the modern bureaucratic state, the wilderness lost its savage image and became safe: a place more of reverie than of revulsion or fear. Meanwhile, its original inhabitants were kept out by dint of force, their earlier uses of the land redefined as inappropriate or even illegal. To this day, for instance, the Blackfeet continue to be accused of "poaching" on the lands of Glacier National Park that originally belonged to them and that were ceded by treaty only with the proviso that they be permitted to hunt there.

The removal of Indians to create an "uninhabited wilderness"—uninhabited as never before in the human history of the place—reminds us just how invented, just how constructed, the American wilderness really is. To return to my opening argument: there is nothing natural about the concept of wilderness. It is entirely a creation of the culture that holds it dear, a product of the very history it seeks to deny. Indeed, one of the most striking proofs of the cultural invention of wilderness is its thoroughgoing erasure of the history from which it sprang. In virtually all of its manifestations, wilderness represents a flight from history. Seen as the original garden, it is a place outside of time, from which human beings had to be ejected before the fallen world of history could properly begin. Seen as the frontier, it is a savage world at the dawn of civilization, whose transformation represents the very beginning of the national historical epic. Seen as the bold landscape of frontier heroism, it is the place of youth and childhood, into which men escape by abandoning their pasts and entering a world of freedom where the constraints of civilization fade into memory. Seen as the sacred sublime, it is the home of a God who transcends history by standing as the One who remains untouched and unchanged by time's arrow. No matter what the angle from which we regard it, wilderness offers us the illusion that we can escape the cares and troubles of the world in which our past has ensnared us.

... Wilderness is the natural, unfallen antithesis of an unnatural civilization that has lost its soul. It is a place of freedom in which we can recover the true selves we have lost to the corrupting influences of our artificial lives. Most of all, it is the ultimate landscape of authenticity. Combining the sacred grandeur of the sublime with the primitive simplicity of the frontier, it is the place where we can see the world as it really is, and so know ourselves as we really are—or ought to be.

But the trouble with wilderness is that it quietly expresses and reproduces the very values its devotees seek to reject. The flight from history that is very nearly the core of wilderness represents the false hope of an escape from responsibility, the illusion that we can somehow wipe clean the slate of our past and return to the tabula rasa that supposedly existed before we began to leave our marks on the world. The dream of an unworked natural landscape is very much the fantasy of people who have never themselves had to work the land to make a living—urban folk for whom food comes from a supermarket or a restaurant instead of a field, and for whom the wooden houses in which they live and work apparently have no meaningful connection to the forests in which trees grow and die. Only people whose relation to the land was already alienated could hold up wilderness as a model for human life in nature, for the romantic ideology of wilderness leaves precisely nowhere for human beings actually to make their living from the land.

This, then, is the central paradox: wilderness embodies a dualistic vision in which the human is entirely outside the natural. If we allow ourselves to believe that nature, to be true, must also be wild, then our very presence in nature represents its fall. The place where we are is the place where nature is not. If this is so—if by definition wilderness leaves no place for human beings, save perhaps as contemplative sojourners enjoying their leisurely reverie in God's natural cathedral—then also by definition it can offer no solution to the environmental and other problems that confront us. To the extent that we celebrate wilderness as the measure with which we judge civilization, we reproduce the dualism that sets humanity and nature at opposite poles. We thereby leave ourselves little hope of discovering what an ethical, sustainable, *honorable* human place in nature might actually look like.

Worse: to the extent that we live in an urban-industrial civilization but at the same time pretend to ourselves that our *real* home is in the wilderness, to just that extent we give ourselves permission to evade responsibility for the lives we actually lead. We inhabit civilization while holding some part of ourselves —what we imagine to be the most precious part—aloof from its entanglements. We work our nine-to-five jobs in its institutions, we eat its food, we drive its cars (not least to reach the wilderness), we benefit from the intricate and all too invisible networks with which it shelters us, all the while pretending that these things are not an essential part of who we are. By imagining that our true home is in the wilderness, we forgive ourselves the homes we actually inhabit. In its flight from history, in its siren song of escape, in its reproduction of the dangerous dualism that sets human beings outside of nature—in all of these ways, wilderness poses a serious threat to responsible environmentalism at the end of the twentieth century.

By now I hope it is clear that my criticism in this [selection] is not directed at wild nature per se, or even at efforts to set aside large tracts of wild land, but rather at the specific habits of thinking that flow from this complex cultural construction called wilderness. It is not the things we label as wilderness that are the problem—for nonhuman nature and large tracts of the natural world *do* deserve protection—but rather what we ourselves mean when we use that label. Lest one doubt how pervasive these habits of thought actually are in contemporary environmentalism, let me list some of the places where wilderness serves as the ideological underpinning for environmental concerns that might otherwise seem quite remote from it. Defenders of biological diversity, for instance, although sometimes appealing to more utilitarian concerns, often point to "untouched" ecosystems as the best and richest repositories of the undiscovered species we must certainly try to protect. Although at first blush an apparently more "scientific" concept than wilderness, biological diversity in fact invokes many of the same sacred values, which is why organizations like the Nature Conservancy have been so quick to employ it as an alternative to the seemingly fuzzier and more problematic concept of wilderness. There is a paradox here, of course. To the extent that biological diversity (indeed, even wilderness itself) is likely to survive in the future only by the most vigilant and self-conscious management of the ecosystems that sustain it, the ideology of wilderness is potentially in direct conflict with the very thing it encourages us to protect.

The most striking instances of this have revolved around "endangered species," which serve as vulnerable symbols of biological diversity while at the same time standing as surrogates for wilderness itself. The terms of the Endangered Species Act in the United States have often meant that those hoping to defend pristine wilderness have had to rely on a single endangered species like the spotted owl to gain legal standing for their case—thereby making the full power of sacred land inhere in a single numinous organism whose habitat then becomes the object of intense debate about appropriate management and use. The ease with which anti-environmental forces like the wise-use movement have attacked such single-species preservation efforts suggests the vulnerability of strategies like these....

Wilderness gets us into trouble only if we imagine that this experience of wonder and otherness is limited to the remote corners of the planet, or that it somehow depends on pristine landscapes we ourselves do not inhabit. Nothing could be more misleading. The tree in the garden is in reality no less other, no less worthy of our wonder and respect, than the tree in an ancient forest that has never known an ax or a saw—even though the tree in the forest reflects a more intricate web of ecological relationships. The tree in the garden could easily have sprung from the same seed as the tree in the forest, and we can claim only its location and perhaps its form as our own. Both trees stand apart from us; both share our common world. The special power of the tree in the wilderness is to remind us of this fact. It can teach us to recognize the wildness we did not see in the tree we planted in our own backyard. By seeing the otherness in that which is most unfamiliar, we can learn to see it too in that which at first seemed merely ordinary. If wilderness can do this—if it can help us perceive and respect a nature we had forgotten to recognize as natural—then it will become part of the solution to our environmental dilemmas rather than part of the problem.

This will only happen, however, if we abandon the dualism that sees the tree in the garden as artificial—completely fallen and unnatural—and the tree in the wilderness as natural—completely pristine and wild. Both trees in some ultimate sense are wild; both in a practical sense now depend on our management and care. We are responsible for both, even though we can claim credit for neither. Our challenge is to stop thinking of such things according to a set of bipolar moral scales in which the human and the nonhuman, the unnatural and the natural, the fallen and the unfallen, serve as our conceptual map for understanding and valuing the world. Instead, we need to embrace the full continuum of a natural landscape that is also cultural, in which the city, the suburb, the pastoral, and the wild each has its proper place, which we permit ourselves to celebrate without needlessly denigrating the others. We need to honor the Other within and the Other next door as much as we do the exotic Other that lives far away—a lesson that applies as much to people as it does to (other) natural things. In particular, we need to discover a common middle ground in which all of these things, from the city to the wilderness, can somehow be encompassed in the word "home." Home, after all, is the place where finally we make our living. It is the place for which we take responsibility, the place we try to sustain so we can pass on what is best in it (and in ourselves) to our children. . . .

Learning to honor the wild—learning to remember and acknowledge the autonomy of the other—means striving for critical self-consciousness in all of our actions. It means that deep reflection and respect must accompany each act of use, and means too that we must always consider the possibility of non-use. It means looking at the part of nature we intend to turn toward our own ends and asking whether we can use it again and again and again—sustainably—without its being diminished in the process. It means never imagining that we can flee into a mythical wilderness to escape history and the obligation to take responsibility for our own actions that history inescapably entails. Most of all, it means practicing remembrance and gratitude, for thanksgiving is the simplest and most basic of ways for us to recollect the nature, the culture, and the history that have come together to make the world as we know it. If wildness can stop being (just) out there and start being (also) in here, if it can start being as humane as it is natural, then perhaps we can get on with the unending task of struggling to live rightly in the world—not just in the garden, not just in the wilderness, but in the home that encompasses them both.

7.3 JOHN VANDERMEER AND IVETTE PERFECTO

Rethinking Rain Forests: Biodiversity and Social Justice

There is no disagreement among ecologists about the importance of rain forests. These tropical ecosystems are home to more than three-quarters of all the species of plants and animals that inhabit the earth. There is also broad agreement that human beings are rapidly destroying this vital natural resource. What is in dispute is the precise cause of this carnage and what can be done to stop it.

John Vandermeer is Alfred Thurneau Professor of Biology, and Ivette Perfecto is an associate professor in the School of Natural Resources and the Environment at the University of Michigan. They have both devoted many years to the study of the ecology of tropical rain forests as well as to the social, political, and economic factors that threaten their existence. They have recently written a book detailing their findings entitled *Breakfast of Biodiversity: The Truth About Rain Forest Destruction,* published by Food First Books. In the following selection from "Rethinking Rain Forests: Biodiversity and Social Justice," *Food First Backgrounder* (Summer 1995), Vandermeer and Perfecto seek to dispel the myths that attribute the assaults on the forests to any one of several contributing factors. Instead, they argue that there is a "web of causality" linking all these factors and that only by designing an action plan that takes into account all the components of this web can the forests be saved.

Key Concept: the interacting factors causing rain forest destruction

The buzz is unmistakable. A huge chain saw cuts effortlessly through the wood of a beautiful rain forest tree, slicing up the trunk it has just felled into smaller bits to be taken away on giant lumber trucks. That image is fixed in our minds. It drives us to the same distraction it has driven so many before us. The rain forests are physically beautiful and contain the vast majority of our relatives on this planet. What sort of person would not be haunted by the sound of chain saws decimating them?

Yet another image is equally haunting. The bulldozed wooden shack, formerly the home of a poor family, constantly reminds us that lives as well as logs are being cut in most areas of tropical rain forests. Hungry children wander among the stumps of once majestic rain forest trees. Their mother cooks over an open fire, and their father fights the onslaught of weeds that continually threaten to choke out the crops the family needs for next year's food. All live in fear that the bulldozers will come again to destroy their present home. What sort of person would not be haunted by the existence of such poverty in a world of plenty?

But for us the power of these two images lies in the way they are connected, a fact we are reminded of every morning we slice up bananas on our breakfast cereal. The banana cannot be grown in the United States, yet it is one of the most popular fruits here. As we all know, it is produced in the world's tropical regions, usually in the same areas where rain forests have flourished in the past. The link between the decimated forest and the hungry children is the banana. That is why it is so easy, as we slice up a banana in Michigan, for our thoughts to wander to the image of the chain saw slicing up the rain forest trees and the children who view the banana as a staple food rather than a luxury.

The majority of life on earth lives in the rain forest. Close to 80% of the terrestrial species of animals and plants are to be found there. And this cradle of life is disappearing at an enormous rate. This is what the popular press has labeled as the "biodiversity" crisis.

Some view the problem from only a utilitarian point of view. It is obvious that we depend on biodiversity for the most elementary aspects of existence —plants convert the sun's energy to a usable form, animals convert unusable plants to a product we can use, bacteria in our stomachs help digest our food. There are a host of other critical functions of life's diversity and furthermore, future utilitarian designs on biodiversity are most likely to follow the patterns of the past—medicines and genes for new crops being the obvious examples.

Yet even if these utilitarian concerns were absent, the spiritual concern that the world's biodiversity is being destroyed should be enough to drive us to action. Less than 50% of the original tropical rain forests of the world are left, and at the present rate of destruction almost all will be gone by 2025. Our families, our memories—indeed a piece of our humanness—will have been destroyed forever. For this reason many have sounded the alarm and called for action.

While we echo this same alarm, we are concerned that the calls for action may not be correctly placed. Indeed, many of these calls are based on one myth or another about what is causing rain forest destruction. We feel that these myths act to mask the true issue. [Here] we present arguments against the five main myths of rain forest destruction and argue that a more complex understanding is necessary to grasp what is causing the destruction of the world's rain forests. So we begin with an analysis of the five myths and conclude with a description of "the causal web," the *true* cause of rain forest destruction.

MYTH ONE: Loggers and logging companies are decimating the rain forest.

Certainly the most immediate and visually spectacular cause of tropical rain forest destruction is logging. Cutting trees is nothing new. The use of rain forest wood has been traditional for most human societies in contact with these ecosystems. But the European invasion of tropical lands accelerated wood cutting enormously, as tropical woods began contributing to the development of the modern industrial society.

The direct consequences of logging, apart from the obvious and dramatic visual effects, are largely unknown. Some facts are deducible from general ecological principles, and a handful of studies have actually measured a few of the consequences, but a detailed knowledge of the direct consequences of logging is lacking.

What can be deduced from ecological principles is not that tropical forests are irreparably damaged by logging, but quite the contrary: tropical forests are potentially quite resilient to disturbance. While this is a debatable deduction, most of the debate centers on how fast a forest will recover after a major disturbance, such as logging, not on whether it will. The process of ecological succession inevitably begins after logging, and the proper question to ask, then, is: how long will it take for the forest to recover?

In analyzing the effects of logging, we cannot assume a uniform process. There are a variety of logging techniques, some likely to lead to rapid forest recovery, others necessitating a longer period for recovery. For example, local residents frequently chop down trees for their own use as fence posts, charcoal, or dugout canoes. Forest recovery after such an intrusion can be thought of as virtually instantaneous, since the removal of a single tree is similar to a tree dying of natural causes, a perfectly natural process that happens regularly in all forests. At the other extreme is clear cutting, the extraction of all trees in an area. Though the physical nature of a clear cut forest is spectacularly different from the mature forest, from other perspectives the damage is not quite as dramatic as it appears. The process of secondary succession that begins immediately after such logging leads rapidly to the establishment of secondary forest. A great deal of biological diversity is contained in a secondary forest. Indeed, a late secondary forest is likely to appear indistinguishable from an old-growth forest to all but the most sophisticated observer, even though it may have been initiated from a clear cut. Large expanses of secondary forest may even contain more biological diversity than similar expanses of old-growth forest. No studies thus far have followed such an area to its return to a "mature" forest again, but a reasonable estimate is that it would take something on the order of 40 to 80 years before the area begins to regain the structure of an old growth forest.

Probably the most common type of commercial logging is not the clear cutting described above but, rather, selective logging. In an area of tropical forest that may contain 400 or more species of trees, only twenty or thirty will be of commercial importance. Thus, a logging company usually seeks out areas with particularly large concentrations of the valuable species and ignores the rest. Often the wood is so valuable that it makes economic sense to build a road to extract just a few trees. Yet these roads offer new access to the forest for hunters, miners, and peasant agriculturists. In most situations this aspect of selective logging contributes most egregiously to deforestation, but it is obviously an *indirect* consequence of the logging operation itself.

A selectively logged forest is damaged, but not destroyed. Even a single year after the selective logging the forest begins taking on the appearance of a "real" forest. If no further cutting occurs, the selectively logged forest may regain the structural features of old growth after ten or twenty years. Although the scars of selective logging will remain for decades to a trained eye, the general structure of the forest may rapidly return. But this is not to say selective logging is, in the end, benign. The roads and partial clearings are obvious entrance points for peasant agriculture, as described below.

MYTH TWO: Peasant farmers are increasing in numbers and cut down rain forests to make farms to feed their families.

This myth is especially popular among neo-Malthusians. The explosive growth in the population of poor people in most tropical countries of the world is seen as a consequence of the basic forces that cause populations to grow generally, and a simple extrapolation suggests that even if this is not the main problem now, it certainly will be if population growth is not somehow curtailed.

Debunking the neo-Malthusian myth is not our purpose here; that has been done well elsewhere. Rather, laying the blame for the destruction of the forest on the peasant farmer is really blaming the victim. Peasant farmers in most rain forest areas are forced to farm under circumstances that are unfavorable, to say the least, from both an ecological and sociopolitical point of view.

At the outset, we must acknowledge the temptation to assume that, in rain forest areas, the potential for agriculture is great. Since there is neither winter nor lack of water, two of the main limiting factors for agriculture in other areas of the world, it is easy to conclude that production might very well be cornucopian. The tremendously lush vegetation of a tropical rain forest only heightens this impression, and indeed this perception may ultimately be valid. The ability to produce for twelve months of the year without worrying about irrigation is definitely a positive aspect to farming in such regions. But, so far at least, the woes are almost insurmountable, as most farmers forced to cultivate in rain forest areas can attest.

The first problem is the soils. Rain forest soils are usually acidic, made up of clay that cannot store nutrients well, and very low in organic matter. Even if nutrients are added to the soil they will be utilized relatively inefficiently because of the acidity, and then they will be washed out of the system because of its low storage capacity.

This problem is actually exacerbated by the forest itself. Because tropical rain forest plants have grown in these poor soil conditions for millions of years, they have evolved mechanisms for storing the system's nutrients in their vegetative matter (leaves, stems, roots, etc.) If they did not, much of the nutrient material would simply wash out of the system and no longer be available to them. This means that a vast majority of the nutrients in the ecosystem are stored in plant material rather than in the soil.

Consequently when a forest is cut down and burned, the nutrients in the vegetation are immediately made available to any crops that have been planted. The crops grow vigorously at first, but any nutrients unused during the first growing season will tend to leach out of the system. The "poverty" of the soil only becomes evident during the second growing season. This pattern is especially invidious when migrant farmers from areas with relatively stable soils arrive in a rain forest area. The first year they may produce a bumper crop, which creates a false sense of security. Then, if the second year is not a complete failure, almost certainly the third or fourth is, and the farmer is forced to move on to cut down another piece of forest.

A second problem is insects, diseases and weeds. The magnitude of the pest problem is often not fully anticipated by farmers or planners, and it is only after problems arise that the surprised agronomists become concerned. This is unfortunate, since one of the few things we can predict with confidence is that when rain forest is converted to agriculture, many pests arrive. The herbivores that used to eat the plants of the rain forest are not eliminated when the forest is cut. They are representatives of the massive biodiversity of tropical rain forests, and the potential number of them is enormous. Herbivores can devastate farmers' fields, and are able to destroy an entire crop in days.

A third problem is that because of the uniformly moist and warm environment, organisms that cause crop diseases find rain forest habitats quite hospitable. Consequently, the potential for losing crops to disease is far greater than in more temperate climates. Finally, just as the hot, wet environment is agreeable for crops, it is also agreeable for competitive plants. Since no two plants can occupy the same space, frequently the crop falls victim to the more aggressive vines and grasses that colonize open areas in tropical rain forest zones. Weeds are thus an especially difficult problem.

These, then, are some of the ecological problems faced by the peasant farmer seeking to establish a farm in a rain forest area. Sociopolitical forces, however, are far more devastating. And most of those sociopolitical forces are associated with a different form of agriculture—modern export agriculture.

When a modern export agricultural operation is set up, it tends to do two things regarding labor. First, it purchases, or sometimes steals, land from local peasant farmers, thus forcing them to move onto more marginal lands, with the kinds of problems we described above. Second, it frequently requires more labor than is locally available, thus acting as a magnet to attract unemployed people from other regions. Indeed, in most rain forest areas this magnet effect is a far more important factor leading to increased local populations than population growth.

But the modern agricultural operation, as detailed in the following section, is subject to dramatic fluctuations in production, since it is usually intimately connected with world agricultural commodity markets. Thus, there is a highly variable need for this labor, which means that today's workers always face the prospect of becoming tomorrow's peasant farmers.

In the contemporary world most peasant farmers find themselves in this precarious position. While it is true that many indigenous groups have lived and farmed in rain forest areas for hundreds of years and certainly deserve the world's attention and support in their attempts at preserving traditional ways of life, the vast majority of poor peasant farmers today are not indigenous. Rather, they are people who have been marginalized by a politico-economic system that needs them to serve as laborers when times are good, and to take care of themselves when times are not. As long as times are good, the banana workers of Central America have jobs. But when economies sour, many of those banana workers suddenly become peasant farmers.

So in the end, the myth of the peasant farmers causing rain forest destruction is perhaps true in the narrow sense that a knitting needle causes yarn to form a sweater. But little understanding of what really drives the process is gained from the simple observation that a peasant's ax can chop a rain forest tree.

MYTH THREE: The transformation of rain forests into large-scale export agriculture is the main factor leading to deforestation.

Given the above description of how peasant agriculture is driven by industrialized agricultural activities, it is no wonder that many have concluded that the modern export agricultural system is the ultimate culprit. Furthermore, the images of large cattle ranchers purposefully burning Amazon rain forests to make cattle pastures fuels this interpretation. Again, there is some merit to this position. However, we feel that it, too, is an inappropriate window through which to view the problem of rain forest destruction.

The direct action of large modern agricultural enterprises is not really as involved in direct rain forest destruction as is popularly believed. Burning Amazon rain forests to replace them with cattle ranches is certainly an example of the direct destruction of rain forests by "big" agriculture. But the vast majority of modern agricultural transformations in tropical areas are confined to areas that had already been converted to agriculture. Developers of expanding banana plantations of Central America claim, for example, to be cutting no primary forest at all. While we doubt their full sincerity, it does seem that about 90% of the current expansion is into areas that had long ago been deforested. Attributing direct deforestation to them is, as they argue, probably quite unfair. On the other hand, their activities are not totally unrelated to the problem, as can be easily seen from a closer examination of their underlying structure.

The basic structure of modern agriculture is frequently misunderstood because of an overly romantic notion of agriculture—the small, independent, family farm, rich with tradition and a work ethic that even a Puritan could be impressed with. Such romanticism is fueled by a confusion between farming and agriculture.

Farming is a resource transformation process in which land, seed, and labor are converted into, for example, peanuts. It is Farmer Brown cultivating the land, sowing the seed, and harvesting the peanuts. Agriculture is the decision to invest money in this year's peanut production; the use of a tractor and cultivator to prepare the land; an automatic seeder for planting; application of herbicides, insecticides, fungicides, nematodes and bactericides to kill unwanted pieces of the ecology; automatic harvest of the commodity; sale of the commodity to a processing company where it is ground up and emulsifiers, taste enhancers, stabilizers and preservatives are added; packing in convenient "pleasing-to-the-consumer" jars; and, finally, marketing under a sexy brand name. In short, while farming is the production of peanuts from the land, agriculture is the production of peanut butter from petroleum. Over the last two hundred years, and especially in the last fifty, much farming has been transformed into agriculture.

The consequence of this evolution is that modern agriculture is remarkably intrusive on local ecologies. Take, for example, the establishment of a banana plantation. When the banana export business began, local peasant farmers grew most of the bananas and sold them to shipping companies. Gradually, the shipping companies turned into the banana producers, with huge areas devoted to the monocultural production of this single crop. To establish a modern banana plantation it is often necessary to construct a complex system of hydrological control wherein the soil is leveled and crisscrossed with drainage channels, significantly altering the physical nature of the soil. Contemporary banana production even includes burying plastic tubing in the ground to eliminate the natural variability in subsurface water depth. Metal monorails hang from braces placed into cement footings to haul the bunches of bananas. To avert fungal diseases, heavy use of fungicides is required, and because of the large scale of the operation chemical methods of pest control are the preferred option. The banana plants create an almost complete shade cover and thus replace all residual vegetation. Pesticide application is sometimes intense, other times almost absent, depending on conditions, but over the long run one can expect an enormous cumulative input of pesticides, the long-term consequences of which are unknown but likely to be unhealthy for both workers and the environment.

A major social transformation is also required. Banana production tends to promote a local "overpopulation crisis" by encouraging a great deal of migration into the area. As the international market for bananas ebbs and flows, workers are alternatively hired and fired. When fired, there is little alternative economic opportunity in banana zones, and displaced workers must either look for a piece of land to farm, or migrate to the cities to join the swelling ranks of shanty town dwellers.

Thus, the direct effect of most modern agricultural activities is not inexorably linked with the cutting and burning of rain forests, despite some obvious and spectacular examples of where it indeed is. More importantly, the overall operation of the modern agricultural system is integrated into a bigger picture. It is that bigger picture that we must examine to understand the causes of rain forest destruction.

MYTH FOUR: Local governments institute policies that cause rain forests to be destroyed.

Probably the most cited example of local government policy that promotes deforestation is that of the infamous transmigration programs of the Indonesian government, in which hundreds of thousands of Javanese farmers have been displaced to the exceedingly poor soils of Kalimantan. However, most local government programs in forestry and agriculture are frequently dictated by very specific economic and political forces that are effectively beyond the control of local governments. Once those forces are understood, it is difficult to lay the full blame on local governments. They may be corrupt, they may be inefficient, but in fact their hands are frequently tied by forces beyond their control.

Given today's global interconnectedness, in order to understand the Third World we must view it as embedded in the modern industrial system. In that system the people who provide the labor in the production processes are not the same people who provide the tools, machines and factories. The former are the workers in the factories, the latter are the owners of the factories. The owners of the machines and tools directly make the management decisions about all production processes. A good manager tries to minimize all production costs, including the cost of labor.

However, the owners of the factories face a complicated and contradictory task. While factory workers constitute a cost of production to be minimized, they also participate, along with the multitudes of other workers in society, in the consumption of products. In trying to maximize profits, factory owners are concerned that their factories' products sell for a high price. This can only happen if workers, in general, are making a lot of money. In contrast to what is desired at the level of the factory, the opposite goal is sought at the level of society. Factory owners must wear two hats, then: one as owners of the factories, and another as members of a social class. Owners wish the laborers to receive as little as possible, but members of the social class benefit if laborers in general receive as much as possible (to enable them to purchase the products produced in the factory). This has long been recognized as one of the classic contradictions of a modern economy.

The situation in much of the Third World appears superficially similar. For the most part we are dealing with agrarian economies in which there are two obvious social classes, those who produce crops for export like cotton, coffee, tea, rubber, bananas, chocolate, beef; and sugar; and those who produce food for their own consumption on their own small farms and, when necessary, provide the labor for export crop producers.

The typical arrangement in the Developed World is an articulated economy, while that in the Third World is disarticulated, in that the two main sectors of the economy are not articulated or connected with one another. The banana company does not really care whether its workers make enough money to buy bananas; that is not its market. The banana company cares that the workers of the Developed World have purchasing power, because those are the people expected to buy most of the bananas. This disarticulation, or dualism, helps to explain the differences between analogous classes in the First and Third Worlds. Flower producers in Colombia do not concern themselves much over the fact that their workers cannot buy their products. On the other hand, the factory owners in the U.S., whether they be private factories or government owned and/or subsidized industries, care quite a lot that the working class has purchasing power. General Motors "cares" that the general population in the U.S. can afford to buy cars. Naturally they aim to pay their own workers as little as possible, but that goal is balanced by their wish for the workers in general to be good consumers.

Seeing this structure at the national level in an underdeveloped country causes one to realize that one of the main, sometimes only, sources of capital to create a civil society is from agricultural exports. Because of the disarticulated nature of the economy the dream of development based on internally derived consumer demand is pie in the sky, and any realist must acknowledge that the only conceivable source of capital to invest in growth must come from exports. And most frequently agricultural exports are the only possibility.

Herein derives the need for Third World governments to continue expanding this export agriculture. This need is an inevitable consequence of the underlying structure of the general world system. Thus to blame local governments for initiating policies that are ultimately damaging to rain forests may be technically correct in that those policies frequently do just what the critics say they do—destroy rain forests. But taking a larger view we see that local governments are effectively constrained to do exactly that. Indeed, we predict that most of today's critics would wind up promoting the very same programs the local governments are currently promoting, if they were suddenly pushed into the same position the local governments currently find themselves.

MYTH FIVE: Decisions made by international agencies cause rain forest destruction.

As before, there is some truth to this position. As well documented, although not yet "retrospected" by Mr. McNamara, the World Bank has left a trail of rain forest destruction in the wake of its many socially and economically destructive programs in the Third World. From the point of view of the decision making agencies, they, along with other agencies involved in the overall problem, seem to be boxed in by circumstances.

The climate for investment is variable in the Developed World. There are times when it is difficult to find profitable investments at home. At such times it is useful to have alternatives to investment. The Third World is one source for those opportunities. The Developed World, because of its basic structure, tends to go through cycles of bust and boom, sometimes severe, other times merely annoying. During low economic times, where is an investor supposed to invest? The Third World provides a sink for investments during rough times in the First World. This is why the dualism of the Third World is a "functional dualism." It functions to provide an escape valve for investors from the Developed World. The West German entrepreneur who started an ornamental plant farm in Costa Rica, on which former peasant farmers work as night watchmen, invested his money in the Third World because opportunities in his native Germany were scarce at the time. What would he have done had there been no peasants willing to work for practically nothing, and no Costa Rica willing to accept his investments at very low taxation? Clearly Costa Rica is, for him, a place to make his capital work until the situation clears up in Germany. Union Carbide located its plant in Bhopal, India, and not in Grand Rapids, Michigan. U.S. pesticide companies export to Third World countries insecticides that have been banned at home. U.S. pharmaceutical companies pollute the ground water in Puerto Rico because they cannot do so (at least not so easily) in the United States. In all cases, Third World people are forced to accept such arrangements, largely because of their extremely underdeveloped economy.

With this analysis, the origin of the Third World as an outgrowth of European expansion (while a correct and useful historical point of view), can be seen as not the only factor to be considered. Even today, the maintenance of the Third World is a consequence of the way our world system operates. The Developed World remains successful at economic development for two reasons. First, because it has an articulated economy, and second, because it is able to weather the storm of economic crisis by seeking investment opportunities in the Third World. The Third World, in contrast, has been so unsuccessful because its economy is disarticulated, lacking the connections that would make it grow in the same way as the Developed World. Yet at a more macro scale, the dualism of the Third World is quite functional, in that it maintains the opportunity for investors from the Developed World to use the Third World as an escape valve in times of crisis. Indeed, it appears that the Developed World remains developed, at least in part, specifically because the Underdeveloped World is underdeveloped.

Given this structure, what really can be expected of international agencies? Their goal is usually stated in very humanitarian rhetoric. But their more basic goal has to be the preservation of the system that gives them their station in life. That is, above and beyond the stated goals of the World Bank or the IMF or the FAO, there must be a commitment to keep the world organized in its current state. Their activities can thus be viewed as trying to solve problems within the context of the current system. They thus become part of what preserves that system.

We should not expect large international agencies to be promoting such causes as land reform for peasant agriculture or labor standard regulations for export agriculture. Indeed, such proposals would be at odds with the manner in which the current world system is functioning, and would represent a legitimate challenge to the existence of the agency itself. Viewed from this perspective, the international agencies are just as boxed in as the local governments. The world system is functioning as well today as it was at the end of World War II—according to the standards adopted by those who benefit from its current structure.

THE WEB OF CAUSALITY

In reading our demythologization of the above five myths, the reader has undoubtedly noted that there really is something valid about each of the myths. Loggers do cut trees down, peasant agriculturists do clear and burn forests, export agriculturists do cut down primary rain forests, local governments do encourage export agriculture, and international agencies do promote programs that destroy rain forests. But our attempt was not to disprove these myths per se, but rather to disprove the idea that any one of them could be the ultimate cause of rain forest destruction. Indeed, in each case we have emphasized not the direct consequences of the agency involved but, rather, the indirect connections that tie each of the agencies into a web of interaction.

We agree with Wallerstein's general assessment that the world is intricately connected, that it no longer makes sense to try understanding isolated pockets, such as nations, and, we add, that isolated thematic pockets are similarly incomprehensible unless embedded in this global framework. For this reason, attempts at understanding tropical rain forest destruction in isolation have largely failed. As should be clear by now, the fate of the rain forest is intimately tied to various agricultural activities, which are embedded in larger structures, some retaining a connection to agriculture, some not. Our position is that there is multiple causation of rain forest destruction, with logging, peasant agriculture, export agriculture, domestic sociopolitical forces, international socio-economic relations, and other factors intricately connected with one another in a "web of causality." This web is key to understanding why we face the problem of rain forest destruction in the first place....

Furthermore, seeing the entire web of causality enables those engaged in highly focused political action to see their actions in relation to other actions, perhaps evoking an analysis of consequences that may be dramatic, even though quite indirect. For example, organizing consumer boycotts can be seen as clearly attacking the connection between consumers and modern export agriculture. But following through the logic of the web also suggests that a successful consumer boycott may likewise reduce the need of modern agriculture for workers, thus creating more peasant farmers, who will likely clear more forest. If a careful analysis of this situation reveals that the loss of jobs will be severe, the political action agenda might then be expanded to form alliances with a local farm worker union calling for job security or a political movement seeking secure land ownership for the increased number of peasant farmers that will surely be created if the boycott is a success.

THE POLITICAL ACTION PLAN

This analysis is meaningless without a program of political action. Political action must focus on the web of causality and eschew single issue foci. Calls for boycotts of tropical timbers or bananas need to be coupled with actions to change investment patterns and international banking pressures.

Above all, political action plans must be formulated so they do not make the situation worse—certainly a conceivable, perhaps even likely, consequence of any action, given the complex nature of the web of causality. It appears obvious that political action needs to be focused not only on rain forests and the subjects traditionally associated with them, but also on social justice. The same peasant farmers who formed the backbone of the Vietnamese liberation forces or the Salvadoran guerrillas are the ones who are forced into the marginal existence that compels them to continually move into the forests. So the same issues that compelled progressive organizers in the past to form solidarity committees and anti-war protests are the issues that must be addressed if the destruction of rain forests is to be stopped.

Just as the most effective political action in the past was organized in conjunction with and to some extent under the leadership of the people for whom social justice was being sought, so today political action should be coordinated with those same people. As that coordination proceeds, the alliances that grow will inevitably lead to the reformulation of goals, which the rain forest conservation activist must acknowledge and respect. Local people, quite obviously, must recognize something about the rain forest that is in their best interest to preserve, and it is the job of the progressive organizer to construct the political action so that such value is evident. In short, the alliance between the people who live in and around the rain forest and those from the outside who seek to stop the tide of rain forest destruction must be a two way alliance. If the people who live around the Lacandon forest in Mexico, for example, have as their major goal the reformulation of the Mexican political system, the rain forest conservationist must join the political movement to change that system —something that many would see as distant from the original goal of preserving rain forests. Political action to preserve rain forests, under the framework of the web of causality, will inevitably involve the serious preservationist in social justice issues, many of which initially may have seemed only marginally associated with the problem of rain forest destruction. Recalling the old slogan, "If you want peace, work for justice," we hope someday to hear, for example, "Save Mexican rain forests, support the Zapatistas," or "If you want to save Cuba's rain forests, break the illegal U.S. blockade."

CHAPTER 8 Pollution

8.1 JOHN EVELYN

Fumifugium: Or the Inconvenience of the Aer and Smoake of London Dissipated

Many environmental courses that include units on acid rain, photochemical smog, ozone depletion, and global warming may leave the impression that serious air pollution problems have only become an issue in the second half of the twentieth century. This is certainly not the case. The earliest reports of noxious smoke in urban areas is associated with the introduction of high-sulfur coal from Newcastle as a fuel by cloth dyers and beer brewers in England's cities in the middle of the thirteenth century. By 1306 King Edward I had introduced restrictions on the burning of coal in London from nearby seashore outcroppings. During the fifteenth and sixteenth centuries increased populations in London and the shortage of firewood resulted in the rapid expansion in the use of coal as a domestic and industrial fuel.

John Evelyn (1620–1706) was a famous English writer whose historical diary is an important chronicle of life in seventeenth-century London. In 1661 Evelyn wrote the tract *Fumifugium: Or the Inconvenience of the*

Aer and Smoake of London Dissipated, from which the following selection has been taken. This tract was addressed to King Charles II. In this work, which has become a much-cited document by historians of environmentalism, Evelyn describes the wretched effects of coal smoke on all aspects of life in the London of his day. His principal proposal for reducing the smoke pollution was to move those trades that burned large quantities of coal to a location five or six miles from London. Although King Charles ignored Evelyn's plea, subsequent government action did drastically reduce the burning of soft coal in London proper. However, by 1933 the industrial revolution had made smoke pollution such a widespread concern that the English National Smoke Abatement Society decided to reprint Evelyn's pamphlet as a means of publicizing their cause.

Key Concept: the long history of urban air pollution

TO THE KINGS MOST SACRED MAJESTY:

SIR,

It was one day, as I was Walking in Your MAJESTIES Palace at WHITE-HALL, (where I have sometimes the honour to refresh myself with the Sight of Your Illustrious Presence, which is the Joy of Your Peoples hearts) that a presumptuous Smoake issuing from one or two Tunnels neer Northumberland-house, and not far from Scotland-yard, did so invade the Court; that all the Rooms, Galleries, and Places about it were filled and infested with it; and that to such a degree, as Men could hardly discern one another for the Clowd, and none could support, without manifest Inconveniency. It was not this which did first suggest to me what I had long since conceived against this pernicious Accident, upon frequent observation; But it was this alone, and the trouble that it must needs procure to Your Sacred Majesty, as well as hazzard to Your Health, which kindled this Indignation of mine, against it, and was the occasion of what it has produced in these Papers.

Your Majesty, who is a Lover of noble Buildings, Gardens, Pictures, and all Royal Magnificences, must needs desire to be freed from this prodigious annoyance; and, which is so great an Enemy to their Lustre and Beauty, that where it once enters there can nothing remain long in its native Splendor and Perfection: Nor must I here forget that Illustrious and divine Princesse, Your Majesties only Sister, the now Dutchesse of Orleans, who at her Highnesse late being in this City, did in my hearing, complain of the Effects of this Smoake both in her Breast and Lungs, whilst She was in Your Majesties Palace. I cannot but greatly apprehend, that Your Majesty (who has been so long accustomed to the excellent Aer of other Countries) may be as much offended at it, in that regard also; especially since the Evil is so Epidemicall; indangering as well the Health of Your Subjects, as it sullies the Glory of this Your Imperial Seat.

Sir, I prepare in this short Discourse, an expedient how this pernicious Nuisance may be reformed. . . .

SIR,

Your Majesties ever Loyal, most obedient Subject and Servant,
J. EVELYN. . . .

PART I.

. . . I shall not here much descant upon the Nature of *Smoakes,* and other Exhalations from things burnt, which have obtained their several *Epithetes,* according to the quality of the Matter consumed, because they are generally accounted noxious and unwholsome; and I would not have it thought, that I doe here *Fumos vendere,* as the world is, or blot paper with insignificant remarks: It was yet haply no inept derivation of that *Critick,* who took our *English,* or rather, *Saxon* appellative, from the Greek word σμυχϖ *corrumpo* and *exuro,* as most agreeable to its destructive effects, especially of what we doe here so much declaim against, since this is certain, that of all the common and familiar materials which emit it, the immoderate use of, and indulgence to *Sea-coale* alone in the City of *London,* exposes it to one of the fowlest Inconveniences and reproaches, than possibly beffall so noble, and otherwise incomparable City: And that, not from the *Culinary* fires, which for being weak, and lesse often fed below, is with such ease dispelled and scattered above, as it is hardly at all discernible, but from some few particular Tunnells and Issues, belonging only to *Brewers, Diers, Lime-burners, Salt,* and *Sope-boylers,* and some other private Trades, *One* of whose *Spiracles* alone, does manifestly infect the *Aer,* more than all the Chimnies of *London* put together besides. And that this is not the least *Hyperbolic,* let the best of Judges decide it, which I take to be our senses: Whilst these are belching it forth their sooty jaws, the City of *London* resembles the face rather of Mount *Ætna,* the *Court of Vulcan, Stromboli,* or the Suburbs of Hell, than an Assembly of Rational Creatures, and the Imperial seat of our incomparable *Monarch.* For when in all other places the *Aer* is most Serene and Pure, it is here Ecclipsed with such a Cloud of Sulphure, as the Sun itself, which gives day to all the World besides, is hardly able to penetrate and impart it here; and the weary *Traveller,* at many Miles distance, sooner smells, than sees the City to which he repairs. This is that pernicious Smoake which sullyes all her Glory, superinducing a sooty Crust or Fur upon all that it lights, spoyling the moveables, tarnishing the Plate, Gildings and Furniture, and corroding the very Iron-bars and hardest Stones with these piercing and acrimonious Spirits which accompany its Sulphure; and executing more in one year, than exposed to the pure *Aer* of the Country it could effect in some hundreds. . . .

It is this horrid Smoake which obscures our Churches, and makes our Palaces look old, which fouls our Clothes, and corrupts the Waters, so as the very Rain, and refreshing Dews which fall in the several Seasons, precipitate this impure vapour, which, with its black and tenacious quality, spots and contaminates whatever is exposed to it....

The Consequences then of all this is, that (as was said) almost one half of them who perish in *London*, dye of *Phthisical* and *Pulmonic* distempers; That the *Inhabitants* are never free from *Coughs* and importunate *Rheumatisms* spitting of *Impostumated* and corrupt matter: for remedy whereof, there is none so infallible, as that, in time, the Patient change his *Aer*, and remove into the *Country*: Such as repair to *Paris* (where it is excellent) and other like Places, perfectly recovering of their health; which is a demonstration sufficient to confirm what we have asserted, concerning the perniciousnesse of that about this City, produc'd only, from this exital and intolerable Accident....

Solutions

PART II.

We know (as the *Proverb* commonly speaks) that, as *there is no Smoake without Fire*; so neither is there hardly any *Fire* without *Smoake*, and the... materials which burn clear are very few, and but comparatively so tearmed: That to talk of serving this vast City (though *Paris* as great, be so supplied) with *Wood*, were madnesse; and yet doubtless it were possible, that much larger porportions of Wood might be brought to *London*, and sold at easier rates, if that were deligently observed, which both our *Laws* enjoyn, as faisible and practised in other places more remote, by Planting and preserving of *Woods* and *Copses*, and by what might by Sea, be brought out of the *Northern Countries*, where it so greatly abounds, and seems inexhaustible. But the *Remedy* which I would propose, has nothing in it of this difficulty, requiring only the Removal of such *Trades*, as are manifest *Nuisances* to the City, which, I would have placed at farther distances; especially, such as in their Works and Fournaces use great quantities of *Sea-Coale*, the sole and only cause of those prodigious Clouds of *Smoake*, which so universally and so fatally infest the *Aer*, and would in no City of *Europe* be permitted, where Men had either respect to Health or Ornament. Such we named to be *Brewers, Diers, Sope* and *Salt-boylers, Lime-burners*, and the like: These I affirm, together with some few others of the same *Classe* removed at competent distance, would produce so considerable (though but partial) a Cure, as Men would even be found to breath a new life as it were, as well as *London* appear a new City, delivered from that, which alone renders it one of the most pernicious and insupportable abodes in the World, as subjecting her Inhabitants to so infamous an *Aer*, otherwise sweet and very healthful: For, (as we said) the *Culinary* fires (and which *charking* would greatly reform) contribute little, or nothing in comparison to these foul mouth'd Issues, and Curles of *Smoake*, which (as the Poet has it) do *Cœlum* subtexere fumo,] and draw a sable Curtain over Heaven. Let any man observe it, upon a *Sunday*, or such time as these Spiracles cease, that the Fires are generally extinguished, and he shall sensibly conclude, by the clearnesse of the Skie, and universal serenity of the *Aer* about it, that all the Chimnies in *London*, do not darken and poyson it so much, as *one* or two of those Tunnels of *Smoake*; and, that, because the most imperceptible transpirations, which *they* send forth, are ventilated, and dispersed with the least breath which is stirring: Whereas the *Columns* and Clowds of *Smoake,* which are belched forth from the sooty Throates of those Works, are so thick and plentiful, that rushing out with great impetuosity, they are capable even to resist the fiercest winds, and being extremely surcharg'd with a fuliginous Body, fall down upon the City, before they can be dissipated, as the more thin and weak is; so as two or three of these *fumid vortices,* are able to whirle it about the whole City, rendring it in a few Moments like the Picture of *Troy* sacked by the *Greeks,* or the approaches of *Mount-Hecla.*

I propose therefore, that by an *Act* of this present *Parliament,* this infernal *Nuisance* be reformed; enjoyning, that all those *Works* be removed five or six miles distant from *London* below the River of *Thames.*

8.2 BEVERLY PAIGEN

Controversy at Love Canal

The potentially disastrous consequences of improper hazardous waste disposal became a public issue in the late 1970s. The problem was dramatized by the evacuation from Niagara Falls, New York, of hundreds of residents, whose health was being threatened by chemicals leaking into their homes, backyards, and local school from the abandoned Love Canal, which had been used for many years as an industrial waste dump. The publicity given to this event led numerous other communities to discover and report local sites where toxic chemicals had been disposed of in open lagoons or were leaking from disintegrating steel drums. Such esoteric chemical names as dioxins and PCBs soon became part of the common lexicon, and numerous local citizens' groups were mobilized to demand the public's protection from exposure to hazardous wastes.

The Love Canal Homeowners Association was fortunate to have both an able leader, Lois Gibbs, who has since become a national organizer of the grassroots Citizens Committee on Hazardous Wastes, and a dedicated volunteer scientific adviser, Beverly Paigen. In 1978 Paigen, a research biologist, was studying differences in susceptibility to chemical toxins at the Roswell Park Memorial Institute of the New York State Department of Health, which is located in Buffalo, New York, a short distance from the Love Canal neighborhood. The developing controversy about the health effects of the chemicals to which the Love Canal residents had been exposed provided an opportunity for Paigen to make use of her professional expertise in a real-world situation. She hoped to help answer the concerns of the homeowners while contributing to her own research interests. As she describes in the following selection from "Controversy at Love Canal," *Hastings Center Report* (June 1982), political and economic factors became serious obstacles in this endeavor and raised a series of important ethical questions about the behavior of various parties with conflicting interests in the resolution of a hazardous waste exposure issue.

Key Concept: the scientific and political dimensions of hazardous waste controversies

During the past four years, I have been involved in a controversy over the impact of hazardous waste buried at Love Canal on the health of the surrounding community. As a scientist employed by a research institute that is a division of the New York State Department of Health, I came to believe that additional studies should be done on the health of Love Canal residents. However, David Axelrod, who became Commissioner of Health in December 1978, maintained that the Health Department's studies were adequate and showed little or no health risk to the community. At the beginning, I thought our differences could be resolved in the traditional scientific manner by examining protocols, experimental design, and statistical analysis. But I was to learn that actual facts made little difference in resolving our disagreements—the Love Canal controversy was predominantly political in nature, and it raised a series of questions that had more to do with values than science.

The Environmental Protection Agency (EPA) estimates that 50,000 hazardous waste sites may exist in the United States, that 90 percent of these pose a potential health threat because they are improperly located or poorly managed, and that 2,000 are currently threatening the health of nearby communities. Thus the hazardous waste problem is large not only because of the number of sites but also because these sites are usually close to the places where people live and work. Because the issues faced at Love Canal will occur again and again, any lessons we can learn may be helpful in preventing or resolving controversy in similar situations and in protecting the public health.

TOXIC WASTES AT LOVE CANAL

In 1942 Hooker Electrochemical Corporation (now Hooker Chemical and Plastics, a subsidiary of Occidental Petroleum Corporation) began to fill an abandoned canal a half-mile long with toxic chemicals from the manufacture of chlorinated hydrocarbons and caustics. More than 21,000 tons of 200 or more chemicals had been deposited in the canal when the Niagara Falls Board of Education approached Hooker Chemical about purchasing the site for a school. Hooker claims that it warned the Board of Education that the site was not appropriate for a school. The company says it sold the property to the Board in 1953 for a token $1.00 only when the Board threatened to take the property by eminent domain. None of the people who were Board members at the time are living to confirm or deny that claim, and the minutes of the meetings do not bear out the claim. The deed transferring the property from Hooker to the Board of Education does contain a clause that, Hooker says, releases the company from liability. It states that the site was filled "with waste products resulting from the manufacturing of chemicals" and that "no claims, suit, action, or demand of any nature whatsoever" could be made against Hooker "for injury to a person or persons, including death resulting therefrom, . . . by reason of the presence of said industrial waste."

An elementary school was built in the center of the site and the north and south portions were sold to developers who built ninety-eight homes along the banks of the former canal. During the ensuing twenty-five years, chemicals from the dump site migrated as a thick black oily mixture through the topsoil into the surrounding community. As early as 1958 Hooker Chemical and city officials were informed that three children had suffered chemical burns from exposed wastes on the surface of the canal. The Niagara Falls Health Department and other local officials took no action on this and many other complaints.

Through the efforts of determined residents and newspaper reporter Michael Brown, the EPA and the New York State Department of Health eventually entered the picture. In 1978, these agencies identified many chemicals in the air of Love Canal homes, and the Department of Health documented an excess frequency of miscarriages in women living in homes immediately adjacent to Love Canal. On August 2, 1978, Robert Whalen, then the New York State Commissioner of Health, declared a health emergency. Shortly thereafter Hugh Carey, the governor, offered to purchase the 239 homes closest to the canal and to assist in relocating the families. A fence was placed around the homes and plans were made to construct trenches to intercept the flow of chemicals from the canal. The Health Department initiated health studies of 850 additional homes in the Love Canal neighborhood by distributing questionnaires and taking blood samples for analysis. In early fall of 1978 the department announced the preliminary results of these studies; officials assured the Love Canal residents that the neighborhood was a safe place to live and that the community beyond the homes that had already been evacuated was not at any increased health risk. This announcement was based on data showing that the miscarriage rate in homes beyond the barrier was no higher than elsewhere.

The community was not reassured, citing visible seepage through basement walls, chemical odors in homes, and odors at storm sewer openings as evidence that chemicals had migrated beyond the fence. The residents also questioned why certain families living three to four blocks from the canal had multiple miscarriages and other illnesses.

HISTORY OF THE CONTROVERSY

At the time I was studying genetic variation in the metabolism of chemicals and trying to determine whether such differences might explain differing susceptibilities to environmental toxins. The exposure of 850 families to low levels of chemicals seemed to me an unusual opportunity to locate susceptible families with several sick members, match these with a healthy neighboring family, and compare the metabolism of chemicals in these two groups. I devised a health questionnaire, and a small group of residents systematically surveyed the neighborhood by telephone, collecting information about the number of people in the family, length of residence, and health.

I planned to plot the illnesses geographically with the following expectations: (1) if illnesses were clustered in families, that would indicate a possible genetic susceptibility to low-level chemical exposure and thus provide families for future study in my laboratory; or (2) if illnesses were geographically clustered, that would probably indicate migration of chemicals from the canal; or (3) if illnesses were randomly distributed, that would indicate no relationship to chemical exposure. Of course I realized that the sample size was small (1,140), so that any findings would only be a signal that further studies needed to be done. Plotting the results on a map revealed a strong geographical clustering of disease that appeared to be related to former stream beds and swales, which are low marshy areas that collect water but do not have a particular direction of flow. These streams or swales no longer exist because they were filled with building rubble as the neighborhood was developed. The predominant soil in the area is clay; thus a low-lying area filled with rubble could provide a permeable conduit for the flow of liquid chemicals from the canal.

The exact locations of these swales, as well as swamps and a pond, were determined by soil scientists under contract to the Health Department. Once these were defined, it was possible to divide the neighborhood into historically wet homes—those bordering streams and swales or built in swamps or the former pond—and historically dry homes. Comparing the incidence of disease in wet and dry homes provided an internal neighborhood control that randomized for reporting bias, occupational exposure and age.

We found a threefold increase in miscarriages in pregnant women who lived in wet homes compared to those who lived in dry homes. Particularly striking were the women in wet homes with three or more miscarriages. The frequency of such habitual aborters reported in the literature is between 0.4 to 0.7 percent. Of the sixty-four women living in wet homes at Love Canal who became pregnant, five (8 percent) had three or more miscarriages. The probability of this occurring by chance is less than 0.001.

Birth defects were also higher in wet homes. Of 120 children born in wet homes, 24 (20 percent) had birth defects compared to 6.8 percent of children born in dry homes. Some of these birth defects were minor, such as club feet, webbed toes, a missing ear, or an extra set of teeth; but many were serious, such as heart defects, missing or nonfunctional kidneys, deafness, absence of diaphragm, and mental retardation. Asthma was 3.5 times more frequent and urinary system disease was 2.8 times more frequent in wet homes than in dry. Various symptoms of central nervous system toxicity were reported by residents in wet homes: these included dizziness, fainting, seizures, blurred vision, depression, hyperactivity in children, suicides, suicide attempts, and nervous breakdowns. Since data included health effects only up to the declaration of the health emergency, the stress and anxiety that may have occurred after that date exerted no influence on the symptoms. Since central nervous system problems are very subjective, I evaluated only those that were severe and could be verified. I chose to group together admissions to a mental hospital and suicide attempts and found that these were five times more frequent in wet areas ($p<.0005$).

The survey, which was conducted without any funds, was adequate to locate sick and healthy families in order to study metabolism of chemicals, but it had never been intended as a full-scale epidemiological study. Once I had the data, however, I found myself in a very uncomfortable position. The data strongly indicated that chemicals might have migrated beyond the fence and that a health risk might be present in some parts of the Love Canal neighborhood. However, Commissioner Whalen was publicly stating that the chemicals had not migrated beyond the fence.

The department's claim arose from two factors. First, the department had analyzed miscarriage frequency by street (99th, 100th, 101st, etc.). Since the rate of miscarriage did not decrease as the streets grew more distant from the canal, the Health Department concluded that the miscarriages were not related to chemical exposure. However, if chemicals had migrated from the canal preferentially along swales, as I suspected, then miscarriage frequency should not necessarily decrease with perpendicular distance from the canal, but instead would be related to the location of the swales. Second, the Health Department's finding that miscarriage frequency in Love Canal women was no higher than elsewhere was based on the Warburton and Fraser study, which the department had selected as a literature control. However, that study differed in several important ways from the Love Canal study conducted by the Health Department. The women in the Warburton and Fraser study had either previously had a child with a birth defect, or in a minority of cases, had given birth to twins. Thus they were not a representative population. In addition, unlike the Love Canal population, the women in the control study were poor and had little or no prenatal care. Finally Warburton and Fraser counted as valid any miscarriages reported by a woman, while the Health Department counted only those that could be independently confirmed by physicians' records. . . .

After reviewing the data analysis by the Health Department and the results of my own study, I felt I needed to convey my concerns to my superiors. On November 1, 1978, I suggested to the Health Department a hypothesis that needed to be tested: adverse pregnancy outcomes were more frequent in wet homes than in dry homes and adverse effects might be occurring in the urinary, respiratory, and central nervous system as well. The department then reevaluated their own data, and on February 8, 1979, the new health commissioner, David Axelrod, announced that women in wet homes were more likely to have miscarriages, babies of low birth weight, and children with congenital abnormalities. At that time families with children under two years of age or with women who could prove they were pregnant were evacuated. But evacuation was to continue only until the youngest child in the family reached the age of two. Women who wished to conceive were concerned because during most of the sensitive first trimester of pregnancy when the fetus is at highest risk they would be involved in proving they were pregnant and in processing their requests for evacuation. These women applied for evacuation but their requests were denied.

At the February 1979 meeting, the health commissioner publicly praised my contributions and promised the residents of Love Canal that studies would follow on the respiratory, urinary, and nervous systems. More than three years have elapsed, but these studies have not been done or even initiated. Still at the time I felt considerable relief. Since my superiors and I now agreed on the essential points, I thought the controversy was over. But I was wrong. The controversy continued for one-and-a-half years with increasing intensity, until the residents of Love Canal were given the opportunity to move out of the neighborhood.

That "scientific" controversy rolled on regardless of scientific facts was a hard lesson for me to learn. Shortly after the February 1979 meeting, I asked a New York state scientist in bewilderment exactly what we were disagreeing about since we both agreed that adverse pregnancy outcomes were elevated and the other diseases had not yet been examined by the state. He replied that the disagreement was now over the exact number of miscarriages and birth defects. Seeking to end that controversy, I said that I would accept their figures. I changed my slides and reports and even wrote to David Rall, chief of the National Institute of Environmental Health Science, that "the diagnosis of diseases in the State's data base is more reliable than those in my data base because the State has checked each disease with physician and hospital records while I have accepted the individual's self-report" (March 13, 1979). The state used this statement against me later, but the controversy I had hoped to end continued even though there was no longer any difference of opinion over facts.

These events caused me to think carefully about the nature of public health controversy, the factors that prevent resolution of controversy, and the ways in which controversy can be reduced or avoided. In discussing these ideas, I will illustrate with examples from Love Canal, but the events are characteristic of many such controversies.

THE ELEMENTS OF CONTROVERSY

The two opposing sides in the Love Canal controversy were the community and the New York State Department of Health. This was somewhat surprising, since the Health Department had declared the health emergency in the first place. However, when the community turned to the agency they regarded as their ally and protector, they felt that the response was inadequate.

Antagonism between a community and a health department is not unique to Love Canal. At a number of other hazardous waste sites, frustrated and angry communities have turned on the local health agency rather than on the industry that created the hazardous waste sites. Perhaps such a response is not so surprising after all. Hooker Chemical was a major industry in the region and employed many people in the Love Canal neighborhood. The community made some allowance for Hooker because the chemicals were buried many years before the chronic toxicity of chemicals was understood and before regulations concerning disposal of toxic waste existed. Hooker Chemical claimed that it had used state-of-the-art technology in burying the waste and that furthermore they had warned the Board of Education not to build a school on the site. The community also understood that the goal of industry is profits and that Hooker was acting in a manner consistent with its goals by using the cheapest method of disposal. So while many in the Love Canal community were inclined to make allowances for Hooker, they did not feel well disposed toward the Health Department, which was acting in the present with full knowledge of chemical toxicity. The stated goal of the department is to protect health, and the salaries of officials came from the community's tax money. So the community viewed the department as acting in a manner inconsistent with its goals and responsibilities when health effects were minimized or ignored.

Once the controversy was under way several factors impeded a resolution:

1. THE FAILURE TO RESOLVE ANY CONTROVERSY MAY BE ADVANTAGEOUS TO ONE SIDE. In this case the state had much to gain from delay. Since

over 600 other hazardous waste sites existed in New York, any action taken at Love Canal would set a precedent. Any state official who recommended positive action at Love Canal would have had to justify spending even more than the $42,000,000 the state had already allowed for construction to prevent further leakage and relocation of the families living closest to the canal. In contrast, an official who delayed was taking very little risk. If the decision to delay was later shown to be wrong, the community would suffer the risk of impaired health and some future official would probably have to worry about the results....

This is not a scientific issue, nor can it be resolved by scientific methods. The issue is ethical, for it is a value judgment to decide whether to make errors on the side of protecting human health or on the side of conserving state resources. In science this problem is called avoiding type I and type II errors. In a type I error, a scientist accepts as true something that is actually false. The custom in science is to insist that we be 95 percent certain that something is true (for example, that the health of Love Canal residents has suffered) before we accept it as a fact. In contrast, a scientist who makes a type II error fails to recognize as true something that actually is true....

Before Love Canal, I also needed to have 95 percent certainty before I was convinced of a result. But seeing this rigorously applied in a situation where the consequences of an error meant that pregnancies were resulting in miscarriages, stillbirths, and children with medical problems, I realized I was making a value judgment. In other issues of public health and safety—bomb threats; possible epidemics, etc.—we do not insist on 95 percent probability of harmful consequences before action is taken. Why is that the criterion in environmental health? This issue should be debated widely in the public health community so that choices can be made not out of habit but with full realization of the values involved.

2. OPPONENTS MAY NOT AGREE ON THE QUESTION THAT NEEDS TO BE ANSWERED. In this case, Commissioner Axelrod informed the residents that the epidemiologist would look for adverse health effects on the human fetus since that was the most vulnerable segment of the population. But the residents assumed that any increase in adverse pregnancy outcomes would indicate that toxic chemicals were present and that the entire population was at risk. They reacted angrily when the commissioner announced that the fetus was indeed at risk, but that the state would evacuate only pregnant women.

The design of a study also has a profound influence on the outcome. For example, in the spring of 1981, the Center for Disease Control (CDC) designed a study to answer the question of health impact at Love Canal. A great deal of money and effort went into answering the question of whether psychological harm occurred. In contrast, few resources went into answering the question of whether there was chromosome damage. The CDC study was canceled due to lack of funds. However, had the design been carried out, after two years and several million dollars, the public would have been informed that Love Canal residents suffered psychological damage but not chromosomal damage—a conclusion dictated by the study design.

The most costly portion of the CDC study called for detailed health histories, physical examinations and laboratory tests for all Love Canal residents. However, these were not to be done on a comparable control population, nor were there plans for data compilation and analysis on the results of the Love Canal residents....

3. IN ANY CONTROVERSY, SINCE THE TYPE AND QUALITY OF INFORMATION GATHERED WILL INFLUENCE THE OUTCOME, NO ONE GROUP SHOULD BE IN COMPLETE CONTROL OF THE INFORMATION-GATHERING PROCESS. At Love Canal the state had the personnel and monies of a very good health department and an additional $1,500,000 specifically allocated by the New York legislature and Congress to gather information about health effects. In contrast, the residents had only their own energy and the help of a few scientists. Of the three small studies done outside the health department, two were done with no money (my survey and a study of nerve conduction by Dr. Stephen Barron) and one with $10,000 (a chromosome study by Dr. Dante Picciano under contract to the EPA). All had flaws in part attributable to the lack of funds. Since the data from these studies were widely available, it was easy for scientists to review and criticize them.

In contrast, the protocols and the data from the well-funded studies of the Health Department were—and still are—secret. According to a fact sheet released by the Department of Health on June 23, 1980, these studies cost $3,292,000 and 205,000 staff hours (122 staff years) and encompassed 4,386 blood samples, 11,138 field interviews, 5,924 soil samples, over 700 air, sump, and water samples, follow-up of 2,000 former residents, and 411 physical examinations of workers involved in remedial construction at Love Canal. With the exception of a provisional draft of the study on adverse pregnancy outcomes, which was released when a newspaper asked for it under the Freedom of Information Act, and the recently published Janerich study, the health studies are not available for scientific review or criticism even though conclusions are frequently quoted publicly....

We must develop ways of providing communities with access to resources and expertise. This was done with remarkable success in one case at Love Canal. Shortly after the health emergency was declared, the New York State Department of Transportation planned some remedial construction work to prevent the further flow of chemicals from the canal into the community. A health and safety plan to protect the community was developed. During negotiations between the department and the community, the lawyer for the community arranged for a sum of money to be set aside for a toxicologist who would be on site during the construction to monitor the state's compliance with the safety plan and to report to the community. The cost was only $10,000 in a project that involved $3 to $4 million....

Agencies that undertake health studies should certainly consider allocating money to the community so that residents can obtain their own expertise. But an even better solution would be to have information gathered and funded by neutral third parties. If a federally based and funded response team were established to deal with many hazardous waste sites, it might be in a position to evaluate the relative seriousness of each situation.

4. Beyond questions of money and expertise is the issue of full expression for dissident and minority opinions. Controversy is stifled, not resolved, by silencing the opposition. At Love Canal, scientists working for New York State who disagreed with the official stance were demoted, transferred, or harassed. For example, William Friedman, regional director of the Department of Environmental Conservation, was demoted to staff engineer in November 1978 (he subsequently left the agency) for "prodd[ing] the Albany hierarchy unrelentingly" (*Buffalo Evening News*, November 12, 1978). And Donald McKensie, a senior sanitary engineer in the regional office, Department of Environmental Conservation, wrote a letter to his superior raising questions about the manner in which the department was handling Love Canal and was promptly transferred from the Love Canal project to air quality projects (*Knickerbocker News*, May 12, 1980).

I too was harassed in a variety of ways. The most important involved my ability to raise money to support my research by grant applications. I first spoke publicly concerning the Love Canal problem in the summer of 1978. That September, the Department of Health administration withdrew one grant I had submitted without even informing me. That winter the administration refused to process papers so I could get funds from another grant that already had been awarded. This denial led the professional staff of Roswell Park Memorial Institute to charge the administration with scientific censorship. In June 1979, I was informed that, because of the "sensitive nature" of my work, all grants and research ideas had to go through a special review process. I was told to outline my research ideas and submit them for review "at the moment of conception before a single experiment was done." My professional mail arrived already opened and scotch-taped shut. My office was entered outside of working hours and my files were searched.

My income tax return was audited by New York State for the first time in over twenty years of filing returns and the auditor's file contained newspaper clippings, including such nonfactual material as "Letters to the Editor," about my role at Love Canal. Later James Tully, Jr., Commissioner of Taxation and Finance, apologized for several "errors in procedure," and wrote that the clippings resulted from a "a misunderstanding" of department policy by a local audit supervisor. Added to these serious problems were a variety of petty harassments, which continued until I announced in 1981 that I was leaving Roswell Park and accepting a position in another state.

In addressing the problem of whistleblowers, the Committee on Scientific Freedom and Responsibility of the American Association for the Advancement of Science (AAAS) stated in 1975 that scientists "must be assured of some form of due process in passing judgment on the issues that they raise. This would call for the presence of independent outside members on any board that passes judgment on the issues, and should also include some right of appeal." My experience showed that no such protection exists.

5. Scientists, who are no strangers to controversy, should follow the social controls on behavior that they have developed for the advancement of knowledge and the detection of error.

These include openness of data, peer review and criticism, publication of data, and replication of experiments.

These norms of scientific behavior were violated at Love Canal. Secrecy of the state's data and protection from peer review and criticism prevailed. Of the repeated attempts to obtain state data, I will mention only a few. The Environmental Defense Fund requested the protocols of the Department of Health studies under the Freedom of Information Act, but this request was denied. On February 8, 1979, a news release from the Health Department made five claims regarding the results of the health studies (no liver disease, no benzene toxicity, no blood problems, no cancer, no epilepsy). My request for data supporting these claims was denied.

A short time later in a face-to-face meeting among the three of us, Governor Hugh Carey asked Commissioner Axelrod to give me the data. Though Dr. Axelrod agreed, he has never done so. Later toxicologist Steven Lester and I were promised by Dr. Axefrod access to the state's computerized data base. On an arranged date, Lester flew to Albany only to be denied access to any of the health data. On another occasion, Dr. Axelrod explained that he wanted to provide health data but could not do so ethically because confidentiality had been promised to the interviewed residents. He suggested that I get signed consent forms from each resident who provided health data to the state, releasing that data to me. The lawyer for the residents drew up the consent form, and the residents organized and obtained hundreds of signed consents. When I sent these to Dr. Axelrod with a cover letter, all I received back was a reprimand for writing my cover letter on Department of Health stationery. (See Archives Library of the State University of New York, Buffalo). The release of data is still "under review." Many more examples could be cited. . . .

6. IN ANY ATTEMPT AT CONTROVERSY RESOLUTION, ALL PARTIES TO THE CONFLICT SHOULD AGREE ON PRECISELY WHAT FACTS NEED RESOLVING; ALL PARTIES SHOULD AGREE ON THE COMPOSITION OF THE BODY CHOSEN TO RESOLVE THE CONTROVERSY; ALL PARTIES SHOULD AGREE ON THE PROCEDURES BY WHICH THAT BODY WILL OPERATE; AND ALL PARTIES MUST AGREE TO ABIDE BY THE DECISIONS. Unfortunately, such a logical procedure rarely occurs.

There were at least two unsuccessful attempts to resolve controversy at Love Canal, and both failed because these requirements were not followed. The first occurred on February 22, 1979, when the local Congressman John LaFalce requested that representatives of the EPA and Department of Health, Education and Welfare meet with Dr. Axelrod and me "so that both sets of data can be reviewed by the federal government. I am hopeful," LaFalce wrote, "that such a meeting will resolve the differences that currently exist and that appropriate actions will then be taken to protect the health and welfare of my people." Dr. David Rall of the NIEHS was appointed chair, and he convened experts in biostatistics and epidemiology. Since Dr. Axelrod refused to have a face-to-face meeting with me, the federal committee met with me on April 12, 1979, and with State Health Department scientists on April 26. Their report agreed with my concerns and recommended that further studies be undertaken. In response to my suggestion that all women of childbearing age who wish more children should be evacuated, they recommended that exposure of Love Canal residents be "minimized to the extent feasible." Those familiar with Love Canal understood that the only way to minimize the exposure was to move the people out of the area. But, as the report pointed out, "The State appears to have a problem in terms of perceived conflict of interest." Thus the report recommended that "the involvement of outside scientists, both in the interpretation of data and the formulation of recommendations to the State policy makers, should be continued. In a manner such as this where there is much public concern it is wise that considerations of findings and of alternatives be conducted openly. To this end it might be wise to include representatives of the local population."

This attempt to resolve controversy failed because the Health Department chose to ignore the report. The department had nothing to gain by resolution of the controversy and had not agreed to abide by the decisions of the group.

The second attempt occurred about a year later. In May 1980, State Senator Thomas Bartosiewicz released a report on the hazardous waste problem in New York State. He then called for an investigation into the handling of Love Canal by the Department of Health. In a letter to Governor Carey and in a Senate resolution, he charged the state agencies (Department of Health and Department of Environmental Conservation) with unethical conduct. The detailed list of charges included:

- Appointment of a Blue Ribbon panel which had secret members and secret recommendations which were withheld from the public:
- Manipulation of health data... to minimize risks;
- Unexplained delays of up to eighteen months before the State was willing to admit a health problem existed;
- Demotion, transfers, and harassment of state employees sympathetic to Love Canal residents;
- An effort by the state to discourage and prevent independent professional health studies....

THE NEXT DECADE

... When controversies arise in communities as they inevitably will, steps can be taken to ease the situation and protect the public health. First, scientists should scrupulously adhere to the norms of their profession such as openness of data, peer review and criticism, and publication of evidence. Second, community involvement should be sought and used at every level of the process. Third, funds should be provided so the community can hire its own experts.

It would also be useful to have a carefully reasoned process for conflict resolution, which could be published by scientific societies and incorporated into the ethical code of behavior for scientists. Such a process would help, even in cases like Love Canal, where the controversy was so political. The standards for conflict resolution should include guidelines for selecting an independent group of fact-finders, the rules of procedure for such a body, an agreement by both parties to abide by the decision, and adequate protection for whistle-blowers. Finally, since in conflicts between bureaucracy and community, the bureaucracy often does not gain by controversy resolution, third parties may have to try to apply pressure in order to get the bureaucracy to participate in such a procedure. . . .

Many of the controversies of Love Canal were stated as scientific issues, but they had their roots in ethical considerations. In such instances the controversy would have become easier to resolve if the ethical considerations and value judgments had been openly stated and understood.

8.3 J. W. MAURITS LA RIVIÈRE

Threats to the World's Water

All the world's creatures (other than those that live in the oceans) are dependent on an adequate supply of uncontaminated freshwater. Unfortunately, this precious resource is in increasingly short supply in many parts of the world. This is the result of inappropriate and wasteful human agricultural, industrial, and domestic water usage that has polluted a large fraction of the earth's lakes, rivers, and aquifers, while placing increasing demand on those that remain relatively pure.

J. W. Maurits la Rivière is a distinguished environmental microbiologist and a former chair of the environmental engineering department at the International Institute for Hydraulic and Environmental Engineering in Delft. The Netherlands. An outspoken leader on water quality issues, he has also served as secretary general of the International Council of Scientific Unions and as president of that organization's Scientific Committee on Problems of the Environment.

In the following selection from "Threats to the World's Water," *Scientific American* (September 1989), la Rivière describes the various factors—including population growth, ignorance, poverty, and unsustainable development—that will result in severe water shortages if corrective action is not taken soon.

Key Concept: water shortage as a threat to the future of the human species

Water is the earth's most distinctive constituent. It set the stage for the evolution of life and is an essential ingredient of all life today; it may well be the most precious resource the earth provides to humankind. One might therefore suppose that human beings would be respectful of water, that they would seek to maintain its natural reservoirs and safeguard its purity. Yet people in countries throughout the world have been remarkably shortsighted and negligent in this respect. Indeed, the future of the human species and many others may be compromised unless there is significant improvement in the management of the earth's water resources.

All the fresh water in the world's lakes and creeks, streams and rivers represents less than .01 percent of the earth's total store of water. Fortunately, this freshwater supply is continually replenished by the precipitation of water vapor from the atmosphere as rain or snow. Unfortunately, much of that precipitation is contaminated on the way down by gases and particles that human activity introduces into the atmosphere.

Fresh water runs off the land and on its way to the ocean becomes laden with particulate and dissolved matter—both natural detritus and the wastes of human society. When the population density in the catchment area is low, waste matter in the water can be degraded by microbes through a process known as natural self-purification. When the self-purifying capacity of the catchment area is exceeded, however, large quantities of these waste substances accumulate in the oceans, where they can harm aquatic life. The water itself evaporates and enters the atmosphere as pure water vapor. Much of it falls back into the ocean; what falls on land is the previous renewable resource on which terrestrial life depends.

The World Resources Institute estimates that 41,000 cubic kilometers of water per year return to the sea from the land, counterbalancing the atmospheric vapor transport from sea to land. Some 27,000 cubic kilometers, however, return to the sea as flood runoff, which cannot be tapped, and another 5,000 cubic kilometers flow into the sea in uninhabited areas. Of the 41,000 cubic kilometers that return to the sea, some amount is retained on land, where it is absorbed by the vegetation, but the precise amount is not known.

This cycle leaves about 9,000 cubic kilometers readily available for human exploitation worldwide. That is a plentiful supply of water, in principle enough to sustain 20 billion people. Yet because both the world's population and usable water are unevenly distributed, the local availability of water varies widely. When evaporation and precipitation balances are worked out for each country, water-poor and water-rich countries can be identified. Iceland, for example, has enough excess precipitation to provide 68,500 cubic meters of water per person per year. The inhabitants of Bahrain, on the other hand, have virtually no access to natural fresh water; they are dependent on the desalinization of seawater. In addition, withdrawal rates per person differ widely from country to country: the average U.S. resident consumes more than 70 times as much water every year as the average resident of Ghana does.

Although the uses to which water is put vary from country to country, agriculture is the main drain on the water supply. Averaged globally, 73 percent of water withdrawn from the earth goes for that purpose. Almost three million square kilometers of land have been irrigated—an area nearly the size of India —and more is being added at the rate of 8 percent a year.

Local water shortages can be solved in two ways. The supply can be increased, either by damming rivers or by consuming capital—by "mining" groundwater. Or known supplies can be conserved, as by increasing the efficiency of irrigation or by relying more on food imports.

In spite of such efforts, there is no doubt that water is becoming increasingly scarce as population, industry and agriculture all expand. Severe shortages occur as demand exceeds supply. Depletion of groundwater is common in, for example, India, China and U.S. In the Soviet Union the water level of both the Aral sea and Lake Baikal is dropping dramatically as a result of agricultural and industrial growth in those areas. Contentious competition for the water of such international rivers as the Nile, the Jordan, the Ganges and the Brahmaputra is a symptom of the increasing scarcity of water.

Another problem brought on by overirrigation is salinization. As water evaporates and is taken up by plants, salt is left behind in the soil. The rate of deposition exceeds the rate at which the salt can be removed by flowing water, and so a residue accumulates. Currently more than a million hectares every year are subject to salinization; in the U.S. alone more than 20 percent of the irrigated land is thus affected.

Human activity in a river basin can often aggravate flood hazards. Deforestation and excessive logging lead not only to increased soil erosion but also to increased runoff; in addition, navigation canals are sometimes dug, which may exacerbate flooding by increasing the amount of water that reaches the floodplain.

Finally, of course, any human activity that accentuates the greenhouse effect and ensuing climatic change must inevitably influence the global water cycle. A projected sea-level rise of between .5 and 1.5 meters in the next century, for instance, not only would pose a coastal flooding problem but also would lead to salinization of water resources, create new wetlands while destroying existing ones and increase the ratio of salt water to fresh water on the globe. Precipitation could rise by between 7 and 15 percent in the aggregate; the geographic variations are not predictable.

Assuring an adequate supply is not the only water problem facing many countries throughout the world: they need to worry about water quality. In its passage through the hydrological cycle, water is polluted by two kinds of waste. There is traditional organic waste: human and animal excreta and agricultural fibrous waste (the discarded parts—often more than half—of harvested plants). And there is waste generated by a wide range of industrial processes and by the disposal, after a brief or long lifetime, of industry's products.

Although organic waste is fully biodegradable, it nonetheless presents a significant problem—and in some places a massive one. Excessive biodegradation can cause oxygen depletion in lakes and rivers. Human excreta contain some of the most vicious contaminants known, including such pathogenic microorganisms as the waterborne agents of cholera, typhoid fever and dysentery.

Industrial waste can include heavy metals and considerable quantities of synthetic chemicals, such as pesticides. These materials are characterized by toxicity and persistence: they are not readily degraded under natural conditions or in conventional sewage-treatment plants. On the other hand, such industrial materials as concrete, paper, glass, iron and certain plastics are relatively innocuous, because they are inert, biodegradable or at least nontoxic.

Wastes can enter lakes and streams in discharges from such point sources as sewers or drainage pipes or from diffuse sources, as in the case of pesticides and fertilizers in runoff water. Wastes can also be carried to lakes and streams along indirect pathways—for example, when water leaches through contaminated soils and transports the contaminants to a lake or river. Indeed, dumps of

FIGURE 1

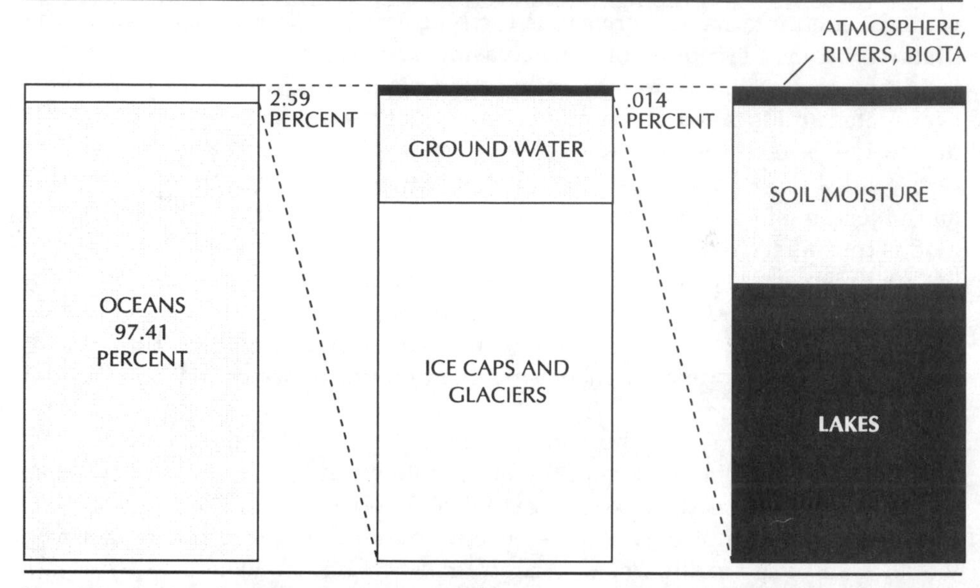

Note: Distribution of water on the planet is highly uneven. Most of it (97.41 percent) is in the oceans; only a small fraction (2.59 percent) is on the land. Even most of the water on land is largely unavailable, because it is sequestered in the form of ice and snow or as groundwater; only a tiny amount (.014 percent) of the earth's total water is readily available to human beings and other organisms.

toxic chemical waste on land have become a serious source of groundwater and surface-water pollution. The metal drums containing the chemicals are nothing less than time bombs that will go off when they rust through. The incidents at Lekerkerk in the Netherlands and at Love Canal in the U.S. are indicators of the pollution of this kind going on worldwide in thousands of chemical-waste dumps.

Some pollutants enter the water cycle by way of the atmosphere. Probably best known among them is the acid that arises from the emission of nitrogen oxides and sulfur dioxide by industry and motor vehicles. Acid deposition, which can be "dry" (as when the gases make direct contact with soil or vegetation) or "wet" (when the acid is dissolved in rain), is causing acidification of low-alkalinity lakes throughout the industrialized world. The acid precipitation also leaches certain positively charged ions out of the soil, and in some rivers and lakes ions can reach concentrations that kill fish.

In areas of intensive animal farming, ammonia released from manure is partly introduced into the atmosphere and partly converted by soil microbes into soluble nitrates in the soil. Since nitrate has high mobility (it is soluble in water and does not bind to soil particles), it has become one of the main pollutants of groundwater, often reaching concentrations that exceed guidelines established by the World Health Organization.

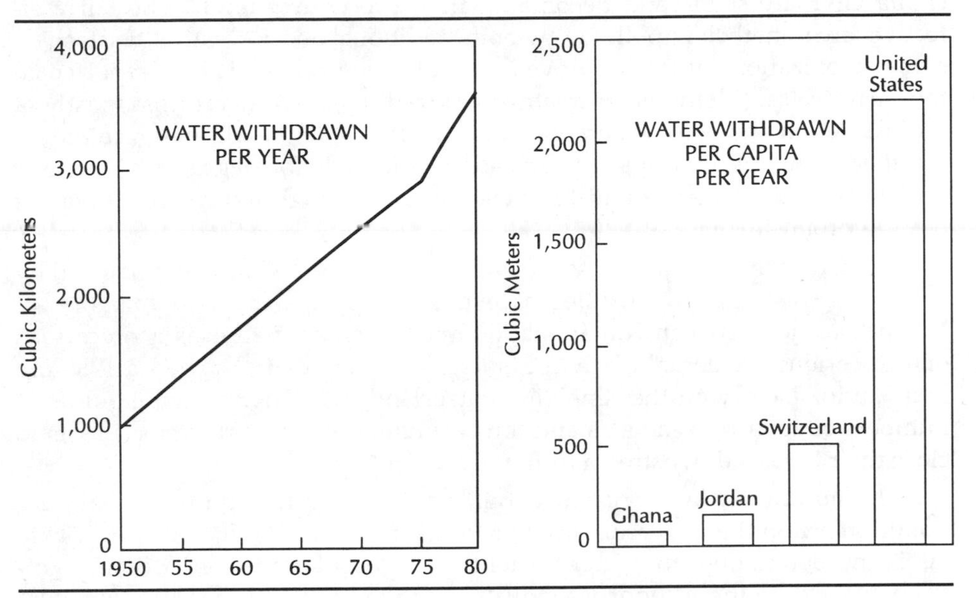

Note: Global water consumption is increasing (*left*), largely in response to a growing population and increasing per capita use by agriculture and industry. Although sufficient fresh water (9,000 cubic kilometers) is currently available, sound water management is necessary to ensure an adequate supply for the future. Per capita consumption rates vary drastically (*right*); the average American, for example, consumes more than 70 times as much water as the average resident of Ghana.

The wind can also carry pollutants—fly ash from coal-burning plants, for example, or sprayed pesticides. These can be carried great distances, eventually to be deposited on the surfaces of lakes or of rivers.

Another recently recognized aspect of water pollution is the accumulation of heavy metals, nutrients and toxic chemicals in the bottom mud in deltas and estuaries of highly polluted rivers, such as the Rhine. Because of their high pollution content, sediments that are dredged up cannot be used for such projects as landfills in populated or agricultural areas. Moreover, there is always the danger that natural processes or human activity will trigger chemical reactions that mobilize the pollutants by rendering them soluble, thus allowing them to spread over great distances.

The quality of inland waters depends not only on the amount of waste generated but also on the decontamination measures that have been put into effect. The degree of success in the battle for water quality differs from country to country, but it can be generalized into a conceptual formula proposed by Werner Stumm and his co-workers of the Swiss Federal Institute for Water Resources and Water Pollution Control in Zurich. The formula holds that the contamination load of a river basin depends on the population in the basin, the per capita gross national product, the effectiveness of decontamination and the amount of river discharge.

Most rivers in the industrialized world, where the population and per capita GNP are stable and decontamination procedures tend to be fairly effective, are nonetheless polluted by both traditional and industrial wastes. Yet some stabilization—if not improvement—of pollution levels was reported in the early 1980's. (Methods for treatment of traditional wastes consist mostly of sedimentation and aerobic and anaerobic microbial degradation, which are intensified forms of natural self-purification.) Methods for degrading inorganic pollutants such as metals and toxic chemicals, although improving, have not been as promising.

Where increasing industrial activity in a river basin has been matched by increasing waste treatment, a decent level of water quality can be maintained. Yet the balance between contamination and decontamination is a precarious one. A serious accidental discharge, such as the one that followed a 1986 fire at a Sandoz factory on the Rhine in Switzerland, is enough to wipe out large numbers of aquatic organisms and force drinking-water purification plants to close their intakes downstream from the accident.

In most newly industrializing countries both organic and industrial river pollution are on the increase, since the annual per capita GNP is rising quickly (as is the population, to a lesser extent) and decontamination efforts are often neglected. In these countries industrialization has had higher priority than reduction of pollution. As a consequence, in some regions (East Asia, for example), degradation of water resources is now considered the gravest environmental problem.

In less developed countries, where the population is growing and where waste treatment is practically non-existent, water pollution by organic wastes is widespread. As a result, millions of people—and children in particular—die each year from water-related diseases that can be prevented by proper sanitation facilities. These countries still suffer from diseases eradicated in the West long ago. Although the United Nations declared the 1980's to be the International Drinking Water Supply and Sanitation Decade and instituted a program to provide safe drinking water and appropriate sanitation for all by 1990, much remains to be done before the program's ambitious goals are met. Some progress has nonetheless been made in several countries, including Mexico, Indonesia and Ghana.

The quality of the water in lakes is comparable to that in rivers. Thousands of lakes, including some large ones, are currently being subjected to acidification or to eutrophication: the process in which large inputs of nutrients, particularly phosphates, lead to the excessive growth of algae. When the overabundant algae die, their microbial degradation consumes most of the dissolved oxygen in the water, vastly reducing the water's capacity to support life. Experience in Europe and North America has shown that the restoration of lakes is possible—at a price—but that the process takes several years. Liming is effective against acidification; flushing out the excess nutrients and restricting the further inflow of nutrients helps to reduce eutrophication.

Although pollution of rivers and lakes is potentially reversible, that is not the case for groundwater. Actually, little is known about the quality of the earth's vast groundwater reserves, except in those instances where particular aquifers are being actively exploited. In Europe and the U.S., where groundwater represents a significant source of fresh water, between 5 and 10 percent of all wells examined are found to have nitrate levels higher than the maximum recommended value of 45 milligrams per liter. Many organic pollutants find their way into groundwater as seepage from waste dumps, leakage from sewers and fuel tanks or as runoff from agricultural land or paved surfaces in proliferating urban and suburban areas.

Because groundwater is cut off from the atmosphere's oxygen supply, its capacity for self-purification is very low: the microbes that normally break down organic pollutants need oxygen to do their job. Prevention of contamination is the only rational approach—particularly for the developing world, where increased reliance on vast groundwater reserves is likely.

The oceans are part of the world's "commons," exploited by many countries and the responsibility of none and therefore all the more difficult to safeguard. More than half of the world's people live on seacoasts, in river deltas and along estuaries and river mouths, and some 90 percent of the marine fish harvest is caught within 320 kilometers of the shore. Every year some 13 billion tons of silt are dumped into coastal zones at the mouths of rivers. Although most of those sediments would have found their way into the ocean anyway, a growing part of the accumulating silt can be attributed to erosion and deforestation caused by human intervention. Depending on the particular agricultural and industrial activities in the catchment area, a coastal zone can be both fertilized and polluted by the silt and dissolved materials that reach it.

The coastal zone is the site of important physicochemical reactions between saltwater and freshwater flows; it is the zone of highest biological productivity, supporting marine life ranging from plankton to fish, turtles and whales. Aquaculture in the coastal zone now produces some 10 percent of the world's fish harvest. The 240,000 square kilometers of coastal mangrove forest are essential habitats for many economically important fish species during part of their life cycle, and they also provide timber and firewood; reed and cypress swamps are other examples of biologically rich coastal wetlands. Finally, of course, coastal zones support a highly profitable tourist industry and include a growing number of protected areas, such as the Great Barrier Reef Marine Park in Australia.

Aside from river discharges, diffuse runoff, atmospheric transport, waste dumping or burning at sea, offshore mining and shipping accidents are the primary ways that some 20 billion tons of dissolved and suspended matter reach the ocean, where they exert their initial effect on the coastal zone.

Polychlorinated biphenyls (PCB's) and other persistent toxic chemicals, including DDT and heavy-metal compounds, have already spread throughout the world's marine ecosystems, in part through gradual accumulation in the food chain. A ban on the use of DDT and PCB's has been enforced for some 10 years in the industrialized countries and has reduced the concentration of such chemicals in the marine life of North American and European coastal waters. The chemicals are, however, still being used and injected into the marine environment in many tropical regions.

Ocean currents are also vehicles for the transport of trash and pollutants. Examples are the nondegradable plastic bottles, pellets and containers that now commonly litter beaches and the ocean's surface. They cause the death of thousands of birds, fish and marine mammals that mistake them for food or get entangled in them. Less spectacular but possibly more serious are the chemical and biological processes (as yet poorly understood) whereby toxic substances such as radioactive wastes are distributed and accumulated.

Excessive sewage discharges from coastal urban areas lead to eutrophication of coastal waters, which can change the composition of plankton populations. The plankton, provided with abundant nutrients in the sewage, may experience rapid population growth, which depletes the supply of available oxygen and so leads to fish kills. Moreover, the presence of pathogenic bacteria in sewage has forced the closing of many kilometers of beaches to swimmers and has led to prohibitions on the harvesting of shellfish, which concentrate the bacteria in their tissues.

About one tenth of 1 percent of the world's total annual oil production —some five million tons a year, or more than one gram per 100 square meters of the ocean's surface—finds its way to the ocean. Large areas of the ocean would be covered with oil accumulated over the past decades were it not for the fact that the oil eventually evaporates or is degraded by bacteria. Although petroleum is almost entirely biodegradable, it takes the microbes that break it down a long time to accomplish the task, because their activity is limited by the low nutrient concentrations in seawater. In the meantime an oil spill's effects are lethal for a variety of plankton, fish larvae and shellfish, as well as for such larger animals as birds and marine mammals.

It is clear that the quality of the water in coastal zones is seriously endangered and that damage to fisheries and marine wildlife is widespread. Regional seas such as the Baltic and the Mediterranean, which have more coastline per square kilometer than the high seas do, suffer more from water pollution. Their poor condition demonstrates what may happen in the future to the larger oceans of the world.

Human activity is clearly responsible for widespread damage to marine ecosystems. What is not firmly established is how quickly toxic substances can accumulate in marine organisms or whether such accumulation is reversible. Nor has it been determined precisely how synthetic chemicals are transported through the oceans and what the likelihood is that toxic substances in bottom sediments will find their way into the human food supply. Yet experience so far dictates utmost caution, the more so because restoration of the oceans is incomparably more difficult than that of lakes and inland seas, if not impossible.

Some management of water resources—of both their quantity and quality—is now widely practiced all over the world, but the results, particularly in quality control, have been inadequate. All signals point to further deterioration in the quality of fresh and marine waters unless aggressive management programs are instituted.

Many of the guiding principles in water management have evolved from past experience and are well known, and yet their application has lagged. Above all, the need for an integrated approach has become apparent. In every river or lake basin, socioeconomic and environmental aspirations must be orchestrated so that human settlements, industry, energy production, agriculture, forests, fisheries and wildlife can coexist. In many cases varied interests are not necessarily in conflict; they can be synergistic. Erosion control, for example, goes hand in hand with reforestation, flood prevention and water conservation.

An integrated approach calls, of course, for closer cooperation at the governmental and intergovernmental level; it goes against the historical allocation of different tasks to different agencies. In many countries water supply and sanitation are handled by separate departments. Departmental budgets are isolated by money-tight walls, making it hard to balance investments made by one department with any resulting gains or savings accrued to another.

Such obstacles are even more formidable in an international setting. A country is unlikely to make significant investments in the decontamination of a river's water if it is other countries, downstream, that are likely to reap the benefits. The less developed countries may actually have a better opportunity to make progress here than the developed ones, where vested interests have entrenched themselves in rigid administrative structures. The United Nations Environmental Program (UNEP), for example, has drawn up an action plan for the Zambezi River based largely on principles of integrated management.

A water-management project should lean toward increasing the efficiency of water consumption rather than toward increasing the supply of water. To increase the supply is often more costly, and in any case it merely postpones a crisis. Indeed, because many countries are already overtaxing their water reserves, increasing efficiency is the only solution in some cases. Irrigation, for example, is terribly inefficient as it is practiced in most countries. Averaged over the world, only about 37 percent of all irrigation water is taken up by agricultural crops; the rest is never absorbed by the plants and can be considered lost. New microirrigation techniques, by which perforated pipes deliver water directly to the plants, provide great opportunities for water conservation, making it possible to expand irrigated fields without building new dams.

The mining of groundwater in order to increase supply should, of course, be avoided at all costs—unless it can be guaranteed that the aquifer from which the water is taken will be replenished. The protection of groundwater quality also deserves special attention. Government officials are more likely to implement pollution-control measures when they (or their constituents) are presented with highly visible signs of pollution, such as rubbish washed onto a beach. Hidden as it is from view, groundwater can therefore become polluted gradually without eliciting an outcry from the public until it is too late to reverse the damage wrought by the pollution.

It has also become apparent that the prevention of pollution, and the restoration of bodies of water that are already polluted, should gradually take precedence over the development of purification technologies. Water-purifying technology is becoming more complex and costly as the number of pollutants in water increases; the money spent on removing contaminants from drinking water would be better spent on preventing the contaminants from entering the water in the first place. The high cost of restoring polluted water bodies also strengthens the appeal of pollution-prevention programs.

For that reason "end of pipe" remedies for industrial water pollution should be replaced by recycling and reuse. Factories designed to minimize water pollution through waste reduction are often more economic than those that construct their own waste-water treatment plants in order to meet environmental standards. Factories that integrate pollution-control techniques are also likely to be more acceptable to an environmentally conscious populace. As Peter Donath of the Ciba-Geigy Corporation, one of the world's largest chemical companies, said at last year's International Rhine Conference, "Only with environmentally sound products and manufacturing processes will the chemical industry be able to maintain social acceptability in the future." As an example of this new trend in chemical engineering, he cited a novel process for the production of naphthalene sulfonic acids that reduces pollution by more than 90 percent.

Pollution of a river or a regional sea is, of course, more easily perceived than the pollution of the oceans, which are much larger; it is not surprising that the UNEP has already established pollution-control programs for 10 regional seas. Although such programs are a good start, they need to be followed up with protection of the oceans in general. A recent step in this direction is an international agreement forbidding the discarding of plastics from ships, which became effective at the beginning of this year. Other existing international conventions regulating marine resources need to be improved by backing them up with better monitoring schemes and enforcement measures.

Parallel with the need for improved water-resources management is the need for more research on the hydrosphere. For example, ecological and toxicologic studies of marine life are badly needed if we are to improve the husbandry of the oceans and gain better understanding of the ecological effects of long-lived pollutants in ocean waters.

Many aspects of the hydrological cycle, including the fluxes between its compartments and the extent of groundwater reserves, are not accurately known. These problems and others are currently being addressed by the International Hydrological Program of the United Nations Educational, Scientific and Cultural Organization. Moreover, major international research programs to study the interactions between climate and the hydrological cycle have recently been launched by the UNEP as well as by the World Health Organization and the nongovernmental International Council of Scientific Unions.

Predicting what is likely to happen if sound principles of water management are not vigorously implemented is all too easy. We have already seen rivers turn into sewers and lakes into cesspools. People die from drinking contaminated water, pollution washes ashore on recreational beaches, fish are poisoned by heavy metals and wildlife habitats are destroyed. A laissez-faire approach to water management will spell more of the same—on a grander scale. One can only hope recognition of that fact will spur governments and people into action.

CHAPTER 9 Global Warming and Ozone Depletion

9.1 MARIO J. MOLINA AND F. S. ROWLAND

Stratospheric Sink for Chlorofluoromethanes: Chlorine Atomc-atalysed Destruction of Ozone

In 1995 the Nobel Prize in chemistry was awarded for work done on the chemistry of the environment for the first time in history. The prize was shared by German chemist Paul Crutzen with two American chemists, Mario J. Molina and F. S. Rowland. These scientists were honored for fundamental work that predicted and led to the discovery of the threat to the stratospheric ozone layer posed by the popular industrial chemicals known as chlorofluorocarbons (CFCs).

In 1973, having earned a Ph.D. in physical chemistry at the University of California, Berkeley, Molina accepted a postdoctoral research position in the university's laboratory of atmospheric chemistry. Rowland, who also received a Ph.D. in physical chemistry from Berkeley, obtained a position at the university's campus in Irvine. There the two chemists proceeded to investigate the atmospheric fate of CFCs, the use of which as refrigerants and spray can propellants was becoming increasingly popular because they are volatile, have low toxicity, and are nonreactive. This work was inspired in part by the work Crutzen had done on chemical reactions that occur in the stratosphere. The next year Molina and Rowland published the article "Stratospheric Sink for Chlorofluoromethanes: Chlorine Atomc-atalysed Destruction of Ozone," *Nature* (June 28, 1974), from which the following selection has been taken. In it, they advanced that CFCs were a potential threat to the stratospheric ozone that protects life on earth from the dangers of high-energy ultraviolet radiation. This conclusion set off a heated controversy: The industrial producers and users of CFCs marshaled their resources to challenge the theory, and in the ensuing years, Rowland, Molina (now a professor at the Massachusetts Institute of Technology), their students, and other scientists had to repeatedly confirm and strengthen their model in the face of a series of scientific objections raised by industry scientists. Rowland devoted much of his energy to attempting to persuade the world's political and scientific leaders about his increasingly firm belief that CFCs represent a serious environmental threat. With the discovery of the Antarctic "ozone hole" in the mid-1980s and its subsequent link to the products of CFC decomposition, most of the skeptics were finally convinced and the international effort to phase out the use of CFCs was set in motion.

Key Concept: chlorofluorocarbons as a threat to stratospheric ozone

Chlorofluoromethanes are being added to the environment in steadily increasing amounts. These compounds are chemically inert and may remain in the atmosphere for 40–150 years, and concentrations can be expected to reach 10 to 30 times present levels. Photodissociation of the chlorofluoromethanes in the stratosphere produces significant amounts of chlorine atoms, and leads to the destruction of atmospheric ozone.

Halogenated aliphatic hydrocarbons have been added to the natural environment in steadily increasing amounts over several decades as a consequence of their growing use, chiefly as aerosol propellants and as refrigerants[1,2]. Two chlorofluoromethanes, CF_2Cl_2 and $CFCl_3$, have been detected throughout the troposphere in amounts (about 10 and 6 parts per 10^{11} by volume, respectively) roughly corresponding to the integrated world industrial production to date[3–5,31]. The chemical inertness and high volatility which make these materials suitable for technological use also mean that they remain in the atmosphere for a long time. There are no obvious rapid sinks for their removal, and they may be useful as inert tracers of atmospheric motions[4–6]. We have attempted to calculate the probable sinks and lifetimes for these molecules. The most important sink for atmospheric $CFCl_3$ and CF_2Cl_2 seems to be stratospheric photolytic dissociation to $CFCl_2 + Cl$ and to $CF_2Cl + Cl$, respectively, at altitudes of 20–40 km. Each of the reactions creates two odd-electron species—one Cl atom and one free radical. The dissociated chlorofluoromethanes can be traced to their ultimate sinks. An extensive catalytic chain reaction leading to the net destruction of O_3 and O occurs in the stratosphere:

$$Cl + O_3 \rightarrow ClO + O_2 \qquad (1)$$

$$ClO + O \rightarrow Cl + O_2 \qquad (2)$$

This has important chemical consequences. Under most conditions in the Earth's atmospheric ozone layer, (2) is the slower of the reactions because there is a much lower concentration of O than of O_3. The odd chlorine chain (Cl, ClO) can be compared with the odd nitrogen chain (NO, NO_2) which is believed to be intimately involved in the regulation of the present level of O_3 in the atmosphere[7–10]. At stratospheric temperatures, ClO reacts with O six times faster than NO_2 reacts with O (refs 11,12). Consequently, the Cl-ClO chain can be considerably more efficient than the NO-NO_2 chain in the catalytic conversion of $O_3 + O \rightarrow 2O_2$ per unit time per reacting chain[13].

PHOTOLYTIC SINK

Both $CFCl_3$ and CF_2Cl_2 absorb radiation in the far ultraviolet[14], and stratospheric photolysis will occur mainly in the 'window' at 1,750–2,200 Å between the more intense absorptions of the Schumann-Runge regions of O_2 and the Hartley bands of O_3. We have extended measurements of absorption coefficients for the chlorofluoromethanes to cover the range 2,000–2,270 Å. Calculations of the rate of photolysis of molecules at a given altitude at these wavelengths is complicated by the intense narrow band structure in the Schumann-Runge region, and the effective rates of vertical diffusion of molecules at these altitudes are also subject to substantial uncertainties. Vertical mixing is frequently modelled through the use of 'eddy' diffusion coefficients[10,15–18], which are presumably relatively insensitive to the molecular weight of the diffusing species. Calculated using a time independent one-dimensional vertical diffusion model with eddy diffusion coefficients of magnitude $K \sim (3 \times 10^3)$-10^4 $cm^2\ s^{-1}$ at altitudes 20–40 km (refs 10,15–18), the atmospheric lifetimes of $CFCl_3$ and CF_2Cl_2 fall into the range of 40–150 yr. The time required for approach toward a steady state is thus measured in decades, and the concentrations of chlorofluoromethanes in the atmosphere can be expected to reach saturation values of 10–30 times the present levels, assuming constant injection at current rates, and no other major sinks. (The atmospheric content is now equivalent to about five years world production at current rates.) Lifetimes in excess of > 10 and > 30 yr can already be estimated from the known industrial production rates and atmospheric concentrations[3,5], and so the stratospheric photochemical sink will be important even if other sinks are discovered.

Our calculation of photodissociation rates is modelled after those of Kockarts[19] and Brinkmann[20], and is globally averaged for diurnal and zenith angle effects. The photodissociation rates at an altitude of 30 km. are estimated to be $3 \times 10^{-7}\ s^{-1}$ for $CFCl_3$ and $3 \times 10^{-8}\ s^{-1}$ for CF_2Cl_2, decreasing for each by about a factor of 10^{-2} at 20 km. The appropriate solar ultraviolet intensities at an altitude of 30 km may be uncertain by a factor of 2 or 3 (ref. 21) and we have therefore calculated lifetimes for photodissociation rates differing from the above by factors of 3 or more. The competition between photodissociation and upward diffusion reduces the relative concentration of chlorofluoromethane at higher altitudes and the concentrations should be very low above 50 km. The peak rate of destruction, and formation of Cl atoms, occurs at 25–35 km, in the region of high ozone concentration. The rates of formation of Cl atoms at different altitudes, and the chlorofluoromethane atmospheric lifetimes are sensitive to the assumed eddy diffusion coefficients, as well as to the photodissociation rates.

The major chain processes in the stratosphere involving species with odd numbers of electrons belong to the H (H, OH, HO_2), N (NO, NO_2), and Cl (Cl, ClO) series. (ClO_2 is rapidly decomposed and its concentration is negligible relative to Cl plus ClO.) These odd-electron chains can only be terminated by interaction with one another, although other reactions can convert one series to another. At most altitudes, the first reaction for converting the Cl-ClO off-electron chain to an even-electron species containing chlorine is the abstraction of H from CH_4, which transfers the odd-electron character to the CH_3 radical:

$$Cl + CH_4 \rightarrow HCl + CH_3 \qquad (3)$$

As stratospheric temperatures the rate constant for Cl atoms[22], for (3) is about 10^{-3} times as fast as (1) and the O_3/CH_4 concentration ratio can make the rate of (3) less than that of (1) by another factor of 10. The Cl atom chain can be renewed by the reaction of OH with HCl (ref. 23):

$$OH + HCl \rightarrow H_2O + Cl \qquad (4)$$

Ultraviolet dissociation by absorption in the range 1,750–2,200 Å can also occur at the higher altitudes. The reaction rate of (4) in the stratosphere depends on the concentration of OH, which is known only roughly. In our estimates, termination of the Cl-ClO chain results from downward diffusion of the longer lived species in the chain (ClO, HCl) and eventual removal by tropospheric processes. The rate of termination thus also depends on diffusion processes and estimates will vary with the choice of eddy diffusion coefficients.

Possible terminations involving the Cl series with itself (for example, Cl + ClO $\rightarrow$ Cl_2O) or with one of the others (for example, Cl + NO $\rightarrow$ NOCl) normally lead to molecules with appreciable absorption coefficients at longer wavelengths, which are very rapidly dissociated again by the much more intense solar fluxes available there. Thus, even if a molecule which temporarily terminates two chains is formed, at least one of which involves the Cl series, the terminating molecule is rapidly photolysed and both chains are regenerated again.

Under most stratospheric conditions, the slow reactions in both the Cl-ClO and NO-NO_2 chains occur between O atoms and ClO and NO_2 molecules. The two chains are interconnected:

$$ClO + NO \rightarrow Cl + NO_2 \qquad (5)$$

The rate of this reaction in the stratosphere is frequently comparable to that of reaction (2). The overall effect is complex and depends on the relative concentrations of ClO_x, NO_x, O_3, O and OH. Reaction (1) is so rapid that the ClO/Cl ratio is usually > 10, even when Cl is produced by both reaction (2) and reaction (5), so that the overall rate of reaction (2) is not directly affected by the occurrence of reaction (5). As soon as Cl is produced, however, HCl can form by reactions (1) or (3), resulting in the temporary termination of the Cl atom chain. Whether or not the chain is then restarted depends primarily on the concentration of OH. There are substantial ranges of stratospheric altitudes in which neither reaction (3) nor reaction (5) seriously impedes the chain process of reactions (1) and (2).

The initial photolytic reaction produces one Cl atom from each of the parent molecules, plus a CX_3 radical (X may be F or Cl). The detailed chemistry of CX_3 radicals in O_2 or air is not completely known, but in the laboratory a phosgene-type molecule, CX_2O, is rapidly produced and another X atom probably Cl (or ClO)—is released from $CFCl_2$ or CF_2Cl[24,25]. CX_2O may also photolyse in the atmosphere to give a third and fourth free halogen atom. Thus, each molecule of $CFCl_3$ initially photolysed probably leads to between two and three Cl atom chains, and CF_2Cl_2 probably produces two Cl atom chains when it is photolysed. Initial calculations suggest that F atom chains will be much shorter than Cl atom chains because the reaction of abstraction from CH_4 is much faster for F atoms[26], whereas the reaction between OH and HF is 17 kcalorie mol^{-1} endothermic and will not occur in the stratosphere. We have not yet attempted to analyse the subsequent reaction paths of HF.

PRODUCTION RATES

The 1972 world production rates for $CFCl_3$ and CF_2Cl_2 are about 0.3 and 0.5 Mton yr^{-1} respectively[1,2,5], and are steadily increasing (by 8.7% per year for total fluorocarbons in the United States from 1961–71) (ref. 1). We have not included any estimates for other chlorinated aliphatic hydrocarbons also found in the atmosphere, such as CCl_4 (refs 3 and 4), $CHCl_3$, C_2Cl_4 and C_2HCl_3 for which there is no evidence for long residence times in the atmosphere[27]. If the stratospheric photolytic sink is the only major sink for $CFCl_3$ and CF_2Cl_2, then the 1972 production rates correspond at steady state to globally averaged destruction rates of about 0.8×10^7 and 1.5×10^7 molecules $cm^{-2}s^{-1}$ and formation rates of Cl atoms of about 2×10^7 and 3×10^7 atoms cm^{-2} s^{-1}, respectively. The total rate of production of 5×10^7 Cl atoms cm^{-2} s^{-1} from the two processes is of the order of the estimated natural flux of NO molecules

($2.5\text{-}15 \times 10^7$ NO molecules $cm^{-2}\ s^{-1}$) involved in the natural ozone cycle[9–12], and of the 5×10^7 NO molecules $cm^{-2}\ s^{-1}$ whose introduction around 25 km from stratospheric aviation is estimated would cause a 6% reduction in the total O_3 column[10].

Photolysis of these chlorofluoromethanes does not occur in the troposphere because the molecules are transparent to wavelengths longer than 2,900 Å. In fact the measured absorption coefficients for $CFCl_3$ and CF_2Cl_2 are falling rapidly at wavelengths longer than 2,000 Å (ref. 14). The reaction between OH and CH_4 is believed to be important in the troposphere[17,28], but the corresponding Cl atom abstraction reaction (for example, $OH + CFCl_3 \rightarrow HOCl + CFCl_2$) is highly endothermic and is negligible under all atmospheric conditions.

Neither $CFCl_3$ nor CF_2Cl_2 is very soluble in water, and they are not removed by rainout in the troposphere. Details of biological interactions of these molecules in the environment are very scarce because they do not occur naturally (except possibly in minute quantities from volcanic eruptions)[29], but rapid biological removal seems unlikely. The relative insolubility in water together with their chemical stability (especially toward hydrolysis)[30] indicates that these molecules will not be rapidly removed by dissolution in the ocean, and the few measurements made so far indicate equilibrium between the ocean surface and air, and therefore a major oceanic sink cannot be inferred[3].

It seems quite clear that the atmosphere has only a finite capacity for absorbing Cl atoms produced in the stratosphere, and that important consequences may result. This capacity is probably not sufficient in steady state even for the present rate of introduction of chlorofluoromethanes. More accurate estimates of this absorptive capacity need to be made in the immediate future in order to ascertain the levels of possible onset of environmental problems.

As with most NO_x calculations, our calculations have been based entirely on reactions in the gas phase, and essentially nothing is known of possible heterogeneous reactions of Cl atoms with particulate matter in the stratosphere. One important corollary of these calculations is that the full impact of the photodissociation of CF_2Cl_2 $CFCl_3$ is not immediately felt after their introduction at ground level because of the delay required for upward diffusion up to and above 25 km. If any Cl atom effect at atmospheric O_3 concentration were to be observed from this source, the effect could be expected to intensify for some time thereafter. A lengthy period (of the order of calculated atmospheric lifetimes) may thus be required for natural moderation, even if the amount of Cl introduced into the stratosphere is reduced in the future.

This research has been supported by the US Atomic Energy Commission. We acknowledge a helpful discussion with Professor H. S. Johnston.

NOTES

1. *Chemical Marketing Reporter,* August 21, 1972.
2. *Chemistry in the Economy* (American Chemical Society, Washington, DC, 1973).
3. Lovelock, J. E., Maggs, R. J., and Wade, R. J., *Nature,* **241,** 194 (1973).

4. Wilkniss, P. E., Lamontagne, R. A., Larson, R. E., Swinnerton, J. W., Dickson, C. R., and Thompson, T., *Nature,* **245,** 45 (1973).
5. Su, C.-W., and Goldberg, E. D., *Nature,* **245,** 27 (1973).
6. Machta, L., Proceedings of the Second IUTAM-IUGG Symposium on Turbulent Diffusion in Environmental Pollution (Charlottesville, 1973).
7. Crutzen, P. J., J. *geophys. Res.,* **30,** 7311 (1971).
8. Johnston, H., *Science,* **173,** 517 (1971).
9. Johnston, H. S., Proceedings of the First Survey Conference, Climatic Impact Assessment Program, US Department of Transport, 90 (1972).
10. McElroy, M. B., Wofsy, S. C., Penner, J. E., and McConnell, J. C., *J. atmos. Sci.,* **31,** 287 (1974).
11. Bemand, P. O., Clyne, M. A. A., and Watson, R. J., *J. Chem. Soc., Faraday I,* **69,** 1356 (1973).
12. Hampson, R., *et al., Chemical Kinetics Data Survey VI,* National Bureau of Standard Interim Report 73–207 (1973).
13. Stolarski, R. S., and Cicerone, R. J., International Association of Geomagnetism and Aeronomy (Kyoto, Japan, 1973); see also *Can. J. Chem.,* (in the press).
14. Doucet, J., Sauvageau, P., and Sandorfy, C., *J. chem. Phys.,* **58,** 3708 (1973).
15. McConnell, J. C., and McElroy, M.B., *J. atmos. Sci.,* **30,** 1465 (1973).
16. Schütz, K., Junge, C., Beck, R., and Albrecht, B., *J. geophys. Res.,* **75,** 2230 (1970).
17. Wofsy, S. C., McConnell, J. C., and McElroy, M. B., *J. geophys. Res.,* **77,** 4477 (1972)
18. Wofsy, S. C., and McElroy, M. B., *J. geophys. Res.,* **78,** 2619 (1973).
19. Kockarts, G., in *Mesospheric Models and Related Experiments* (edit. by Fiocco, G.), 168, Reidel, Dodrrecht (1971).
20. Brinkmann, R., *ibid.,* 89.
21. Hudson, R. D., and Mahle, S. H., *J. geophys. Res.,* **77,** 2902 (1972).
22. Clyne, M. A. A., and Walker, R. F., *J. Chem, Soc., Faraday I,* **69,** 1547 (1973).
23. Takacs, G. A., and Glass, G. P., *J. phys. Chem.,* **77,** 1948 (1973).
24. Marsh, D., and Heicklen, J., *J. phys. Chem.,* **69,** 4410 (1965).
25. Heicklen, J., *Adv. Photochem.,* **7,** 57 (1969).
26. Homann, K. H., Solomon, W. C., Warnatz, J., Wagner, H. G., and Zetzch, C., *Ber. Bunsenges. phys. Chem.,* **74,** 585 (1970).
27. Murray, A. J., and Riley, J. P., *Nature,* **242,** 37 (1973).
28. Levy, H., *Planet. Space Sci.,* **20,** 919 (1972); **21,** 575 (1973).
29. Stoiber, R. E., Leggett, D. C., Jenkins, T. F., Murrmann, R. P., and Rose, W. J., *Bull. geol. Soc. Am.,* **82,** 2299 (1971).
30. Hudlicky, M., *Chemistry of Organic Fluorine Compounds,* 340 (MacMillan, New York, 1962).
31. Lovelock, J. E., *Nature,* **230,** 379 (1971).

9.2 RICHARD ELLIOT BENEDICK

Ozone Diplomacy: New Directions in Safeguarding the Planet

By 1987 the world had been shocked by the discovery of the Antarctic "ozone hole," a recurring and precipitous decrease in the protective stratospheric ozone concentration in a huge and expanding area surrounding the South Pole. Furthermore, most scientists studying this problem agreed that this phenomenon is related to and confirms the theory that chlorofluorocarbons (CFCs) pose a threat to the ozone layer, which was first proposed 13 years earlier by physical chemists F. S. Rowland and Mario J. Molina. In 1987 an international conference of political and scientific delegates from both developed and developing nations, meeting in Montreal, Canada, reached historic agreement that few political analysts had thought was possible: the Montreal Protocol, which committed the signatory nations to significantly reducing the use of CFCs and related industrial chemicals. In subsequent years, as evidence of the destruction of stratospheric ozone resulting from atmospheric contamination by these substances mounted, this initial agreement was strengthened, and by 1996 the production of CFCs by developed nations had been banned.

Richard Elliot Benedick is a seasoned diplomat with many years of experience in the U.S. Foreign Service. As deputy assistant secretary of state for environment, health, and natural resource issues, he was assigned to be the chief U.S. negotiator at the Montreal meeting. In *Ozone Diplomacy: New Directions in Safeguarding the Planet* (Harvard University Press, 1991), he describes the lessons he learned from that experience. In the following selection, which has been taken from the first and last chapters of *Ozone Diplomacy,* Benedick explains how these lessons can be applied in the quest for international agreements on other important global environmental problems. In particular, he emphasizes the need for action on the more complex and difficult issue of reducing the potentially disastrous consequences of climate change due to atmospheric accumulation of greenhouse gases.

Key Concept: strategies for negotiating international environmental agreements

LESSONS FROM HISTORY

On September 16, 1987, a treaty was signed that was unique in the annals of international diplomacy. Knowledgeable observers had long believed that this particular agreement would be impossible to achieve because the issues were so complex and arcane and the initial positions of the negotiating parties so widely divergent. Those present at the signing shared a sense that this was not just the conclusion of another important negotiation, but rather a historic occasion. It was hailed as "the most significant international environmental agreement in history," "a monumental achievement," and "unparalleled as a global effort."

The Montreal Protocol on Substances That Deplete the Ozone Layer mandated significant reductions in the use of several extremely useful chemicals. At the time of the treaty's negotiation, chlorofluorocarbons (CFCs) and halons were rapidly proliferating compounds with wide applications in thousands of products, including refrigeration, air conditioning, aerosol sprays, solvents, transportation, plastics, insulation, pharmaceuticals, computers, electronics, and fire fighting. Scientists suspected, however, that as these substances were released into the atmosphere and diffused to its upper reaches, they might cause future damage to a remote gas—the stratospheric ozone layer—that shields life on Earth from potentially disastrous levels of ultraviolet radiation.)

By their action, the signatory countries sounded the death knell for an important part of the international chemical industry, with implications for billions of dollars in investment and hundreds of thousands of jobs in related sectors. The protocol did not simply prescribe limits on these chemicals based on "best available technology," which had been a traditional way of reconciling environmental goals with economic interests. Rather, the negotiators established target dates for replacing products that had become synonymous with modern standards of living, even though the requisite technologies did not yet exist.

"Politics," Lord Kennet stated in special hearings on the accord in the British House of Lords a year later, "is the art of taking good decisions on insufficient evidence." Perhaps the most extraordinary aspect of the treaty was its imposition of substantial short-term economic costs to protect human health and the environment against unproved future dangers—dangers that rested on scientific theories rather than on firm data.

At the time of the negotiations and signing, no measurable evidence of damage existed. Thus, unlike environmental agreements of the past, the treaty was not a response to harmful developments or events but rather a preventive action on a global scale. Indeed, the Montreal Protocol added a new dimension to the 1972 Declaration of the United Nations Conference on the Human Environment at Stockholm, which appealed to nations "to ensure that activities within their jurisdiction of control do not cause damage to the environment of other States"—a responsibility that had been more often an ideal than a reality.

As the international community and national political leaders grapple with other environmental problems involving scientific uncertainty, long-term risks, and multilateral cooperation, the model of ozone diplomacy offers a valuable precedent....

The Montreal Protocol was a response to a new type of problem facing the modern world. In the past two decades there has been a growing realization that ecological dangers can imperil the security of all peoples. Ozone layer depletion, climate change, destruction of tropical rain forests, toxic wastes, pollution of oceans and fresh water, massive loss of biological diversity—these issues are moving to the top of the world's agendas, and they will increasingly dominate international relations in the 1990s. The drafting of the Montreal Protocol was only the first of a series of international events in the late 1980s that brought together national leaders to consider these global threats. Symbolic of the awakening political awareness was the unprecedented concern over the deteriorating environment expressed at the 1989 and 1990 economic summit meetings of the major industrial democracies. Twenty years after the Stockholm conference, world leaders will convene in June 1992 in Brazil at the United Nations Conference on Environment and Development, which will generate crucial guidelines for international approaches to these problems over the coming decades.

This new generation of ecological issues exemplifies the interconnectedness of life and its natural support systems on Earth. Modern scientific discoveries are revealing that localized activities can have global consequences and that dangers can be slow and perhaps barely perceptible in their development, yet with long-term and virtually irreversible effects. The concept is not obvious: a perfume spray in Paris helps to destroy an invisible gas in the stratosphere and thereby contributes to skin cancer deaths and species extinction half a world away and several generations in the future.

Neither traditional environmental law nor traditional diplomacy offers guidelines for confronting such situations. Environmental problems of the past were normally localized or regional, and their effects were self-evident. Such events could be addressed largely through unilateral actions, national legislation, and occasional international treaties, all based on unmistakable evidence of damage.

However, if the international community is to respond effectively to the new environmental challenges, governments must undertake coordinated actions while some major questions remain unresolved-and before damage becomes tangible and thereby possibly irremediable. The negotiators of the Montreal Protocol dealt with dangers that could touch every nation and all life on Earth, over periods far beyond the normal time horizons of politicians. But the potentially grave consequences could be neither measured nor predicted with any certainty at the time.

The Montreal Protocol thus became a prototype for an evolving new form of international cooperation. In achieving the treaty, consensus was forged and decisions were made on a balancing of probabilities. And the risks of waiting for more complete evidence were finally deemed to be too great.

As U.S. delegation head, I summed up the case for action in a plenary address at one of the early negotiating rounds:

> When we build a bridge, we build it to withstand much stronger pressures than it is ever likely to confront. And yet, when it comes to protecting the global atmo-

sphere, where the stakes are so much higher, the attitude [of some people] seems to be equivalent to demanding certainty that the bridge will collapse as a justification for strengthening it. If we are to err in designing measures to protect the ozone layer, then let us, conscious of our responsibility to future generations, err on the side of caution.

Challenges and Leadership

Ironically, as time has passed, and as disturbing new scientific discoveries have vindicated the concerns of the protocol's designers, the events at Montreal have acquired an aura of inevitability. It all seems easy in retrospect. But memories are short.

Both before and during the negotiations, major uncertainties surrounded such fundamental questions as the possible degree of future damage to stratospheric ozone, the extent to which CFCs and other chemicals were responsible, the prospective growth of demand for these chemicals, the significance of any adverse effects from ozone layer depletion, and the length of time before serious harm might occur. Many governments whose cooperation was needed displayed attitudes ranging from indifference to outright hostility. International industry was strongly opposed to regulatory action. And the gaps separating the negotiating parties often seemed unbridgeable.

The problem of protecting the stratospheric ozone layer presented an unusual challenge to diplomacy. Military strength was irrelevant to the situation. Economic power also was not decisive. Neither great wealth nor sophisticated technology was necessary to produce large quantities of ozone destroying chemicals. Traditional notions of national sovereignty became questionable when local decisions and activities could affect the well-being of the entire planet. Because of the nature of ozone depletion, no single country or group of countries, however powerful, could effectively solve the problem. Without far-ranging cooperation, the efforts of some nations to protect the ozone layer would be undermined.

In this unique and complex situation, a small and previously little-known UN agency, the United Nations Environment Programme (UNEP), together with the U.S. government, assumed primary leadership roles. They faced an uphill battle all the way. How the consensus for international action was achieved is a case study in modern diplomacy. . . .

LOOKING AHEAD: A NEW GLOBAL DIPLOMACY

Perhaps the most poignant image of our time is that of Earth as seen by the space voyagers: a blue sphere, shimmering with life and light, alone and unique in the cold vastness of the cosmos. From this perspective, the maps of geopolitics vanish, and the underlying interconnectedness of all the components of this extraordinary living system—animal, plant, water, land, and atmosphere —becomes strikingly evident.

Experimenting With Planet Earth

Humanity has learned that the activities of modern industrial economies, driven by consumer demands and burgeoning populations, can alter delicate natural balances. We can no longer pretend that nothing will happen as the planet is subjected to billions of tons of pollutants. It is not that Earth itself is necessarily fragile. It may be, rather, that our own tenure turns out to be less secure and less inevitable as a consequence of the planet's responses to unnatural conditions created by human actions.

The Antarctic ozone hole conveyed a warning. Nature is capable of producing unpleasant surprises: even seemingly small interferences—in this case, an increase in stratospheric chlorine concentrations of a little more than one part per billion—can trigger dramatic and sudden reactions. Recent experience with the forests of Europe and North America indicates that other areas may also have unforeseen thresholds beyond which natural processes are unable to absorb the assaults of contemporary society. The world may not have the luxury of early warning signals before an irreversible collapse occurs in some other part of the planet's ecosystem.

A 1986 report by the National Aeronautics and Space Administration observed that "we are conducting one giant experiment on a global scale by increasing the concentrations of trace gases in the atmosphere without knowing the environmental consequences." This statement echoed almost verbatim the warning three decades earlier of Roger Revelle and Hans Suess, two scientists at the Scripps Institution of Oceanography. Concerned about the implications for global warming of the rapid accumulation of carbon dioxide in the atmosphere resulting from fossil fuel combustion, they concluded in 1957 that "human beings are now carrying out a large scale geophysical experiment of a kind that could not have happened in the past."

Lessons for a New Diplomacy

The international community was successful in its approach to the problem of protecting the stratospheric ozone layer. The experience gained from this process suggests several elements of a new kind of diplomacy for addressing such similar global ecological threats as greenhouse warming.

Scientists must play an unaccustomed but critical role in international environmental negotiations. Without modern science and technology, the world would have remained unaware of the danger to ozone until it was too late. Science became the driving force behind ozone policy. The formation of a commonly accepted body of data and analyses and the narrowing of ranges of uncertainty were prerequisites to a political solution among negotiating parties initially far apart. In effect, a community of scientists from many nations, committed to scientific objectivity, developed through their research an interest in protecting the planet's ozone layer that transcended divergent national interests. In this process, the scientists had to assume some responsibility for relating the implications of their findings to alternative remedial strategies. Close collaboration between scientists and key government officials who also became convinced of the long-term dangers ultimately prevailed over the more parochial and short-run interests of some national politicians and industrialists.

Governments may have to act while there is still scientific uncertainty, responsibly balancing the risks and costs of acting or not acting. By the time the evidence on such issues as the ozone layer and climate change is beyond all dispute, the damage may be irreversible, and it may be too late to forestall serious harm to human life and draconian costs to society. Politicians must therefore resist a tendency to lend too much credence to self-serving economic interests that demand scientific certainty, maintain that dangers are remote and unlikely, and insist that the costs of changing their ways are astronomical. The signatories at Montreal knowingly imposed substantial short-run economic dislocations even though the evidence was incomplete; the prudence of their decision was demonstrated when the scientific models turned out to have underestimated the effects of CFCs on ozone. Governments must sponsor the needed research and act responsibly on the basis of often equivocal results. Unfortunately, the current tools of economic analysis are inadequate aids in this task and can even be deceptive indicators; they are in urgent need of reform.

Educating and mobilizing public opinion are essential to generate pressure on hesitant governments and private companies. The interest of the media in the ozone issue and the collaboration with television and press by diplomats, environmental groups, and legislators had a major influence on governmental decisions and on the international negotiations. Concerned consumers brought about the collapse of the CFC aerosol market. And in their educational efforts, the proponents of ozone layer protection generally avoided invoking apocalypse and resisted temptations to overstate their case in order to capture public attention; exaggerated pronouncements could have damaged credibility and provided ammunition to those interest groups that wanted to delay action.

Multilateral diplomacy, involving coordinated negotiations among many governments, is essential when the issues have planetary consequences. The manifold activities of an international organization—the United Nations Environment Programme—were crucial in promoting a global approach to the protection of stratospheric ozone. UNEP coordinated research, informed governments and world public opinion, and played an indispensable catalytic and mediating role during the negotiation and implementation of the protocol.

Strong leadership by a major country can be a significant force for developing international consensus. The U.S. government reflected its concerns over the fate of the ozone layer through stimulating and supporting both American and international scientific research. Then, convinced of the dangers, it undertook extensive diplomatic and scientific initiatives to promote an ozone protection plan to other countries, many of which were initially hostile or indifferent to the idea. As the largest emitter of both ozone-destroying chemicals and greenhouse gases, the United States has enormous potential to influence the policy considerations of other governments in favor of environmental protection. In fact, because of the geographic size and population of the United States, its economic and scientific strength, and its international interests and influence, progress in addressing global environmental problems can probably not be achieved without American leadership.

It may be desirable for a leading country or group of countries to take preemptive environmental protection measures in advance of a global agreement. When influential governments make such a commitment, they legitimate change and thereby undercut the arguments of those who insist that change is impossible. Preemptive actions can also support moral suasion in encouraging future participation by other countries. In addition, action by major countries can slow dangerous trends and hence buy time for future negotiations and for development of technological solutions. The 1978 U.S. ban on aerosols relieved pressure on the ozone layer and lent greater authority to the U.S. government when it subsequently campaigned for even more stringent worldwide measures. Although environmental controls might conceivably affect a country's international competitiveness in the short run, they might also, by stimulating research into alternative technologies, give that country's industry a head start on the future.

The private sector—including citizens' groups and industry—is very much involved in the new diplomacy. The activities of both environmental organizations and private industry in undertaking research, lobbying governments, and influencing public opinion definitely affected the international debate on the ozone issue. A major by-product of the ozone negotiations was the development of closer relations among hitherto separate environmental groups around the world, reflected in their new and unaccustomed cooperation at conferences on the ozone layer and climate change in 1989 and 1990. Environmental organizations can also play an informal future watchdog role in monitoring compliance by industry with internationally agreed-upon commitments. For their part, industrialists are becoming aware that their corporate image is increasingly affected by environmental issues. The intellectual and financial resources of the corporate sector are, moreover, essential for developing the required technological solutions. These groups will expect to be involved in establishing policies and negotiating international accords on climate and other emerging global ecological issues.

Economic and structural inequalities among countries must be adequately reflected in any international regulatory regime. In the longer run, the developing countries, with their huge and growing populations, could undermine efforts to protect the global environment. As a consequence of the ozone issue, the richer nations for the first time acknowledged a responsibility to help developing countries to implement needed environmental policies without sacrificing aspirations for improved standards of living. The Montreal Protocol broke new ground with the creation of the multilateral ozone fund and the commitments for technology transfer, while illuminating the issues that must be considered in arriving at realistic and equitable solutions to future global problems.

The effectiveness of a regulatory agreement is enhanced when it employs market incentives to stimulate technological innovation. Technology is dynamic and not, as some industrialists have seemed to imply, a static element. But left completely on its own, the market does not necessarily bring forth the right technologies to protect the environment. The ozone protocol set targets that were initially beyond the reach of the existing best-available technology. They appeared difficult but were in fact achievable for most of industry—and thereby averted monolithic industrial opposition that might have delayed international agreement. The treaty actually stimulated collaboration among otherwise competing companies in research and testing, thereby saving both time and money in the development of replacement technologies and products. By expeditiously getting the protocol established in international law even with a 50 percent reduction target, the negotiators effectively signaled the marketplace that research into solutions would now be profitable—thus setting the stage for the later decisions for phaseout.

The signing of a treaty is not necessarily the decisive event in a negotiation; the process before and after signing is critical. It was extremely important to separate the complicated ozone-protection issue into manageable components. The informal fact-finding efforts during 1985 and 1986 and again during 1989 and 1990—workshops, conferences, consultations—established an environment conducive to building personal relationships and generating creative ideas, and thereby facilitated the formal negotiations. During the negotiations themselves, the use of small working groups and a single basic "chairman's text" aided in the gradual emergence of consensus. The developments following the 1987 signing illustrated the wisdom of designing the treaty as a flexible instrument. By providing for periodic integrated assessments—the first of which was advanced from 1990 to 1989 in response to the rapidly changing science—the negotiators made the accord adaptable to evolving circumstances. In effect, the protocol became a continuing process rather than a static solution.

Firmness and pragmatism combined are important ingredients of diplomatic success. The proponents of strong controls refrained from extreme positions but never relented in their pressure for a meaningful treaty. They did not insist on perfect solutions that might have unnecessarily prolonged the negotiations. Nor did they wait for universal participation, or even for agreement among all potential major players. Instead, they achieved an interim solution with built-in flexibility that could serve as a springboard for future action.

Individuals can make a surprisingly significant difference. UNEP's Mostafa Tolba provided overall personal leadership, initiating critical consultations with key governments, private interest groups, and international organizations. During the negotiations, he moved from group to group, arguing for flexibility, applying pressure, often floating his own proposals as a stimulus to the participants. Individual scientists, negotiators, environmentalists, and industry officials also provided ideas, decisions, and actions that proved vital to the final outcome.

Toward Action on Climate Change

The relevance of the experience with the ozone treaties has not been lost on the international community as it addresses greenhouse warming. The establishment in late 1988 and the subsequent functioning of the Intergovernmental Panel on Climate Change—with its multiple scientific, economic, and policy workshops and varied participation from public and private sectors—parallel the fact-gathering and analytical phases of the ozone history. UNEP and the World Meteorological Organization are again actively involved in the process.

Similarly, many governments announced their support during 1989 for a framework agreement on climate change, comparable to the 1985 Vienna Convention on Protecting the Ozone Layer. Such a climate convention need not be a complicated undertaking, and it should be achieved at the earliest possible date. The existence of gaps in scientific and economic knowledge should not become an excuse for delaying the negotiations.

Ideally, a framework convention would enable governments to formalize their agreement in principle on the dimensions of the climate problem and the scope of possible responses. Governments would undertake general obligations for policies and actions to mitigate and adapt to global warming. They would also agree on coordinated research and monitoring to develop additional data as guidelines for future measures.

It might be useful to go beyond the Vienna precedent and try to build into a climate convention some general targets and timetables. However, it would probably be questionable for advocates of stringent greenhouse gas controls to attempt to load a convention with overly detailed and still-controversial commitments. Premature insistence on optimal solutions could have the unintended effect of bogging down the negotiations and prolonging the entire process. On the other hand, an early convention—even if general in its terms—would itself establish a political momentum for further concrete actions.

The framework convention would provide the legal and logistical structure for the crucial step corresponding to the Montreal Protocol: agreement on specific international measures. Work on such measures could well begin even before completion of the framework convention. Because of the complexity of the climate issue, it would not be realistic to attempt to achieve an ideal set of responses at a single stroke; here again, the quest for perfection might only serve to delay action. Instead, it might be more effective—and realistic—to think in terms of incremental stages and partial solutions, following the examples of the Vienna Convention and the Montreal Protocol.

Thus, governments could negotiate a number of separate implementing protocols, each one containing specific measures for dealing with a different aspect of the climate problem. One example would be a treaty mandating greater energy efficiency in the transportation sector, which should be feasible because it need involve only a relative handful of major producers of transportation equipment. The ozone accord itself exemplifies interim progress on the climate problem by means of a constituent protocol. Such partial solutions can be significant; a NASA study estimated that if CFCs had continued to increase at the growth rates of the 1970s, they would by 1989 have surpassed carbon dioxide in their greenhouse impact.

It might be useful to establish standing negotiations under a permanent secretariat, similar to arrangements for the Geneva disarmament talks. By this means, individual protocols could simultaneously be in process of development, each at its own pace.

A climate convention and protocols need not strive right away for universal membership—that could be an unnecessarily complicating factor. In actuality, the overwhelming proportion of carbon emissions from fossil fuels and deforestation is concentrated in a relatively small number of industrialized and developing nations.

Indeed, the major industrialized countries, which are primarily responsible for the world's current precarious ecological condition, could make a vital contribution by agreeing on preemptive actions even before a broader climate treaty is negotiated. The United States and Canada, the Soviet Union, the European Community, and Japan together account for about 60 percent of carbon emissions from fossil fuels. By not delaying actions to increase energy efficiency and reduce carbon dioxide emissions, these countries could significantly slow the warming trend. Doing this would buy time for innovation in energy-efficient technologies and renewable energy sources that could be shared with developing and Eastern European countries to aid them in assuming their own responsibility.

Global Stewardship

Mostafa Tolba has described the Montreal Protocol as "the beginning of a new era of environmental statesmanship." The history of the ozone treaty reflects a new reality: nations must work together in the face of global threats, because if some major actors do not participate, the efforts of others will be vitiated. The process of arriving at the agreement, and the developments that followed its signing, represented new directions for diplomacy, involving unusual emphasis on science and technology, on market forces, on equity, and on flexibility. For all of this, the Montreal Protocol should prove to be a lasting model of international cooperation.

In the realm of international relations, there will always be resistance to change, and there will always be uncertainties—political, economic, scientific, psychological. The ozone protocol's greatest significance, in fact, may be as much in the domain of ethics as environment: it may signal a shift in attitude among critical segments of society in the face of uncertain but potentially grave threats that require coordinated action by sovereign states. The treaty showed that, even in the real world of ambiguity and imperfect knowledge, the international community is capable of undertaking difficult cooperative actions for the benefit of future generations. The Montreal Protocol may thus be the forerunner of an evolving global diplomacy, through which nations accept common responsibility for stewardship of the planet.

9.3 THE INTERGOVERNMENTAL PANEL ON CLIMATE CHANGE (IPCC)

Summary for Policymakers: The Science of Climate Change – IPCC Working Group I

The idea that the heat trapping ability of infrared absorbing gases in the atmosphere is similar to that of the glass panes in a greenhouse was first proposed by the French mathematical physicist Jean-Baptiste-Joseph Fourier in 1827. In 1896 the Swedish chemist Svante Arrhenius, who later won the 1903 Nobel prize in chemistry, predicted that if atmospheric carbon dioxide (CO_2) levels doubled due to the burning of fossil fuels, the resulting increase in the average temperature at the Earth's surface would amount to four to six degrees Celsius (seven to ten degrees Fahrenheit).

The Arrhenius prediction about global warming was all but forgotten for more than half a century until direct observations and historical data demonstrated that by 1960, atmospheric CO_2 levels had risen to 315 ppm from the preindustrial level of 280 ppm. Careful measurements since that discovery have shown that the present CO_2 level is approaching 370 ppm. The Arrhenius prediction that the average temperature on Earth will rise four to six degrees Celsius is likely to occur before the year 2075 if present fossil fuel use and forest destruction trends continue. Most atmospheric scientists agree that a rise of more than two degrees Celsius would cause major changes in the world's weather patterns and a significant increase in sea levels. However, the Arrhenius prediction has remained controversial due to the difficulties of developing accurate models that take into account all of the complex atmospheric interactions that determine global temperature fluctuations. In 1995, for the first time, an international organization of scientists reached a consensus that signs of greenhouse gas-induced global warming have probably been observed already.

The Intergovernmental Panel on Climate Change (IPCC)

In 1988, due to concern about the serious disruptive effects that would result from significant, short-term changes in world climate, the United Nations Environment Programme joined with the World Meteorological Organization to establish the Intergovernmental Panel on Climate Change (IPCC) to assess the available scientific, technical, and socioeconomic information concerning greenhouse gas-induced climate change. In 1995 the IPCC, which includes most of the world's meteorologists and other atmospheric scientists who are studying the global warming problem, issued *Climate Change 1995: IPCC Second Assessment Report* (The Intergovernmental Panel on Climate Change, April 12, 1999). The following selection is taken from the section of that report "Summary for Policy Makers: The Science of Climate Change–IPCC Working Group I." (Working Group I is the IPCC panel charged with assessing available information on the science of climate arising from human activities.) This report included for the first time the statement that "the balance of evidence suggests a discernible human influence on global climate." This conclusion has provided considerable support to those scientists and environmental activists who have been pressing for a worldwide effort to reduce the effects of global warming by curtailing greenhouse gas emissions. The accumulating evidence of increasing average world temperatures and other related atmospheric phenomena since 1995 has reduced the group of scientists who continue to reject the global warming predictions to a small minority, most of whom are either employed by, or funded by, industries that have a financial stake in resisting proposals for significant reductions in the release of greenhouse gases.

Key Concept: causes and effects of greenhouse gas-induced climate change

Considerable progress has been made in the understanding of climate change[1] science since 1990 and new data and analyses have become available.

1. GREENHOUSE GAS CONCENTRATIONS HAVE CONTINUED TO INCREASE

Increases in greenhouse gas concentrations since pre-industrial times (i.e., since about 1750) have led to a positive radiative forcing[2] of climate, tending to warm the surface and to produce other changes of climate.

- The atmospheric concentrations of greenhouse gases, inter alia, carbon dioxide (CO_2), methane (CH_4) and nitrous oxide (N_2O) have grown significantly: by about 30%, 145%, and 15%, respectively (values for 1992). These trends can be attributed largely to human activities, mostly fossil-fuel use, land-use change and agriculture.

- The growth rates of CO_2, CH_4 and N_2O concentrations were low during the early 1990s. While this apparently natural variation is not yet fully explained, recent data indicate that the growth rates are currently comparable to those averaged over the 1980s.
- The direct radiative forcing of the long-lived greenhouse gases (2.45 Wm-2) is due primarily to increases in the concentrations of CO_2 (1.56 Wm-2), CH_4 (0.47 Wm-2) and N_2O (0.14 Wm-2) (values for 1992).
- Many greenhouse gases remain in the atmosphere for a long time (for $C0_2$ and N_20, many decades to centuries), hence they affect radiative forcing on long time-scales.
- The direct radiative forcing due to the CFCs [chlorofluorocarbons] and HCFCS [a group of chemicals that are often used as an interim replacement for CFCs] combined is 0.25 Wm-2. However, their net radiative forcing is reduced by about 0.1 Wm-2 because they have caused stratospheric ozone depletion which gives rise to a negative radiative forcing.
- Growth in the concentration of CFCs, but not HCFCs, has slowed to about zero. The concentrations of both CFCs and HCFCs, and their consequent ozone depletion, are expected to decrease substantially by 2050 through implementation of the Montreal Protocol and its Adjustments and Amendments.
- At present, some long-lived greenhouse gases (particularly HFCs (a CFC substitute), PFCs and SF_6) contribute little to radiative forcing but their projected growth could contribute several per cent to radiative forcing during the 21st century.
- If carbon dioxide emissions were maintained at near current (1994) levels, they would lead to a nearly constant rate of increase in atmospheric concentrations for at least two centuries, reaching about 500 ppmv (approaching twice the pre-industrial concentration of 280 ppmv) by the end of the 21st century.
- A range of carbon cycle models indicates that stabilization of atmospheric CO_2 concentrations at 450, 650 or 1000 ppmv could be achieved only if global anthropogenic CO_2 emissions drop to 1990 levels by, respectively, approximately 40, 140 or 240 years from now, and drop substantially below 1990 levels subsequently.
- Any eventual stabilized concentration is governed more by the accumulated anthropogenic CO_2 emissions from now until the time of stabilization than by the way those emissions change over the period. This means that, for a given stabilized concentration value, higher emissions in early decades require lower emissions later on. Among the range of stabilization cases studied, for stabilization at 450, 650 or 1000 ppmv, accumulated anthropogenic emissions over the period 1991 to 2100 are 630 GtC[3], 1030 GtC and 1410 GtC, respectively (approximately 15% in each case). For comparison the corresponding accumulated emissions for IPCC IS92 emission scenarios range from 770 to 2190 GtC.
- Stabilization of CH_4 and N_2O concentrations at today's levels would involve reductions in anthropogenic emissions of 8% and more than 50% respectively.

- There is evidence that tropospheric ozone concentrations in the Northern Hemisphere have increased since pre-industrial times because of human activity and that this has resulted in a positive radiative forcing. This forcing is not yet well characterized, but it is estimated to be about 0.4 Wm-2 (15% of that from the long-lived greenhouse gases). However, the observations of the most recent decade show that the upward trend has slowed significantly or stopped.

2. ANTHROPOGENIC AEROSOLS TEND TO PRODUCE NEGATIVE RADIATIVE FORCINGS

- Tropospheric aerosols (microscopic airborne particles) resulting from combustion of fossil fuels, biomass burning and other sources have led to a negative direct forcing of about 0.5 Wm-2, as a global average, and possibly also to a negative indirect forcing of a similar magnitude. While the negative forcing is focused in particular regions and subcontinental areas, it can have continental to hemispheric scale effects on climate patterns
- Locally, the aerosol forcing can be large enough to more than offset the positive forcing due to greenhouse gases.
- In contrast to the long-lived greenhouse gases, anthropogenic aerosols are very short-lived in the atmosphere, hence their radiative forcing adjusts rapidly to increases or decreases in emissions.

3. CLIMATE HAS CHANGED OVER THE PAST CENTURY

At any one location, year-to-year variations in weather can be large, but analyses of meteorological and other data over large areas and over periods of decades or more have provided evidence for some important systematic changes.

- Global mean surface air temperature has increased by between about 0.3 and 0.6°C since the late 19th century; the additional data available since 1990 and the re-analyses since then have not significantly changed this range of estimated increase.
- Recent years have been among the warmest since 1860, i.e., in the period of instrumental record, despite the cooling effect of the 1991 Mt Pinatubo volcanic eruption.
- Night-time temperatures over land have generally increased more than daytime temperatures.

- Regional changes are also evident. For example, the recent warming has been greatest over the mid-latitude continents in winter and spring, with a few areas of cooling, such as the North Atlantic ocean. Precipitation has increased over land in high latitudes of the Northern Hemisphere, especially during the cold season.
- Global sea level has risen by between 10 and 25 cm over the past 100 years and much of the rise may be related to the increase in global mean temperature.
- There are inadequate data to determine whether consistent global changes in climate variability or weather extremes have occurred over the 20th century. On regional scales there is clear evidence of changes in some extremes and climate variability indicators (e.g., fewer frosts in several widespread areas; an increase in the proportion of rainfall from extreme events over the contiguous states of the USA). Some of these changes have been toward greater variability; some have been toward lower variability.
- The 1990 to mid-1995 persistent warm-phase of the El Niño-Southern Oscillation (which causes droughts and floods in many areas) was unusual in the context of the last 120 years.

4. THE BALANCE OF EVIDENCE SUGGESTS A DISCERNIBLE HUMAN INFLUENCE ON GLOBAL CLIMATE

Any human-induced effect on climate will be superimposed on the background "noise" of natural climate variability, which results both from internal fluctuations and from external causes such as solar variability or volcanic eruptions. Detection and attribution studies attempt to distinguish between anthropogenic and natural influences. "Detection of change" is the process of demonstrating that an observed change in climate is highly unusual in a statistical sense, but does not provide a reason for the change. "Attribution" is the process of establishing cause and effect relations, including the testing of competing hypotheses.

Since the 1990 IPCC Report, considerable progress has been made in attempts to distinguish between natural and anthropogenic influences on climate. This progress has been achieved by including effects of sulphate aerosols in addition to greenhouse gases, thus leading to more realistic estimates of human-induced radiative forcing. These have then been used in climate models to provide more complete simulations of the human-induced climate-change "signal". In addition, new simulations with coupled atmosphere-ocean models have provided important information about decade to century time-scale natural internal climate variability. A further major area of progress is the shift of focus from studies of global-mean changes to comparisons of modelled and observed spatial and temporal patterns of climate change.

The most important results related to the issues of detection and attribution are:

- The limited available evidence from proxy climate indicators suggests that the 20th century global mean temperature is at least as warm as any other century since at least 1400 A.D. Data prior to 1400 are too sparse to allow the reliable estimation of global mean temperature.
- Assessments of the statistical significance of the observed global mean surface air temperature trend over the last century have used a variety of new estimates of natural internal and externally-forced variability. These are derived from instrumental data, palaeodata, simple and complex climate models, and statistical models fitted to observations. Most of these studies have detected a significant change and show that the observed warming trend is unlikely to be entirely natural in origin.
- More convincing recent evidence for the attribution of a human effect on climate is emerging from pattern-based studies, in which the modelled climate response to combined forcing by greenhouse gases and anthropogenic sulphate aerosols is compared with observed geographical, seasonal and vertical patterns of atmospheric temperature change. These studies show that such pattern correspondences increase with time, as one would expect, as an anthropogenic signal increases in strength. Furthermore, the probability is very low that these correspondences could occur by chance as a result of natural internal variability only. The vertical patterns of change are also inconsistent with those expected for solar and volcanic forcing.
- Our ability to quantify the human influence on global climate is currently limited because the expected signal is still emerging from the noise of natural variability, and because there are uncertainties in key factors. These include the magnitude and patterns of long-term natural variability and the time-evolving pattern of forcing by, and response to, changes in concentrations of greenhouse gases and aerosols, and land surface changes. Nevertheless, the balance of evidence suggests that there is a discernible human influence on global climate.

5. CLIMATE IS EXPECTED TO CONTINUE TO CHANGE IN THE FUTURE

The IPCC has developed a range of scenarios, IS92a-f of future greenhouse gas and aerosol precursor emissions based on assumptions concerning population and economic growth, land-use, technological changes, energy availability and fuel mix during the period 1990 to 2100. Through understanding of the global carbon cycle and of atmospheric chemistry, these emissions can be used to project atmospheric concentrations of greenhouse gases and aerosols and the perturbation of natural radiative forcing. Climate models can then be used to develop projections of future climate.

- The increasing realism of simulations of current and past climate by coupled atmosphere-ocean climate models has increased our confidence in their use for projection of future climate change. Important uncertainties remain, but these have been taken into account in the full range of projections of global mean temperature and sea-level change.
- For the mid-range IPCC emission scenario, IS92a, assuming the "best estimate" value of climate sensitivity[4] and including the effects of future increases in aerosol, models project an increase in global mean surface air temperature relative to 1990 of about 2°C by 2100. This estimate is approximately one-third lower than the "best estimate" in 1990. This is due primarily to lower emission scenarios (particularly for CO_2 and the CFCs), the inclusion of the cooling effect of sulphate aerosols, and improvements in the treatment of the carbon cycle. Combining the lowest IPCC emission scenario (IS92c) with a "low" value of climate sensitivity and including the effects of future changes in aerosol concentrations leads to a projected increase of about 1°C by 2100. The corresponding projection for the highest IPCC scenario (IS92e) combined with a "high" value of climate sensitivity gives a warming of about 3.5°C. In all cases the average rate of warming would probably be greater than any seen in the last 10,000 years, but the actual annual to decadal changes would include considerable natural variability. Regional temperature changes could differ substantially from the global mean value. Because of the thermal inertia of the oceans, only 50–90% of the eventual equilibrium temperature change would have been realized by 2100 and temperature would continue to increase beyond 2100, even if concentrations of greenhouse gases were stabilized by that time.
- Average sea level is expected to rise as a result of thermal expansion of the oceans and melting of glaciers and ice-sheets. For the IS92a scenario, assuming the "best estimate" values of climate sensitivity and of ice-melt sensitivity to warming, and including the effects of future changes in aerosol, models project an increase in sea level of about 50 cm from the present to 2100. This estimate is approximately 25% lower than the "best estimate" in 1990 due to the lower temperature projection, but also reflecting improvements in the climate and ice-melt models. Combining the lowest emission scenario (IS92c) with the "low" climate and ice-melt sensitivities and including aerosol effects gives a projected sea-level rise of about 15 cm from the present to 2100. The corresponding projection for the highest emission scenario (IS92e) combined with "high" climate and ice-melt sensitivities gives a sea-level rise of about 95 cm from the present to 2100. Sea level would continue to rise at a similar rate in future centuries beyond 2100, even if concentrations of greenhouse gases were stabilized by that time, and would continue to do so even beyond the time of stabilization of global mean temperature. Regional sea-level changes may differ from the global mean value owing to land movement and ocean current changes.

- Confidence is higher in the hemispheric-to-continental scale projections of coupled atmosphere-ocean climate models than in the regional projections, where confidence remains low. There is more confidence in temperature projections than hydrological changes.
- All model simulations, whether they were forced with increased concentrations of greenhouse gases and aerosols or with increased concentrations of greenhouse gases alone, show the following features: greater surface warming of the land than of the sea in winter; a maximum surface warming in high northern latitudes in winter, little surface warming over the Arctic in summer; an enhanced global mean hydrological cycle, and increased precipitation and soil moisture in high latitudes in winter. All these changes are associated with identifiable physical mechanisms.
- In addition, most simulations show a reduction in the strength of the north Atlantic thermohaline circulation and a widespread reduction in diurnal range of temperature. These features too can be explained in terms of identifiable physical mechanisms.
- The direct and indirect effects of anthropogenic aerosols have an important effect on the projections. Generally, the magnitudes of the temperature and precipitation changes are smaller when aerosol effects are represented, especially in northern mid-latitudes. Note that the cooling effect of aerosols is not a simple offset to the warming effect of greenhouse gases, but significantly affects some of the continental scale patterns of climate change, most noticeably in the summer hemisphere. For example, models that consider only the effects of greenhouse gases generally project an increase in precipitation and soil moisture in the Asian summer monsoon region, whereas models that include, in addition, some of the effects of aerosols suggest that monsoon precipitation may decrease. The spatial and temporal distribution of aerosols greatly influences regional projections, which are therefore more uncertain.
- A general warming is expected to lead to an increase in the occurrence of extremely hot days and a decrease in the occurrence of extremely cold days.
- Warmer temperatures will lead to a more vigorous hydrological cycle; this translates into prospects for more severe droughts and/or floods in some places and less severe droughts and/or floods in other places. Several models indicate an increase in precipitation intensity, suggesting a possibility for more extreme rainfall events. Knowledge is currently insufficient to say whether there will be any changes in the occurrence or geographical distribution of severe storms, e.g., tropical cyclones.
- Sustained rapid climate change could shift the competitive balance among species and even lead to forest dieback, altering the terrestrial uptake and release of carbon. The magnitude is uncertain, but could be between zero and 200 GtC over the next one to two centuries, depending on the rate of climate change.

6. THERE ARE STILL MANY UNCERTAINTIES

Many factors currently limit our ability to project and detect future climate change. In particular, to reduce uncertainties further work is needed on the following topics:

- Estimation of future emissions and biogeochemical cycling (including sources and sinks) of greenhouse gases, aerosols and aerosol precursors and projections of future concentrations and radiative properties. Representation of climate processes in models, especially feedbacks associated with clouds, oceans, sea ice and vegetation, in order to improve projections of rates and regional patterns of climate change.
- Systematic collection of long-term instrumental and proxy observations of climate system variables (e.g., solar output, atmospheric energy balance components, hydrological cycles, ocean characteristics and ecosystem changes) for the purposes of model testing, assessment of temporal and regional variability, and for detection and attribution studies.

Future unexpected, large and rapid climate system changes (as have occurred in the past) are, by their nature, difficult to predict. This implies that future climate changes may also involve "surprises". In particular, these arise from the non-linear nature of the climate system. When rapidly forced, non-linear systems are especially subject to unexpected behaviour. Progress can be made by investigating non-linear processes and sub-components of the climatic system. Examples of such non-linear behaviour include rapid circulation changes in the North Atlantic and feedbacks associated with terrestrial ecosystem changes.

NOTES

1. Climate change in IPCC Working Group I usage refers to any change in climate over time whether due to natural variability or as a result of human activity. This differs from the usage in the UN Framework Convention on Climate Change where "climate change" refers to a change of climate which is attributed directly or indirectly to human activity that alters the composition of the global atmosphere and which is in addition to natural climate variability observed over comparable time periods.
2. A simple measure of the importance of a potential climate change mechanism. Radiative forcing is the perturbation to the energy balance of the Earth-atmosphere system (in Watts per square metre [Wm-2]).
3. 1 GtC = 1 billion tonnes of carbon.
4. In IPCC reports, climate sensitivity usually refers to the long-term (equilibrium) change in global mean surface temperature following a doubling of atmospheric equivalent CO_2 concentration. More generally, it refers to the equilibrium change in surface air temperature following a unit change in radiative forcing (°C/Wm-2).

PART FOUR

Population Issues and the Environment

On the Internet . . .

Sites appropriate to Part Four

This is the endangered species home page of the United States Fish & Wildlife Service Web site.

http://www.fws.gov/r9endspp/endspp.html

The virtual library of ecology, biodiversity, and the environment is located at this Web site.

http://conbio.rice.edu/vl/

The Population Council, a nonprofit, nongovernmental organization seeks to improve the well-being and reproductive health of current and future generations. The council also seeks to achieve a humane, equitable balance between people and resources. This is the home page of the council.

http://www.popcouncil.org/default.html

This Web site contains information about human population growth and a link to the Natural Resources Defense Council home page.

http://www.nrdc.org/nrdc/bkgrd/pogrow.html

CHAPTER 10 Species Extinction and the Loss of Biodiversity

10.1 E. O. WILSON

The Current State of Biological Diversity

It may properly be said that the extinction of species of flora and fauna is a natural phenomenon that has been occurring since the beginning of life on earth. Why then is there a growing concern among biologists about endangered species to the point where many consider it a crisis worthy of immediate international attention? The answer is that, although there is insufficient information to deduce the precise rate of species extinction, there is broad agreement among experts that the rate has been accelerating rapidly as the direct result of human developmental activities.

The ability of species to evolve and adapt to changing conditions requires the continued existence of a diverse pool of genetic material, which dwindles as species disappear. The long term maintenance of ecosystem stability requires an appropriate balance between the evolution of new and the extinction of old species.

Harvard University entomology professor E. O. Wilson has developed a worldwide reputation as an expert on evolution and species diversity through his pioneering studies of ants and his founding of the controversial field of sociobiology. In recent years he has become one of the most eloquent spokespeople for the need to curb the environmentally destructive, unsustainable practices that threaten a massive reduction in the diversity of the world's biological endowment. The following selection is from "The Current State of Biological Diversity," in Wilson's edited book *Biodiversity* (National Academy Press, 1988).

Key Concept: the urgency of maintaining biodiversity

Biological diversity must be treated more seriously as a global resource, to be indexed, used, and above all, preserved. Three circumstances conspire to give this matter an unprecedented urgency. First, exploding human populations are degrading the environment at an accelerating rate, especially in tropical countries. Second, science is discovering new uses for biological diversity in ways that can relieve both human suffering and environmental destruction. Third, much of the diversity is being irreversibly lost through extinction caused by the destruction of natural habitats, again especially in the tropics. Overall, we are locked into a race. We must hurry to acquire the knowledge on which a wise policy of conservation and development can be based for centuries to come.

To summarize the problem, I review some current information on the magnitude of global diversity and the rate at which we are losing it. I concentrate on the tropical moist forests, because of all the major habitats, they are richest in species and because they are in greatest danger.

THE AMOUNT OF BIOLOGICAL DIVERSITY

Many recently published sources, especially the multiauthor volume *Synopsis and Classification of Living Organisms,* indicate that about 1.4 million living species of all kinds of organisms have been described. Approximately 750,000 are insects, 41,000 are vertebrates, and 250,000 are plants (that is, vascular plants and bryophytes). The remainder consists of a complex array of invertebrates, fungi, algae, and microorganisms. Most systematists agree that this picture is still very incomplete except in a few well-studied groups such as the vertebrates and flowering plants. If insects, the most species-rich of all major groups, are included, I believe that the absolute number is likely to exceed 5 million. Recent intensive collections made by Terry L. Erwin and his associates in the canopy of the Peruvian Amazon rain forest have moved the plausible upper limit much higher. Previously unknown insects proved to be so numerous in these samples that when estimates of local diversity were extrapolated to include all rain forests in the world, a figure of 30 million species was obtained. In an even earlier stage is research on the epiphytic plants, lichens, fungi, roundworms, mites, protozoans, bacteria, and other mostly small organisms that abound in the treetops. Other major habitats that remain poorly explored include the coral reefs, the floor of the deep sea, and the soil of tropical forests and savannas. Thus, remarkably, we do not know the true number of species on Earth, even to the nearest order of magnitude. My own guess, based on the described fauna and flora and many discussions with entomologists and other specialists, is that the absolute number falls somewhere between 5 and 30 million.

A brief word is needed on the meaning of species as a category of classification. In modern biology, species are regarded conceptually as a population or series of populations within which free gene flow occurs under natural conditions. This means that all the normal, physiologically competent individuals at a given time are capable of breeding with all the other individuals of the opposite sex belonging to the same species or at least that they are capable of being linked genetically to them through chains of other breeding individuals. By definition they do not breed freely with members of other species.

This biological concept of species is the best ever devised, but it remains less than ideal. It works very well for most animals and some kinds of plants, but for some plant and a few animal populations in which intermediate amounts of hybridization occur, or ordinary sexual reproduction has been replaced by self-fertilization or parthenogenesis, it must be replaced with arbitrary divisions.

New species are usually created in one or the other of two ways. A large minority of plant species came into existence in essentially one step, through the process of polyploidy. This is a simple multiplication in the number of gene-bearing chromosomes—sometimes within a preexisting species and sometimes in hybrids between two species. Polyploids are typically not able to form fertile hybrids with the parent species. A second major process is geographic speciation and takes much longer. It starts when a single population (or series of populations) is divided by some barrier extrinsic to the organisms, such as a river, a mountain range, or an arm of the sea. The isolated populations then diverge from each other in evolution because of the inevitable differences of the environments in which they find themselves. Since all populations evolve when given enough time, divergence between all extrinsically isolated populations must eventually occur. By this process alone the populations can acquire enough differences to reduce interbreeding between them should the extrinsic barrier between them be removed and the populations again come into contact. If sufficient differences have accumulated, the populations can coexist as newly formed species. If those differences have not yet occurred, the populations will resume the exchange of genes when the contact is renewed.

Species diversity has been maintained at an approximately even level or at most a slowly increasing rate, although punctuated by brief periods of accelerated extinction every few tens of millions of years. . . .

Each species is the repository of an immense amount of genetic information. The number of genes range from about 1,000 in bacteria and 10,000 in some fungi to 400,000 or more in many flowering plants and a few animals. A typical mammal such as the house mouse (*Mus musculus*) has about 100,000 genes. This full complement is found in each of its myriad cells, organized from four strings of DNA, each of which comprises about a billion nucleotide pairs. (Human beings have genetic information closer in quantity to the mouse than to the more abundantly endowed salamanders and flowering plants; the difference, of course, lies in what is encoded.) If stretched out fully, the DNA would be roughly 1-meter long. But this molecule is invisible to the naked eye because it is only 20 angstroms in diameter. If we magnified it until its width equalled that of wrapping string, the fully extended molecule would be 960 kilometers long. As we traveled along its length, we would encounter some 20 nucleotide pairs or "letters" of genetic code per inch, or about 50 per centimeter. The full information contained therein, if translated into ordinary-size letters of printed text, would just about fill all 15 editions of the *Encyclopaedia Britannica* published since 1768.

The number of species and the amount of genetic information in a representative organism constitute only part of the biological diversity on Earth. Each species is made up of many organisms. For example, the 10,000 or so ant species have been estimated to comprise 10^{15} living individuals at each moment of time. Except for cases of parthenogenesis and identical twinning, virtually no two members of the same species are genetically identical, due to the high levels of genetic polymorphism across many of the gene loci. At still another level, wide-ranging species consist of multiple breeding populations that display complex patterns of geographic variation in genetic polymorphism. Thus, even if an endangered species is saved from extinction, it will probably have lost much of its internal diversity. When the populations are allowed to expand again, they will be more nearly genetically uniform than the ancestral populations. The bison herds of today are biologically not quite the same—not so interesting—as the bison herds of the early nineteenth century. . . .

HOW MUCH DIVERSITY IS BEING LOST?

No precise estimate can be made of the numbers of species being extinguished in the rain forests or in other major habitats, for the simple reason that we do not know the numbers of species originally present. However, there can be no doubt that extinction is proceeding far faster than it did prior to 1800. The basis for this statement is not the direct observation of extinction. To witness the death of the last member of a parrot or orchid species is a near impossibility. With the exception of the showiest birds, mammals, or flowering plants, biologists are reluctant to say with finality when a species has finally come to an end. There is always the chance (and hope) that a few more individuals will turn up in some remote forest remnant or other. But the vast majority of species are not monitored at all. Like the dead of Gray's "Elegy Written in a Country Churchyard," they pass from the Earth without notice. . . .

Using the area-species relationship, Simberloff has projected ultimate losses due to the destruction of rain forests in the New World tropical mainland. If present levels of forest removal continue, the stage will be set within a century for the inevitable loss of 12% of the 704 bird species in the Amazon basin and 15% of the 92,000 plant species in South and Central America.

As severe as these regional losses may be, they are far from the worst, because the Amazon and Orinoco basins contain the largest continuous rain forest tracts in the world. Less extensive habitats are far more threatened. An extreme example is the western forest of Ecuador. This habitat was largely undisturbed until after 1960, when a newly constructed road network led to the swift incursion of settlers and clear-cutting of most of the area. Now only patches remain, such as the 0.8-square-kilometer tract at the Rio Palenque Biological Station. This tiny reserve contains 1,033 plant species, perhaps one-quarter of which are known only to occur in coastal Ecuador. Many are known at the present time only from a single living individual.

In general, the tropical world is clearly headed toward an extreme reduction and fragmentation of tropical forests, which will be accompanied by a massive extinction of species. At the present time, less than 5% of the forests are protected within parks and reserves, and even these are vulnerable to political and economic pressures. For example, 4% of the forests are protected in Africa, 2% in Latin America, and 6% in Asia. Thus in a simple system as envisioned by the basic models of island biogeography, the number of species of all kinds of organisms can be expected to be reduced by at least one-half—in other words, by hundreds of thousands or even (if the insects are as diverse as the canopy studies suggest) by millions of species. In fact, the island-biogeographic projections appear to be conservative for two reasons. First, tropical species are far more localized than those in the temperate zones. Consequently, a reduction of 90% of a tropical forest does not just reduce all the species living therein to 10% of their original population sizes, rendering them more vulnerable to future extinction. That happens in a few cases, but in many others, entire species are eliminated because they happened to be restricted to the portion of the forest that was cut over. Second, even when a portion of the species survives, it will probably have suffered significant reduction in genetic variation among its members due to the loss of genes that existed only in the outer portions.

The current reduction of diversity seems destined to approach that of the great natural catastrophes at the end of the Paleozoic and Mesozoic eras—in other words, the most extreme in the past 65 million years. In at least one important respect, the modern episode exceeds anything in the geological past. In the earlier mass extinctions, which some scientists believe were caused by large meteorite strikes, most of the plants survived even though animal diversity was severely reduced. Now, for the first time, plant diversity is declining sharply. . . .

WHAT CAN BE DONE?

The biological diversity most threatened is also the least explored, and there is no prospect at the moment that the scientific task will be completed before a large fraction of the species vanish. Probably no more than 1,500 professional systematists in the world are competent to deal with the millions of species found in the humid tropic forests. . . .

The problem of tropical conservation is thus exacerbated by the lack of knowledge and the paucity of ongoing research. In order to make precise assessments and recommendations, it is necessary to know which species are present (recall that the great majority have not even received a scientific name) as well as their geographical ranges, biological properties, and possible vulnerability to environmental change.

It would be a great advantage, in my opinion, to seek such knowledge for the entire biota of the world. Each species is unique and intrinsically valuable. We cannot expect to answer the important questions of ecology and other branches of evolutionary biology, much less preserve diversity with any efficiency, by studying only a subset of the extant species.

I will go further: the magnitude and control of biological diversity is not just a central problem of evolutionary biology; it is one of the key problems of science as a whole. At present, there is no way of knowing whether there are 5, 10, or 30 million species on Earth. There is no theory that can predict what this number might turn out to be. With reference to conservation and practical applications, it also matters why a certain subset of species exists in each region of the Earth, and what is happening to each one year by year. Unless an effort is made to understand all of diversity, we will fall far short of understanding life in these important respects, and due to the accelerating extinction of species, much of our opportunity will slip away forever.

Lest this exploration be viewed as an expensive Manhattan Project unattainable in today's political climate, let me cite estimates I recently made of the maximum investment required for a full taxonomic accounting of all species: 25,000 professional lifetimes (4,000 systematists are at work full or part time in North America today); their final catalog would fill 60 meters of library shelving for each million species. Computer-aided techniques could be expected to cut the effort and cost substantially. In fact, systematics has one of the lowest cost-to-benefit ratios of all scientific disciplines.

It is equally true that knowledge of biological diversity will mean little to the vast bulk of humanity unless the motivation exists to use it. Fortunately, both scientists and environmental policy makers have established a solid linkage between economic development and conservation. The problems of human beings in the tropics are primarily biological in origin: overpopulation, habitat destruction, soil deterioration, malnutrition, disease, and even, for hundreds of millions, the uncertainty of food and shelter from one day to the next. These problems can be solved in part by making biological diversity a source of economic wealth. Wild species are in fact both one of the Earth's most important resources and the least utilized. We have come to depend completely on less than 1% of living species for our existence, the remainder waiting untested and fallow. In the course of history, according to estimates made by Myers, people have utilized about 7,000 kinds of plants for food; predominant among these are wheat, rye, maize, and about a dozen other highly domesticated species. Yet there are at least 75,000 edible plants in existence, and many of these are superior to the crop plants in widest use. Others are potential sources of new pharmaceuticals, fibers, and petroleum substitutes. In addition, among the insects are large numbers of species that are potentially superior as crop pollinators, control agents for weeds, and parasites and predators of insect pests. Bacteria, yeasts, and other microorganisms are likely to continue yielding new medicines, food, and procedures of soil restoration. Biologists have begun to fill volumes with concrete proposals for the further exploration and better use of diversity, with increasing emphasis on the still unexplored portions of the tropical biota....

In response to the crisis of tropical deforestation and its special threat to biological diversity, proposals are regularly being advanced at the levels of policy and research. For example, Nicholas Guppy, noting the resemblance of the lumbering of rain forests to petroleum extraction as the mining of a nonrenewable resource for short-term profit, has recommended the creation of a cartel, the Organization of Timber-Exporting Countries (OTEC). By controlling production and prices of lumber, the organization could slow production while encouraging member states to "protect the forest environment in general and gene stocks and special habitats in particular, create plantations to supply industrial and fuel wood, benefit indigenous tribal forest peoples, settle encroachers, and much else." In another approach, Thomas Lovejoy has recommended that debtor nations with forest resources and other valuable habitats be given discounts or credits for undertaking conservation programs. Even a small amount of forgiveness would elevate the sustained value of the natural habitats while providing hard currency for alternatives to their exploitation.

Another opportunity for innovation lies in altering somewhat the mode of direct economic assistance to developing countries. A large part of the damage to tropical forests, especially in the New World, has resulted from the poor planning of road systems and dams. For example, the recent settlement of the state of Rondonia and construction of the Tucurui Dam, both in Brazil, are now widely perceived by ecologists and economists alike as ill-conceived. Much of the responsibility of minimizing environmental damage falls upon the international agencies that have the power to approve or disapprove particular projects.

The U.S. Congress addressed this problem with amendments to the Foreign Assistance Act in 1980, 1983, and 1986, which call for the development of a strategy for conserving biological diversity. They also mandate that programs funded through the U.S. Agency for International Development (USAID) include an assessment of environmental impact. In implementing this new policy, USAID has recognized that "the destruction of humid tropical forests is one of the most important environmental issues for the remainder of this century and, perhaps, well into the next," in part because they are "essential to the survival of vast numbers of species of plants and animals." In another sphere, The World Bank and other multinational lending agencies have come under increasing pressure to take a more active role in assessing the environmental impact of the large-scale projects they underwrite....

In the end, I suspect it will all come down to a decision of ethics—how we value the natural worlds in which we evolved and now, increasingly, how we regard our status as individuals. We are fundamentally mammals and free spirits who reached this high a level of rationality by the perpetual creation of new options. Natural philosophy and science have brought into clear relief what might be the essential paradox of human existence. The drive toward perpetual expansion—or personal freedom—is basic to the human spirit. But to sustain it we need the most delicate, knowing stewardship of the living world that can be devised. Expansion and stewardship may appear at first to be conflicting goals, but the opposite is true. The depth of the conservation ethic will be measured by the extent to which each of the two approaches to nature is used to reshape and reinforce the other. The paradox can be resolved by changing its premises into forms more suited to ultimate survival, including protection of the human spirit. I recently wrote in synecdochic form about one place in South America to give these feelings more exact expression:

> To the south stretches Surinam eternal, Surinam serene, a living treasure awaiting assay. I hope that it will be kept intact, that at least enough of its million-year history will be saved for the reading. By today's ethic its value may seem limited, well beneath the pressing concerns of daily life. But I suggest that as biological knowledge grows the ethic will shift fundamentally so that everywhere, for reasons that have to do with the very fiber of the brain, the fauna and flora of a country will be thought part of the national heritage as important as its art, its language, and that astonishing blend of achievement and farce that has always defined our species.[1]

REFERENCES

1. Wilson, E. O. *Biophilia* (Harvard University Press, Cambridge, Mass) 1984.

10.2 STEPHEN JAY GOULD

The Golden Rule: A Proper Scale for Our Environmental Crisis

Harvard University professor, Stephen Jay Gould, a world-renowned paleontologist and evolutionary theorist, is also one of America's foremost science essayists. Since 1974 Gould has written a monthly essay aimed at the educated nonscientist entitled "This View of Life" for *Natural History* magazine. He has also published similar essays in other periodicals, such as *The New York Review of Books, Discover,* and *Nature.*

Gould's most significant scientific contribution is the theory of punctuated equilibrium, which he and fellow paleontologist, Niles Eldredge, developed in the early 1970s. Gould and Eldredge rejected the excepted dogma that evolution is a continual, gradual process. They presented evidence to support the theory that evolution is characterized by long, stable periods with little change, punctuated by "moments" of relatively massive change during which many new species evolve.

There are several recurrent themes in Gould's essays. He repeatedly explains the boundaries that distinguish science from other forms of human activity, and stresses his opposition to the teaching of the religious doctrine of creationism as science. He stresses the highly unpredictable, nonlinear course taken by evolutionary development to make the point that Homo sapiens are neither the inevitable, ultimate result of natural selection nor the pinacle of a simple ascending evolutionary ladder. Gould asserts that modern humans are just one of millions of nature's species, and that each species is unique and valuable in its own way.

The following selection is from Gould's essay "The Golden Rule: A Proper Scale for Our Environmental Crisis," which appears in his book *Eight Little Piggies: Reflections in Natural History* (W. W. Norton, 1993). In this essay Gould stresses that scientific analyses require the use of appropriate scales of space and time. He states that ignoring this requirement discredits the importance of protecting endangered species and perpetrates a fatal logical fallacy.

Key Concept: species extinction and other environmental crises are valid concerns on the time scale of human existence

Patience enjoys a long pedigree of favor. Chaucer pronounced it "an heigh vertu, certeyn" ("The Franklin's Tale"), while the New Testament had already made a motto of the Old Testament's most famous embodiment: "Ye have heard of the patience of Job" James 5:11). Yet some cases seem so extended in diligence and time that another factor beyond sheer endurance must lie behind the wait. When Alberich, having lost the Ring of the Niebelungen fully three operas ago, shows up in Act 2 of *Götterdämmerung* to advise his son Hagen on strategies for recovery, we can hardly suppress a flicker of admiration for this otherwise unlovable character. (I happen to adore Wagner, but I do recognize that a wait through nearly all of the Ring cycle would be, to certain unenlightened folks, the very definition of eternity in Hades.)

Patience of this magnitude usually involves a deep understanding of a fundamental principle, central to my own profession of geology but all too rarely grasped in daily life—the effects of scale. Phenomena unfold on their own appropriate scales of space and time and may be invisible in our myopic world of dimensions assessed by comparison with human height and times metered by human lifespans. So much of accumulating importance at earthly scales—the results of geological erosion, evolutionary changes in lineages—is invisible by the measuring rod of a human life. So much that matters to particles in the microscopic world of molecules—the history of a dust grain subject to Brownian motion, the fate of shrunken people in *Fantastic Voyage* or *Inner Space*—either averages out to stability at our scale or simply stands below our limits of perception.

It takes a particular kind of genius or deep understanding to transcend this most pervasive of all conceptual biases and to capture a phenomenon by grasping a proper scale beyond the measuring rods of our own world. Alberich and Wotan know that pursuit of the Ring is dynastic or generational, not personal. William of Baskerville (in Umberto Eco's *Name of the Rose*) solves his medieval mystery because he alone understands that, in the perspective of centuries, the convulsive events of his own day (the dispute between papacies of Rome and Avignon) will be forgotten, while the only surviving copy of a book by Aristotle may influence millennia. Architects of medieval cathedrals had to frame satisfaction on scales beyond their own existence, for they could not live to witness the completion of their designs.

May I indulge in a personal anecdote on the subject of scale? I loved to memorize facts as a child, but rebelled at those I deemed unimportant (baseball stats were in, popes of Rome and kings of England out). In sixth grade, I had to memorize the sequence of land acquisitions that built America. I could see the rationale behind learning the Louisiana Purchase and the Mexican Cession, for they added big chunks to our totality. But I remember balking, and publicly challenging the long-suffering Ms. Stack, at the Gadsden Purchase of 1853. Why did I have to know about a sliver of southern Arizona and New Mexico?

Now I am finally hoist on my own petard (blown up by my own noxious charge according to the etymologies). After a lifetime of complete nonimpact by the Gadsden Purchase, I have become unwittingly embroiled in a controversy about a tiny bit of territory within this smallest of American growing points. A little bit of a little bit—so much for effects of scale and the penalties of blithe ignorance.

The case is a classic example of a genre (environmentalists vs. developers) made familiar in recent struggles to save endangered populations—the snail darter of a few years back, the northern spotted owl vs. timber interests. The University of Arizona, with the backing of an international consortium of astronomers, wishes to build a complex of telescopes atop Mount Graham in southeastern Arizona (part of the Gadsden Purchase). But the old-growth spruce-fir habitat on the mountaintop provides the central range for *Tamiasciurus hudsonicus grahamensis*, the Mount Graham Red Squirrel—a distinctive subspecies that lives nowhere else, and that forms the southernmost population of the entire species. The population has already been reduced to some one hundred survivors, and destruction of 125 acres of spruce-fir growth (to build the telescopes) within the 700 or so remaining acres of best habitat might well administer a coup de grâce to this fragile population.

I cannot state an expert opinion on details of this controversy (I have already confessed my ignorance about everything involving the Gadsden Purchase and its legacy). Many questions need to be answered. Is the population already too small to survive in any case? If not, could the population, with proper management, coexist with the telescopes in the remaining habitat?

I do not think that, practically or morally, we can defend a policy of saving every distinctive local population of organisms. I can cite a good rationale for the preservation of species, for each species is a unique and separate natural object that, once lost, can never be reconstituted. But subspecies are distinctive local populations of species with broader geographical ranges. Subspecies are dynamic, interbreedable, and constantly changing; what then are we saving by declaring them all inviolate? Thus, I confess that I do not agree with all arguments advanced by defenders of the Mount Graham Red Squirrel. One leaflet, for example, argues: "The population has been recently shown to have a fixed, homozygous allele which is unique in Western North America." Sorry folks. I will stoutly defend species, but we cannot ask for the preservation of every distinctive gene, unless we find a way to abolish death itself (for many organisms carry unique mutations).

No, I think that for local populations of species with broader ranges, the brief for preservation must be made on a case by case basis, not on a general principle of preservation (lest the environmental movement ultimately lose popular support for trying to freeze a dynamic evolutionary world *in statu quo*). On this proper basis of individual merit, I am entirely persuaded that the Mount Graham Red Squirrel should be protected, for two reasons.

First, the squirrel itself: The Mount Graham Red is an unusually interesting local population within an important species. It is isolated from all other populations and forms the southernmost extreme of the species's range. Such peripheral populations, living in marginal habitats, are of special interest to students of evolution.

Second, the habitat: Environmentalists continually face the political reality that support and funding can be won for soft, cuddly, and "attractive" animals, but not for slimy, grubby, and ugly creatures (of potentially greater evolutionary interest and practical significance) or for habitats. This situation had led to the practical concept of "umbrella" or "indicator" species—surrogates for a larger ecological entity worthy of preservation Thus, the giant panda (really quite a boring and ornery creature despite its good looks) raises money to save the remaining bamboo forests of China (and a plethora of other endangered creatures with no political clout); the northern spotted owl has just rescued some magnificent stands of old-growth giant cedars, Douglas fir, and redwoods (and I say Hosanna); and the Mount Graham Red Squirrel may save a rare and precious habitat of extraordinary evolutionary interest.

The Pinaleno Mountains, reaching 10,720 feet at Mount Graham, are an isolated fault block range separated from others by alluvial and desert valleys that dip to less than 3,000 feet in elevation. The high peaks of the Pinalenos contain an important and unusual fauna for two reasons. First, they harbor a junction of two biogeographic provinces: the Nearctic or northern by way of the Colorado Plateau and the Neotropical or southern via the Mexican Plateau. The Mount Graham Red Squirrel (a northern species) can live this far south because high elevations reproduce the climate and habitat found nearer sea level in the more congenial north. Second, and more important to evolutionists, the old growth spruce-fir habitats on the high peaks of the Pinalenos are isolated "sky islands"—10,000-year-old remnants of a habitat more widely spread over the region of the Gadsden Purchase during the height of the last ice age. In evolutionary terms, these isolated pieces of habitat are true islands—patches of more northern microclimate surrounded by southern desert. They are functionally equivalent to bits of land in the ocean. Consider the role that islands (like the Galápagos) have played both in developing the concepts of evolutionary theory and in acting as cradles of origin (through isolation) or vestiges of preservation for biological novelties.

Thus, whether or not the telescopes will drive the Mount Graham Red Squirrel to extinction (an unsettled question well outside my area of expertise), the sky islands of the Pinalenos are precious habitats that should not be compromised. Let the Mount Graham Red Squirrel, so worthy of preservation in its own right, also serve as an indicator species for the unique and fragile habitat that it occupies.

But why should I, a confirmed eastern urbanite who has already disclaimed all concern for the Gadsden Purchase, choose to involve myself in the case of the Mount Graham Red Squirrel? The answer, unsurprisingly, is that I have been enlisted—involuntarily and on the wrong side to boot. I am fighting mad, and fighting back.

The June 7, 1990, *Wall Street Journal* ran a pro-development, anti-squirrel opinion piece by Michael D. Copeland (identified as "executive director of the Political Economy Research Center in Bozeman, Montana") under the patently absurd title: "No Red Squirrels? Mother Nature May Be Better Off." (I can at least grasp, while still rejecting, the claim that nature would be no worse off if the squirrels died, but I am utterly befuddled at how anyone could devise an argument that the squirrels inflict a positive harm upon the mother of us all!) In any case, Mr. Copeland misunderstood my writings in formulating a supposedly scientific argument for his position.

Now scarcely a day goes by when I do not read a misrepresentation of my views (usually by creationists, racists, or football fans in order of frequency).

My response to nearly all misquotation is the effective retort of preference: utter silence. (Honorable intellectual disagreement should always be addressed; misquotation should be ignored, when possible and politically practical). I make an exception in this case because Copeland cited me in the service of a classic false argument—the standard, almost canonical misuse of my profession of paleontology in debates about extinction. We have been enlisted again and again, in opposition to our actual opinions and in support of attitudes that most of us regard as anathema, to uphold arguments by developers about the irrelevance (or even, in this case, the benevolence) of modern anthropogenic extinction. This standard error is a classic example of failure to understand the importance of scale—and thus I return to the premise and structure of my introductory paragraphs (did you really think that I waffled on so long about scale only so I could talk about the Gadsden Purchase?).

Paleontologists do discuss the inevitability of extinction for all species—in the long run, and on the broad scale of geological time. We are fond of saying that 99 percent or more of all species that ever lived are now extinct. (My colleague Dave Raup often opens talks on extinction with a zinging one-liner: "To a first approximation, all species are extinct.") We do therefore identify extinction as the normal fate of species. We also talk a lot—more of late since new data have made the field so exciting—about mass extinctions that punctuate the history of life from time to time. We do discuss the issue of eventual "recovery" from the effects of these extinctions, in the sense that life does rebuild or surpass its former diversity several million years after a great dying. Finally, we do allow that mass extinctions break up stable faunas and, in this sense, permit or even foster evolutionary innovations well down the road (including the dominance of mammals and the eventual origin of humans, following the death of dinosaurs).

From these statements about extinction in the fullness of geological time (on scales of millions of years), some apologists for development have argued that extinction at any scale (even of local populations within years or decades) poses no biological worry but, on the contrary, must be viewed as a comfortable part of an inevitable natural order. Or so Copeland states:

> Suppose we lost a species. How devastating would that be? "Mass extinctions have been recorded since the dawn of paleontology," writes Harvard Paleontologist Stephen Gould.... The most severe of these occurred approximately 250 million years ago... with an estimated 96 percent extinction of species, says Mr. Gould.... There is general agreement among scientists that today's species represent a small proportion of all those that have ever existed—probably less than 1 percent. This means that more than 99 percent of all species ever living have become extinct.

From these facts, largely irrelevant to red squirrels on Mount Graham, Copeland makes inferences about the benevolence of extinction in general (though the argument only applies to geological scales):

> Yet, in spite of these extinctions, both Mr. Gould and University of Chicago paleontologist Jack Sepkoski say that the actual number of living species has probably increased over time. [True, but not as a result of mass extinctions, despite

> Copeland's next sentence.] The "niches" created by extinctions provide an opportunity for a vigorous development of new species.... Thus, evolutionary history appears to have been characterized by millions of species extinctions and subsequent increases in species numbers. Indeed, by attempting to preserve species living on the brink of extinction, we may be wasting time, effort and money on animals that will disappear over time, regardless of our efforts.

But all will "disappear over time, regardless of our efforts"—millions of years from now for most species if we don't interfere. The mean lifespan of marine invertebrate species lies between 5 and 10 million years; terrestrial vertebrate species turn over more rapidly, but still average in the low millions. By contrast, *Homo sapiens* may be only 200,000 years old or so and may enjoy a considerable future if we don't self-destruct. Similarly, recovery from mass extinction takes its natural measure in millions of years—as much as 10 million or more for fully rekindled diversity after major catastrophic events.

These are the natural time scales of evolution and geology on our planet. But what can such vastness possibly mean for our legitimately parochial interest in ourselves, our ethnic groups, our nations, our cultural traditions, our blood lines? Of what conceivable significance to us is the prospect of recovery from mass extinction 10 million years down the road if our entire species, not to mention our personal lineage, has so little prospect of surviving that long?

Capacity for recovery at geological scales has no bearing whatever upon the meaning of extinction today. We are not protecting Mount Graham Red Squirrels because we fear for global stability in a distant future not likely to include us. We are trying to preserve populations and environments because the comfort and decency of our present lives, and those of fellow species that share our planet, depend upon such stability. Mass extinctions may not threaten distant futures, but they are decidedly unpleasant for species caught in the throes of their power. At the appropriate scale of our lives, we are just a species in the midst of such a moment. And to say that we should let the squirrels go (at our immediate scale) because all species eventually die (at geological scales) makes about as much sense as arguing that we shouldn't treat an easily curable childhood infection because all humans are ultimately and inevitably mortal. I love geological time—a wondrous and expansive notion that sets the foundation of my chosen profession—but such vastness is not the proper scale of my personal life.

The same issue of scale underlies the main contribution that my profession of paleontology might make to our larger search for an environmental ethic. This decade, a prelude to the millennium, is widely and correctly viewed as a turning point that will lead either to environmental perdition or stabilization. We have fouled local nests before and driven regional faunas to extinction, but we were never able to unleash planetary effects before this century's concern with nuclear fallout, ozone holes, and putative global warming. In this context, we are searching for proper themes and language to express our environmental worries.

I don't know that paleontology has a great deal to offer, but I would advance one geological insight to combat a well-meaning, but seriously flawed (and all too common), position and to focus attention on the right issue at the proper scale. Two linked arguments are often promoted as a basis for an environmental ethic:

1. We live on a fragile planet now subject to permanent derailment and disruption by human intervention;
2. Humans must learn to act as stewards for this threatened world.

Such views, however well intentioned, are rooted in the old sin of pride and exaggerated self-importance. We are one among millions of species, stewards of nothing. By what argument could we, arising just a geological microsecond ago, become responsible for the affairs of a world 4.5 billion years old, teeming with life that has been evolving and diversifying for at least three-quarters of this immense span? Nature does not exist for us, had no idea we were coming, and doesn't give a damn about us. Omar Khayyam was right in all but his crimped view of the earth as battered, when he made his brilliant comparison of our world to an eastern hotel:

> Think, in this battered caravanserai
> Whose portals are alternate night and day,
> How sultan after sultan with his pomp
> Abode his destined hour, and went his way.

This assertion of ultimate impotence could be countered if we, despite our late arrival, now held power over the planet's future. But we don't, despite popular misperception of our might. We are virtually powerless over the earth at our planet's own geological timescale. All the megatonnage in all our nuclear arsenals yields but one ten-thousandth the power of the 10 km asteroid that might have triggered the Cretaceous mass extinction. Yet the earth survived that larger shock and, in wiping out dinosaurs, paved a road for the evolution of large mammals, including humans. We fear global warming, yet even the most radical model yields an earth far cooler than many happy and prosperous times of a prehuman past. We can surely destroy ourselves, and take many other species with us, but we can barely dent bacterial diversity and will surely not remove many million species of insects and mites. On geological scales, our planet will take good care of itself and let time clear the impact of any human malfeasance.

People who do not appreciate the fundamental principle of appropriate scales often misread such an argument as a claim that we may therefore cease to worry about environmental deterioration,just as Copeland argued falsely that we need not fret about extinction. But I raise the same counterargument. We cannot threaten at geological scales, but such vastness has no impact upon us. We have a legitimately parochial interest in our own lives, the happiness and prosperity of our children, the suffering of our fellows. The planet will recover from nuclear holocaust, but we will be killed and maimed by billions, and our cultures will perish. The earth will prosper if polar icecaps melt under a global greenhouse, but most of our major cities, built at sea level as ports and harbors, will founder, and changing agricultural patterns will uproot our populations.

We must squarely face an unpleasant historical fact. The conservation movement was born, in large part, as an elitest attempt by wealthy social leaders to preserve wilderness as a domain for patrician leisure and contemplation (against the image, so to speak, of poor immigrants traipsing in hordes through the woods with their Sunday picnic baskets). We have never entirely shaken this legacy of environmentalism as something opposed to immediate human needs, particularly of the impoverished and unfortunate. But the Third World expands and contains most of the pristine habitat that we yearn to preserve. Environmental movements cannot prevail until they convince people that clean air and water, solar power, recycling, and reforestation are best solutions (as they are) for human needs at human scales—and not for impossibly distant planetary futures.

I have a decidedly unradical suggestion to make about an appropriate environmental ethic—one rooted, with this entire essay, in the issue of appropriate human scale vs. the majesty, but irrelevance, of geological time. I have never been much attracted to the Kantian categorical imperative in searching for an ethic—to moral laws that are absolute and unconditional, and do not involve any ulterior motive or end. The world is too complex and sloppy for such uncompromising attitudes (and God help us if we embrace the wrong principle and then fight wars, kill, and maim in our absolute certainty). I prefer the messier "hypothetical imperatives" that invoke desire, negotiation, and reciprocity. Of these "lesser," but altogether wiser and deeper principles, one has stood out for its independent derivation, with different words but to the same effect, in culture after culture. I imagine that our various societies grope towards this principle because structural stability (and basic decency necessary for any tolerable life) demand such a maxim. Christians call this principle the "golden rule"; Plato, Hillel, and Confucius knew the same maxim by other names. I cannot think of a better principle based on enlightened self-interest. If we all treated others as we wish to be treated ourselves, then decency and stability would have to prevail.

I suggest that we execute such a pact with our planet. She holds all the cards, and has immense power over us—so such a compact, which we desperately need but she does not at her own timescale, would be a blessing for us and an indulgence for her. We had better sign the papers while she is still willing to make a deal. If we treat her nicely, she will keep us going for a while. If we scratch her, she will bleed, kick us out, bandage up, and go about her business at her own scale. Poor Richard told us that "necessity never made a good bargain," but the earth is kinder than human agents in the "art of the deal." She will uphold her end; we must now go and do likewise.

CHAPTER 11 Population Control Controversies

11.1 BETSY HARTMANN

Reproductive Rights and Wrongs: The Global Politics of Population Control

Biologists Paul R. Ehrlich and Garret Hardin have made cases for aggressive population control as a primary, essential strategy to prevent what they see as an otherwise inevitable future of global food and resource shortages and environmental devastation. Other experts, such as Barry Commoner, Frances M. Lappé, Joseph Collins, and Cary Fowler, have focused on inappropriate technology and political systems that promote inequitable distribution of resources as the causes of food shortages and environmental problems in the present and immediate future. This latter perspective has been elaborated on in recent years by women's organizations and representatives of developing nations.

Betsy Hartmann, an author and researcher, is currently director of the Population and Development Program at Hampshire College in Amherst, Massachusetts. She has written that "the threat to livelihoods, democracy and the environment posed by the fertility of poor women hardly compares to that posed by the consumption patterns of the rich or the ravages of militaries." Hartmann has lived in villages in India and Bangladesh and has made significant contributions to the theories and strategies supported by the international women's health and reproductive rights movement.

In the following selection from her book *Reproductive Rights and Wrongs: The Global Politics of Population Control,* rev. ed. (South End Press, 1995), Hartmann explains the opposition of women's organizations to the policies of the "population establishment." The selection includes the statement of an alternative program supported by the international Committee on Women, Population and the Environment.

Key Concept: feminist rejection of the establishment's population control policies

In my wildest dreams I wish "population" could be dropped all together from the development lexicon and replaced by concern for real people, real environments, not the fixed images of dark babies as bombs, women as wombs, statistical manipulations as absolute truth. Human welfare, yes: education, employment, health care, social and economic justice, reproductive rights, and the long-awaited peace dividend. Environmental protection, yes: curbs on pollution and waste, demilitarization, support for alternative technologies, farming systems and values which strengthen a sense of democratic community, not crass materialism.

POPULATION CONTROL, NO!

This was the slogan of the Women's International Tribunal and Meeting on Reproductive Rights held in Amsterdam in 1984, where 400 women from around the world came together in a powerful condemnation of both population control and antiabortion forces. That meeting solidified my own activism in the international women's health movement. It was held at a time when the movement, particularly its reproductive rights wing, was coming into its own. On the local level feminist initiatives were defining birth control in very different ways than the population establishment. They were responding to women's needs not only through the provision of more comprehensive reproductive health services and sexuality education, but through grassroots organizing around women's economic and political rights.

On the national level, in countries such as Brazil, the Philippines, and India, coalitions of women's groups were exerting pressure on the state to reform health policy, while internationally, the movement was expanding. The centralization of power in Northern countries was eroding, as the movement became more democratic and geographically diverse.

At the same time feminists within the population establishment were engaged in their own reform efforts, mainly through articulating a quality of care agenda. As the antiabortion movement gained strength, finding a powerful political ally in U.S. President Ronald Reagan, other members of the population establishment recognized the value of a strategic alliance with the international women's health movement. In the mid-1980s, dialogues began between representatives of the two groups, often organized by the New York-based International Women's Health Coalition (IWHC) which acted as a political broker. IWHC maintained that it was possible to "balance the scales" between population control and women's health. The first stones had been laid in the foundation of the Cairo consensus.

Who has gained most from the deal is still an open question. Without the participation of the women's movement, the Cairo consensus would no doubt be less attentive to women's rights and more Malthusian than it already is. But have women's organizations given up too much in the process? In particular, should they accept working within a framework which still blames population growth disproportionately for economic and environmental problems, sets targets for population stabilization and scarcely addresses the much more salient issues of unequal terms of trade, debt and structural adjustment, income distribution, and arms control?

This question now divides the movement. A number of groups have agreed to work within a population policy framework, although they have tried in the process to expand it. According to the Women's Declaration on Population Policies prepared in advance of Cairo:

> Population policies... need to address a wide range of conditions that affect the reproductive health and rights of women and men. These include unequal distribution of material and social resources among individuals and groups, based on gender, age, race, religion, social class, rural-urban residence, nationality and other social criteria; changing patterns of sexual and family relationships; political and economic policies that restrict girls' and women's access to health services and methods of fertility regulation; and ideologies, laws and practices that deny women's basic rights.

Despite this opening, the Declaration essentially sets forth a reproductive health agenda, with calls for more "women decision-makers" and financial resources for meeting program requirements. There is no fundamental challenge to the population paradigm, and government and international agencies, despite mention of their past shortcomings, are perceived as basically benign.

There are two key reasons women's organizations have agreed to work within the population framework. Some reproductive rights activists genuinely believe in the urgency of slowing population growth. Marge Berer, for example, writes that the women's movement should "acknowledge that the world cannot sustain an unlimited number of people, just as women's bodies cannot sustain unlimited pregnancies." Likening the planet to a woman's body, however, is a comparison fraught with peril.

For others it is more of a strategic choice—they would argue that accepting the legitimacy of a population framework allows women greater access to decision-makers, or in some cases, the opportunity to be decision-makers themselves. It reflects the movement's growing sophistication and professionalization, with some kind of compromise as the inevitable price of success.

But does influencing and interacting with the establishment necessarily depend on articulating women's concerns within a population framework? No, write Judith Richter and Loes Keysers, who set forth an alternative agenda of engagement. They note that more liberal people in the establishment may welcome support from the women's movement in their struggle against coercive population control, on the one hand, and the Vatican and fundamentalists, on the other.

> Women's advocates are thus in a relative position of strength. Could the splits in the population field not be used differently? Why not—instead of embracing a population agenda—enroll the population soft liners and family planners, for example, into a people's alliance for reproductive rights and health? This would allow feminists to set the terms of reference. It would create space for a shift of paradigm, rather than a shift from hard to soft population control.

The Declaration of People's Perspectives on "Population," issued by women from 23 countries meeting in 1993 in Comilla, Bangladesh, strongly rejects the population framework:

> Women's basic needs of food, education, health, work, social and political participation, a life free of violence and oppression should be addressed on their own merit. Meeting women's needs should be de-linked from population policy including those expressed as apparent humanitarian concerns for women.

The Comilla Declaration also makes clear its support for women's access to safe contraception and legal abortion as part of general health care. "Our resistance to population control policies must never be confused with the opposition of the religious and political right to the same policies."

Nevertheless, women who insist on remaining outside of the population framework are often accused of playing into the hands of the Vatican. There are also debates on which is the greater enemy of reproductive rights: religious fundamentalism or population control. Clearly, this depends on one's specific situation—whether one is fighting restrictive abortion legislation in Latin America and Eastern Europe, for example, or sterilization abuse in India and China. In many places, including the United States, both are enemies, both must be confronted simultaneously.

So far the movement has managed to maintain an uneasy unity and some sense of common identity—its strength is its members' ability to discuss and debate openly, to accept difference and heterogeneity, to insist on democratic processes and leadership. But the road ahead will be difficult. Besides the debate over population, there are many other unresolved issues: If it is politically pragmatic for some members to work within mainstream institutions, how can they remain accountable to the outside movement? Power doesn't necessarily corrupt, but it often separates and isolates. Does participation in official processes, such as U.N. conferences, siphon too much energy away from grassroots organizing? How can the movement, especially at the international level, maintain any grassroots authenticity? In an era of scarce resources, will the agenda conform too much to the funders' priorities? Where will funding come from?

I myself would argue for a strategy of principled pragmatism. While it makes political sense to dialogue and interact with the establishment periodically, as well as to have sympathetic women in positions of power within it, I believe the international women's health movement should not accept—and does not need to accept—the population framework in its efforts to reform health and family planning policy. When the movement does accept the framework, it loses its critical edge, dulls its tools of analysis, and ends up endorsing narrow technocratic agendas, rather than a broader politics of social and economic transformation. It divorces itself further from the poor women it is supposed to represent and places too much faith in official rhetoric. After Cairo, when the consensus snake begins to shed its skin, its progressive trappings will probably be the first to go. Already, Bangladesh and Indonesia are being put forward as the models which other countries should follow.

Stripped of all the economic arguments, political justifications, and soft-sell marketing, population control at heart is a philosophy without a heart, in which human beings become objects to be manipulated. It is a philosophy of domination, for its architects must necessarily view people of different sex, race, and class as inferior, less human than themselves, or otherwise they could not justify the double standards they employ.

Population control profoundly distorts our world view, and negatively affects people in the most intimate areas of their lives. Instead of promoting ethics, empathy, and true reproductive choice, it encourages us to condone coercion. And even on the most practical level, it is no solution to the serious economic, political, and environmental problems we face at the end of the century.

In saying no to population control from a prochoice, feminist perspective, one often feels like a voice in the wilderness, especially in the United States where Malthusianism is a popular religion, a veritable article of faith. But inhabiting the political wilderness is preferable to accepting conventional wisdom that is unwise. And if one listens closely, one can hear many other voices raised in protest and one can join them until collectively, they are too loud and clear to ignore.

The Committee on Women, Population and the Environment is an alliance of women activists, community organizers, health practitioners, and scholars of diverse races, cultures, and countries of origin working for women's empowerment and reproductive freedom, and against poverty, inequality, racism and environmental degradation. Issued in 1992, their statement, "Women, Population and the Environment: Call for a New Approach" continues to gather individual and organizational endorsements from around the world.

CALL FOR A NEW APPROACH

We are troubled by recent statements and analyses that single out population size and growth as a primary cause of global environmental degradation.

We believe the major causes of global environmental degradation are:

- Economic systems that exploit and misuse nature and people in the drive for short-term and short-sighted gains and profits.

- The rapid urbanization and poverty resulting from migration from rural areas and from inadequate planning and resource allocation in towns and cities.
- The displacement of small farmers and indigenous peoples by agribusiness, timber, mining, and energy corporations, often with encouragement and assistance from international financial institutions, and with the complicity of national governments.
- The disproportionate consumption patterns of the affluent the world over. Currently, the industrialized nations, with 22 percent of the world's population, consume 70 percent of the world's resources. Within the United States, deepening economic inequalities mean that the poor are consuming less, and the rich more.
- Technologies designed to exploit but not to restore natural resources.
- Warmaking and arms production which divest resources from human needs, poison the natural environment and perpetuate the militarization of culture, encouraging violence against women.

Environmental degradation derives thus from complex, interrelated causes. Demographic variables can have an impact on the environment, but reducing population growth will not solve the above problems. In many countries, population growth rates have declined yet environmental conditions continue to deteriorate.

Moreover, blaming global environmental degradation on population growth helps to lay the groundwork for the re-emergence and intensification of top-down, demographically driven population policies and programs which are deeply disrespectful of women, particularly women of color and their children.

In Southern countries, as well as in the United States and other Northern countries, family planning programs have often been the main vehicles for dissemination of modern contraceptive technologies. However, because so many of their activities have been oriented toward population control rather than women's reproductive health needs, they have too often involved sterilization abuse; denied women full information on contraceptive risks and side effects; neglected proper medical screening, follow-up care, and informed consent; and ignored the need for safe abortion and barrier and male methods of contraception. Population programs have frequently fostered a climate where coercion is permissible and racism acceptable.

Demographic data from around the globe affirm that improvements in women's social, economic, and health status and in general living standards, are often keys to declines in population growth rates. We call on the world to recognize women's basic right to control their own bodies and to have access to the power, resources, and reproductive health services to ensure that they can do so.

National governments, international agencies, and other social institutions must take seriously their obligation to provide the essential prerequisites for women's development and freedom. These include:

1. Resources such as fair and equitable wages, land rights, appropriate technology, education and access to credit.

2. An end to structural adjustment programs, imposed by the IMF [International Monetary Fund], the World Bank, and repressive governments, which sacrifice human dignity and basic needs for food, health, and education to debt repayment and 'free market', male-dominated models of unsustainable development.
3. Full participation in the decisions which affect our own lives, our families, our communities, and our environment, and incorporation of women's knowledge systems and expertise to enrich these decisions.
4. Affordable, culturally appropriate, and comprehensive health care and health education for women of all ages and their families.
5. Access to safe, voluntary contraception and abortion as part of broader reproductive health services which also provide pre- and post-natal care, infertility services, and prevention and treatment of sexually transmitted diseases including HIV and AIDS.
6. Family support services that include child-care, parental leave and elder care.
7. Reproductive health services and social programs that sensitize men to their parental responsibilities and to the need to stop gender inequalities and violence against women and children.
8. Speedy ratification and enforcement of the UN Convention on the Elimination of All Forms of Discrimination Against Women as well as other UN conventions on human rights.

People who want to see improvements in the relationship between the human population and natural environment should work for the full range of women's rights; global demilitarization; redistribution of resources and wealth between and within nations; reduction of consumption rates of polluting products and processes and of non-renewable resources; reduction of chemical dependency in agriculture; and environmentally responsible technology. They should support local, national, and international initiatives for democracy, social justice, and human rights.

The Ultimate Resource

The late Julian L. Simon was a professor of economics and business administration at the University of Maryland at College Park and the author of *Population Matters: People, Resources, Environment, and Immigration* (Transaction Publishers, 1990). He was also perhaps the best known of the small but vocal group of scholars that are referred to as ecological and environmental optimists, or (more pejoratively) by their opponents as "cornucopians." Simon rejected population control policies based not on political concerns but rather on the contention that environmental conditions and standards of living are likely to improve as a function of an increasing global population. He asserted that his optimistic view about the increased future availability of natural resources was held by a majority of resource economists.

The following selection is from Simon's book *The Ultimate Resource* (Princeton University Press, 1981). The book's title refers to the human imagination coupled with the human spirit. Simon cites data to support his claim that virtually all measures of human well-being have improved since the eighteenth century, when a decline in the death rate resulted in rapid growth of worldwide human population. He saw no reason why such trends should not continue into the indefinite future.

Simon's critics maintain that he supported his argument with a careful selection of data and inappropriate long-term extrapolations of short-term trends. He was also accused of displaying ignorance of the basic principles of ecology.

Key Concept: human beings as the ultimate resource

INTRODUCTION

What Are the *Real* Population and Resource Problems?

Is there a natural-resource problem now? Certainly there is—just as there has always been. The problem is that natural resources are scarce, in the sense that it costs us labor and capital to get them, though we would prefer to get them for free.

Are we now "entering an age of scarcity"? You can see anything you like in a crystal ball. But almost without exception, the best data—the long-run economic indicators—suggest precisely the opposite. The relevant measures of scarcity—the costs of natural resources in human labor, and their prices relative to wages and to other goods—all suggest that natural resources have been becoming *less* scarce over the long run, right up to the present.

How about pollution? Is this not a problem? Of course pollution is a problem—people have always had to dispose of their waste products so as to enjoy a pleasant and healthy living space. But on the average we now live in a less dirty and more healthy environment than in earlier centuries.

About population now: Is there a population "problem"? Again, of course there is a population problem, just as there has always been. When a couple is about to have a baby, they must prepare a place for the child to sleep safely. Then, after the birth of the child, the parents must feed, clothe, guard, and teach it. All of this requires effort and resources, and not from the parents alone. When a baby is born or a migrant arrives, a community must increase its municipal services—schooling, fire and police protection, and garbage collection. None of these are free.

Beyond any doubt, an additional child is a burden on people other than its parents—and in some ways even on them—for the first fifteen or twenty-five years of its life. Brothers and sisters must do with less of everything except companionship. Taxpayers must cough up additional funds for schooling and other public services. Neighbors have more noise. During these early years the child produces nothing, and the income of the family and the community is spread around more thinly than if the baby were not born. And when the child grows up and first goes to work, jobs are squeezed a bit, and the output and pay per working person go down. All this clearly is an economic loss for other people.

Almost equally beyond any doubt, however, an additional person is also a boon. The child or immigrant will pay taxes later on, contribute energy and resources to the community, produce goods and services for the consumption of others, and make efforts to beautify and purify the environment. Perhaps most significant of all for the more-developed countries is the contribution that the average person makes to increasing the efficiency of production through new ideas and improved methods.

The real population problem, then, is *not* that there are too many people or that too many babies are being born. It is that others must support each additional person before that person contributes in turn to the well-being of others.

Which is more weighty, the burden or the boon? That depends on the economic conditions. . . . But also, to a startling degree, the decision about whether the overall effect of a child or migrant is positive or negative depends on the values of whoever is making the judgment—your preference to spend a dollar now rather than to wait for a dollar-plus-something in twenty or thirty years, your preferences for having more or fewer wild animals alive as opposed to more or fewer human beings alive, and so on. Population growth is a problem but not *just* a problem; it is a boon, but not just a boon. So your values are all-important in judging the net effect of population growth, and whether there is "overpopulation" or "underpopulation."

An additional child is, from the economic point of view, like a laying chicken, a cacao tree, a new factory, or a new house. A baby is a durable good in which someone must invest heavily long before the grown adult begins to provide returns on the investment. But whereas "Travel now, pay later" is inherently attractive because the pleasure is immediate and the piper will wait, "Pay now, benefit from the child later" is inherently problematic because the sacrifice comes first.

You might respond that additional children will *never* yield net benefits, because they will use up irreplaceable resources.... [A]dditional persons do, in fact, produce more than they consume, and... natural resources are not an exception. But let us agree that there is still a population problem, just as there is a problem with all good investments: Long before there are benefits, we must tie up capital that could otherwise be used for immediate consumption.

Please notice that I have restricted the discussion to the *economic* aspect of investing in children—that is, to a child's effect on the material standard of living. If we also consider the non-economic aspects of children—what they mean to parents and to others who enjoy a flourishing of humanity—then the case for adding children to our world becomes even stronger. And if we also keep in mind that most of the costs of children are borne by their parents rather than by the community, whereas the community gets the lion's share of the benefits later on, especially in developed countries, the essential differences between children and other investments tend to strengthen rather than weaken the case for having more children....

CONCLUSION

The Ultimate Resource

... In the short run, all resources are limited.... In the short run, a greater use of any resource means pressure on supplies and a higher price in the market, or even rationing. Also in the short run there will always be shortage crises because of weather, war, politics, and population movements. The results that an individual notices are sudden jumps in taxes, inconveniences and disruption, and increases in pollution.

The longer run, however, is a different story. The standard of living has risen along with the size of the world's population since the beginning of recorded time. And with increases in income and population have come less severe shortages, lower costs, and an increased availability of resources, including a cleaner environment and greater access to natural recreation areas. And there is no convincing economic reason why these trends toward a better life, and toward lower prices for raw materials (including food and energy), should not continue indefinitely.

Contrary to common rhetoric, there are no meaningful limits to the continuation of this process.... There is no physical or economic reason why human resourcefulness and enterprise cannot forever continue to respond to impending shortages and existing problems with new expedients that, after an adjustment period, leave us better off than before the problem arose. Adding more people will cause us more such problems, but at the same time there will be more people to solve these problems and leave us with the bonus of lower costs and less scarcity in the long run. The bonus applies to such desirable resources as better health, more wilderness, cheaper energy, and a cleaner environment.

This process runs directly against Malthusian reasoning and against the apparent common sense of the matter, which can be summed up as follows: The supply of any resource is fixed, and greater use means less to go around. The resolution of this paradox is not simple. Fuller understanding begins with the idea that the relevant measure of scarcity is the cost or price of a resource, not any physical measure of its calculated reserves. And the appropriate way for us to think about extracting resources is not in physical units, pounds of copper or acres of farmland, but rather in the services we get from these resources—the electrical transmission capacity of copper, or the food values and gastronomic enjoyment the farmland provides. Following on this is the fact that economic history has not gone as Malthusian reasoning suggests. The prices of all goods, and of the services they provide, have fallen in the long run, by all reasonable measures. And this irrefutable fact must be taken into account as a fundamental datum that can reasonably be projected into the future, rather than as a fortuitous chain of circumstances that cannot continue.

Resources in their raw form are useful and valuable only when found, understood, gathered together, and harnessed for human needs. The basic ingredient in the process, along with the raw elements, is human knowledge. And we develop knowledge about how to use raw elements for our benefit only in response to our needs. This includes knowledge for finding new sources of raw materials such as copper, for growing new resources such as timber, for creating new quantities of capital such as farmland, and for finding new and better ways to satisfy old needs, such as successively using iron or aluminum or plastic in place of clay or copper. Such knowledge has a special property: It yields benefits to people other than the ones who develop it, apply it, and try to capture its benefits for themselves. Taken in the large, an increased need for resources usually leaves us with a permanently greater capacity to get them, because we gain knowledge in the process. And there is no meaningful physical limit—even the commonly mentioned weight of the earth—to our capacity to keep growing forever.

Perhaps the most general matter at issue here is what Gerald Holton calls a "thema." The thema underlying the thinking of most writers who have a point of view different from mine is the concept of fixity or finiteness of resources in the relevant system of discourse. This is found in Malthus, of course. But the idea probably has always been a staple of human thinking because so much of our situation must sensibly be regarded as fixed in the short run—the bottles of beer in the refrigerator, our paycheck, the amount of energy parents have to play basketball with their kids. But the thema underlying my thinking about resources (and the thinking of a minority of others) is that the relevant system of discourse has a long enough horizon that it makes sense to treat the system as not fixed, rather than finite in any operational sense. We see the resource system as being as unlimited as the number of thoughts a person might have, or the number of variations that might ultimately be produced by biological evolution. That is, a key difference between the thinking of those who worry about impending doom, and those who see the prospects of a better life for more people in the future, apparently is whether one thinks in closed-system or open-system terms. For example, those who worry that the second law of thermodynamics dooms us to eventual decline necessarily see our world as a closed system with respect to energy and entropy; those who view the relevant universe as unbounded view the second law of thermodynamics as irrelevant to this discussion. I am among those who view the relevant part of the physical and social universe as open for most purposes. Which thema is better for thinking about resources and population is not subject to scientific test. Yet

it profoundly affects our thinking. I believe that here lies the root of the key difference in thinking about population and resources.

Why do so many people think in closed-system terms? There are a variety of reasons. (1) Malthusian fixed-resources reasoning is simple and fits the isolated facts of our everyday lives, whereas the expansion of resources is complex and indirect and includes all creative human activity—it cannot be likened to our own larders or wallets. (2) There are always immediate negative effects from an increased pressure on resources, whereas the benefits only come later. It is natural to pay more attention to the present and the near future compared with the more distant future. (3) There are often special-interest groups that alert us to impending shortages of particular resources such as timber or clean air. But no one has the same stake in trying to convince us that the long-run

prospects for a resource are better than we think. (4) It is easier to get people's attention (and television time and printer's ink) with frightening forecasts than with soothing forecasts. (5) Organizations that form in response to temporary or nonexistent dangers, and develop the capacity to raise funds from public-spirited citizens and governments that are aroused to fight the danger, do not always disband when the danger evaporates or the problem is solved. (6) Ambition and the urge for profit are powerful elements in our successful struggle to satisfy our needs. These motives, and the markets in which they work, often are not pretty, and many people would prefer not to depend on a social system that employs these forces to make us better off. (7) Associating oneself with environmental causes is one of the quickest and easiest ways to get a wide reputation for high-minded concern; it requires no deep thinking and steps on almost no one's toes.

The apparently obvious way to deal with resource problems—have the government control the amounts and prices of what consumers consume and suppliers supply—is inevitably counter-productive in the long run because the controls and the price fixing prevent us from making the cost-efficient adjustments that we would make in response to the increased short-run costs, adjustments that eventually would more than alleviate the problem. Sometimes governments must play a crucial role to avoid short-run disruptions and disaster, and to ensure that no group consumes public goods without paying the real social cost. But the appropriate times for governments to play such roles are far fewer than the times they are called upon to do so by those inclined to turn to authority to tell others what to do, rather than allow each of us to respond with self-interest and imagination.

I do not say that all is well. Children are hungry and sick; people live out lives of physical and intellectual poverty, and lack of opportunity; war or some new pollution may do us all in. What I *am* saying is that for most of the relevant economic matters I have checked, the *trends* are positive rather than negative. And I doubt that it does the troubled people of the world any good to say that things are getting worse though they are really getting better. And false prophecies of doom can damage us in many ways.

Is a rosy future guaranteed? Of course not. There always will be temporary shortages and resource problems where there are strife, political blundering, and natural calamities—that is, where there are people. But the natural world allows, and the developed world promotes through the marketplace, responses to human needs and shortages in such manner that one backward step leads to 1.0001 steps forward, or thereabouts. That's enough to keep us headed in a life-sustaining direction. The main fuel to speed our progress is our stock of knowledge, and the brake is our lack of imagination. The ultimate resource is people—skilled, spirited, and hopeful people who will exert their wills and imaginations for their own benefit, and so, inevitably, for the benefit of us all.

11.3 JOEL E. COHEN

Human Carrying Capacity: An Overview

Joel E. Cohen has been professor of populations and head of the Laboratory of Populations at The Rockefeller University, New York, since 1975. In 1995, he was jointly appointed professor of populations in the Department of International and Public Affairs at Columbia University, New York, where he also serves on the faculty of the Center for Environmental Research and Conservation.

His latest book *How Many People Can the Earth Support?* (W. W. Norton, 1995), from which the following selection is taken, has been called an "admirable tour de force on human population" by the distinguished paleontologist and curator of the American Museum of Natural History, Niles Eldredge. On the controversial question referred to in the book's title, Cohen takes the middle ground between the position of deep ecologists, such as David Foreman, who assert that the earth's population has already far exceeded its sustainable limit and optimistic economists who echo Julian Simon's view that the larger the population, the better. Cohen's readers will not find a definitive response to the dispute about the earth's carrying capacity, but they should become convinced that it is a highly complex and subjective issue. Cohen maintains that any analysis that results in definitive answers to questions about the relationships among population, resources, and the environment should be assessed with a highly critical eye. One must be careful when contemplating value-laden assumptions. Nevertheless, Cohen warns against passivity in the face of irrefutable evidence that uncontrolled population growth and environmentally inappropriate technological developments pose serious threats to the future of human civilization.

Key Concept: the earth's carrying capacity as a complex but vital issue

The question of how many people the world can support is unanswerable in a finite sense. What do we want?

Are there global limits, absolute limits beyond which we cannot go without catastrophe or overwhelming costs? There are, most certainly.

—George Woodwell 1985

Joel E. Cohen

CASE STUDY: EASTER ISLAND

The constraints on the Earth's human carrying capacity are just as real as the wide range of choices within those boundaries. The history of Easter Island provides a case study of human choices and natural constraints in a small world. While exotic in location and culture, Easter Island is of general interest as one example of the many civilizations that undercut their own ecological foundations.

The island is one of the most isolated bits of land on the Earth. The inhabited land nearest to Easter Island is Pitcairn Island, 2,250 kilometers northwest; the nearest continental place, Concepción, Chile, is 3,747 kilometers southeast. The island is roughly triangular in plan, with sides of 16, 18 and 22 kilometers and an area of 166.2 square kilometers (a bit larger than Staten Island, a borough of the city of New York). About 2.5 million years old, the volcanic island rose from the sea floor 2,000 meters below sea level. A plateau occupies the middle of the island and a peak rises nearly 1,000 meters above sea level.

Radiocarbon dating suggests that people, almost certainly Polynesians, occupied the island by A.D. 690 at the latest; scattered earlier radiocarbon dates from the fourth and fifth centuries are uncertain. The first arrivals found an island covered by a rainforest of huge palms. The islanders were probably isolated from outside human contact until the island was spotted by Dutch sailors in 1722.

During this millennium or millennium and a half of isolation, a fantastic civilization arose. Its most striking material remains are 800 to 1,000 giant statues, or *moai*, two to ten meters high, carved in volcanic tuff and scattered over the island. Many are probably still buried by rubble and soil. The largest currently known is 20 meters (65 feet) long and weighs about 270 tonnes. It was left unfinished.

According to pollen cores recently taken from volcanic craters on the island, a tree used for rope was originally dominant on the island. At different times, depending on the site, between the eighth and the tenth centuries, forest pollen began to decline. Forest pollen reached its lowest level around A.D. 1400, suggesting that the last forests were destroyed by then. The deforestation coincided with soil erosion, visible in soil profiles. The Polynesian rat introduced for food by the original settlers consumed the seeds of forest trees, preventing regeneration. Freshwater supplies on the island diminished. In the 1400s or 1500s, large, stemmed obsidian flakes used as daggers and spearheads appeared for the first time; previously obsidian had been used only for tools.

While early visitors in 1722 and 1770 do not mention fallen *moai*, Captain Cook in 1774 reported that many statues had fallen next to their platforms and that the statues were not being maintained. Something drastic, probably some variant of intergroup warfare, probably happened between 1722, when the Dutch thought the statue cult was still alive, and 1774, when Cook thought it finished. A visitor in 1786 observed that the island no longer had a chief.

The population history of the island is full of uncertainties. The prehistory is based on the dating of sites of agricultural and human occupation. The best current estimate is that the population began with a boatload of settlers in the first half millennium after Christ, perhaps around A.D. 400. The population remained low until about A.D. 1100. Growth then accelerated and the population doubled every century until around 1400. Slower growth continued until at most 6,000 to 8,000 people occupied the island around 1600. The maximum population may have reached 10,000 people in A.D. 1680. A decline then set in. Jean François de Galaup, Comte de La Pérouse, who visited the island in 1786, estimated a population of 2,000, and this estimate is now accepted as roughly correct. Smallpox swept the island in the l860s, introduced by returning survivors among the islanders who had been enslaved and taken to Peru. The population numbered 111 by 1877. In 1888, the island was attached to Chile. Since then, Chileans have added to the population. The present population of 2,100 includes 800 children.

The plausibility of these numbers can be checked from the annual rates of population growth or decline that they imply. An increase from 50 people in A.D. 400 to 10,000 people in A.D. 1680 requires an annual increase of 0.41 percent. If the number of original settlers were 100 instead of 50, the implied population growth rate would be 0.36 percent; if 25 instead of 50, 0.47 percent. The long-term growth rate of around 0.4 percent per year is within the historical experience of developing countries before the post–World War II public health evolution. If the population declined from 10,000 in 1680 to 111 in 1877 (including removals by Peruvian slave traders), the annual rate of decline was 2.3 percent. If the population maximum in 1680 was only 6,000 instead of 10,000, then the annual rate of decline was 2.0 percent. A population decline from 10,000 in 1680 to the 2,000 reported by La Pérouse in 1786 requires an annual decline of 1.5 percent. In round numbers, Easter Island's human population seems to have increased by about 0.4 percent per year for about 13 centuries, then to have declined by about 2 percent per year for about two centuries before resuming a rise in the twentieth century.

Paul Bahn, a British archeologist and writer, and John Flenley, an ecologist and geographer in New Zealand, synthesized the archeological and historical data in an interpretive model.

> Forest clearance for the growing of crops would have led to population increase, but also to soil erosion and decline of soil fertility. Progressively more land would have had to be cleared. Trees and shrubs would also be cut down for canoe building, firewood, house construction, and for the timbers and ropes needed in the movement and erection of statues. Palm fruits would be eaten, thus reducing regeneration of the palm. Rats, introduced for food, could have fed on the palm fruits, multiplied rapidly and completely prevented palm regeneration. The over-exploitation of prolific sea bird resources would have eliminated these from all but the offshore islets. Rats could have helped in this process by eating eggs. The abundant food provided by fishing, sea birds and rats would have encouraged rapid initial human population growth. Unrestrained human population increase

would later put pressure on availability of land, leading to disputes and eventually warfare. Non-availability of timber and rope would make it pointless to carve further statues. A disillusionment with the efficacy of the statue religion in providing the wants of the people could lead to the abandonment of this cult. Inadequate canoes would restrict fishing to inshore waters, leading to further decline in protein supplies. The result could have been general famine, warfare and the collapse of the whole economy, leading to a marked population decline.

Of course, most of this is hypothesis. Nevertheless, there is evidence, as we have seen, that many features of this model did in fact occur. There certainly was deforestation, famine, warfare, collapse of civilization and population decline.

Supposing you accept this summary of the island's history, would you accept the following conclusion of Bahn and Flenley? "We consider that Easter Island was a microcosm which provides a model for the whole planet." Easter Island shares important features with the whole planet and differs in others. Draw your own conclusion.

LIVING IN THE LAND OF LOST ILLUSIONS: HUMAN CARRYING CAPACITY AS AN INDICATOR

A number or range of numbers, presented as a constraint independent of human choices, is an inadequate answer to the question "How many people can the Earth support?" While trying to answer this question, I learned to question the question.

If an absolute numerical upper limit to human numbers on the Earth exists, it lies beyond the bounds that human beings would willingly tolerate. Human physical requirements for bare minimal subsistence are very modest, closer to the level of Auschwitz than to the modest comforts of the Arctic Inuit or the Kalahari bushmen. For most people of the world, expectations of well-being have risen so far beyond subsistence that human choices will prevent human numbers from coming anywhere near absolute upper limits. If human choices somehow failed to prevent population size from approaching absolute upper limits, then gradually worsening conditions for human and other life on the Earth would first prompt and eventually enforce human choices to stop such an approach. As different people have different expectations of well-being, some people would be moved to change their behavior sooner than others. Social scientists focus on the choices and minimize the constraints; natural scientists do the reverse. In reality, neither choices nor constraints can be neglected.

An ideal tool for estimating how many people the Earth can support would be a model, simple enough to be intelligible, complicated enough to be potentially realistic and empirically tested enough to be credible. The model would require users to specify choices concerning technology, domestic and international political institutions, domestic and international economic arrangements (including recycling), domestic and international demographic arrangements, physical, chemical and biological environments, fashions, tastes, moral values, a desired typical level of material well-being and a distribution of well-being among individuals and areas. Users would specify how much they wanted each characteristic to vary as time passes and what risk they would tolerate that each characteristic might go out of the desired range of variability. Users would state how long they wanted their choices to remain in effect. They would specify the state of the world they wished to leave at the end of the specified period. The model would first check all these choices for internal consistency, detect any contradictions and ask users to resolve them or to specify a balance among contradictory choices. The model would then attempt to reconcile the choices with the constraints imposed by food, water, energy, land, soil, space, diseases, waste disposal, nonfuel minerals, forests, biological diversity, biologically accessible nitrogen, phosphorus, climatic change and other natural constraints. The model would generate a complete set of possibilities, including human population sizes, consistent with the choices and the constraints....

The speed registered on the speedometer of a car, the current total fertility rate of a population and the gross national or domestic product of an economy are, every one, indirect and incomplete summaries of more complicated realities: they are summary indicators, approximate but useful. Likewise, estimates of the human carrying capacity of the Earth are indicators. They indicate the population that can be supported under various assumptions about the present or future. Estimates of the Earth's human carrying capacity are conditional on current choices and on natural constraints, all of which may change as time passes. This view of estimates of human carrying capacity as conditional and changing differs sharply from a common view that there is one right number (perhaps imperfectly known) for all time.

Human carrying capacity is more difficult to estimate than some of the standard demographic indicators, like expectation of life or the total fertility rate, because human carrying capacity depends on populations and activities around the world. The expectation of life of a country can be determined entirely from the mortality experienced by the people within the country. But that same country's human carrying capacity depends not only on its soils and natural resources and population and culture and economy, but also on the prices of its products in world markets and on the resources and products other countries can and are willing to trade. When the world consisted of largely autonomous localities, it may have made sense to think of the Earth's human carrying capacity as the sum of local human carrying capacities; but no longer.

BEYOND EQUILIBRIUM

Think of a man engaged in four activities: lying on his back on the floor with his arms and legs relaxed; standing erect but at ease; walking at a comfortable pace; and running. When lying on his back, the man is at a passive equilibrium. If you push him gently on one side, he may rock a bit but will roll back to his original position. If you push him hard enough, he may switch from a passive supine equilibrium to a passive prone equilibrium. Whether he is supine or prone, he can remain in his present equilibrium without effort.

Standing is a much more complicated matter. Opposing muscles are constantly adjusting their tension to maintain upright posture, under the guidance of the body's sensory and nervous systems for maintaining balance. The man may not appear to be working, but his oxygen consumption increases and he will fall to the floor if he relaxes completely. If pushed hard, he may not be able to stay standing. His apparent equilibrium is dynamically maintained by constant control.

Walking is controlled falling. The man's center of gravity moves forward from his area of support and he puts one foot forward with just the right timing and placement to catch himself. He then pivots over the forward foot and continues to fall forward with just-in-time support from alternating legs. Anyone who has watched a child learn to walk, or an adult learn to walk with crutches, appreciates the complex sequential coordination, muscular strength and balance required to walk. The equilibrium of walking is not a stationary state at all, but a sustained motion.

Finally, running is more than an acceleration of walking because the runner may have both legs off the ground at once. The effort required, the speed of the motion and the vulnerability to collapse increase compared to walking. On a rocky mountain ridge or a crowded city street, running is impossible; simplification and control of the environment are required to sustain the equilibrium of steady running.

The purpose of this foray into kinesiology is to find some old and new analogies for the situation of humans on the Earth. If the population size of the human species was ever in a passive equilibrium regulated by the environment, it must surely have been before people gained control of fire. By using fire, early peoples massively reshaped their environment to their own advantage, with the effect of increasing their own population size. For example, they periodically burned grasslands to encourage plants desirable to themselves and to the game they hunted. After the mastery of fire, people moved from a supine equilibrium to a controlled balance analogous to standing. With the invention of shifting cultivation some ten or so millennia ago, and then settled cultivation, the human species initiated a form of controlled forward falling analogous to walking. Humans invented cities and farms and learned to coordinate them. Where agriculture failed, civilizations collapsed. In the last four centuries, surpluses of food released huge numbers of people from being tied to the land and enabled them to make machines and technologies that further loosened their ties to the land. Machines for handling energy, materials and information released people from old work, imposed new work and transformed much of the natural world. More than ever before, the land that still supported people became a partly human creation. For humans now, the notion of a static, passive equilibrium is inappropriate, useless. So is the notion of a static "human carrying capacity" imposed by the natural world on a passive human species. There is no choice but to try to control the direction, speed, risks, duration and purposes of our falling forward.

← cc is dynamic

*

PART FIVE

Human Health and the Environment

On the Internet . . .

Sites appropriate to Part Five

This is the home page for the Institute for Food and Development Policy, also known as Food First. This organization publishes newsletters, reports, and books on all aspects of agricultural policies. Food First seeks to establish food as a basic human right.

http://www.foodfirst.org

The Rachel Carson Homestead Association was created to preserve and restore Rachel Carson's childhood home; to design and implement education programs in keeping with her environmental ethic; and to serve as an international learning center about her life and legacy. This Web site contains information about Rachel Carson and about the Rachel Carson Homestead Association.

http://www.rachelcarson.org

This Web site is part of an educational service entitled "Environmental Concepts Made Easy" and focuses on environmental estrogens and other hormones. This Web site is maintained by scientists at Tulane University.

http://www.mcl.tulane.edu/cbr/ECME/EEHome/default.html

CHAPTER 12 Food

12.1 WENDELL BERRY

The Agricultural Crisis as a Crisis of Culture

Wendell Berry, author, poet, essayist, novelist, farmer, and professor, is one of America's profound ecological thinkers. From the vantage point of his Kentucky farm, Berry offers a biting critique of contemporary culture based on his belief that human abuse of nature is a result of the fact that people no longer live with the land, as they did in times past. Berry's philosophy is that we should "act locally, think globally," rather than the reverse prescription, which is a slogan supported by many environmental activists.

In the following selection from *The Unsettling of America: Culture and Agriculture* (Sierra Club Books, 1977), Berry analyzes the many economic and ecological problems that are associated with modern American agriculture and finds that their roots are in the substitution of mindless technology and "bigness" for the culturally complex, thoughtful communities that thrived in the era of the family farm. He argues that economic stability and environmentally sustainable practices require a farming culture that promotes interdependence and responsible cooperation rather than an agricultural system based on a simplistic, narrow, profit-based definition of efficiency.

Key Concept: the cultural dimensions of sustainable agriculture

In my boyhood, Henry County, Kentucky, was not just a rural county, as it still is—it was a *farming* county. The farms were generally small. They were farmed by families who lived not only upon them, but within and *from* them.

These families grew gardens. They produced their own meat, milk, and eggs. The farms were highly diversified. The main money crop was tobacco. But the farmers also grew corn, wheat, barley, oats, hay, and sorghum. Cattle, hogs, and sheep were all characteristically raised on the same farms. There were small dairies, the milking more often than not done by hand. Those were the farm products that might have been considered major. But there were also minor products, and one of the most important characteristics of that old economy was the existence of markets for minor products. In those days a farm family could easily market its surplus cream, eggs, old hens, and frying chickens. The power for field work was still furnished mainly by horses and mules. There was still a prevalent pride in workmanship, and thrift was still a forceful social ideal. The pride of most people was still in their homes, and their homes looked like it.

This was by no means a perfect society. Its people had often been violent and wasteful in their use of the land and of each other. Its present ills had already taken root in it. But I have spoken of its agricultural economy of a generation ago to suggest that there were also good qualities indigenous to it that might have been cultivated and built upon.

That they were not cultivated and built upon—that they were repudiated as the stuff of a hopelessly outmoded, unscientific way of life—is a tragic error on the part of the people themselves; and it is a work of monstrous ignorance and irresponsibility on the part of the experts and politicians, who have prescribed, encouraged, and applauded the disintegration of such farming communities all over the country.

In the decades since World War II the farms of Henry County have become increasingly mechanized. Though they are still comparatively diversified, they are less diversified than they used to be. The holdings are larger, the owners are fewer. The land is falling more and more into the hands of speculators and professional people from the cities, who—in spite of all the scientific agricultural miracles—still have much more money than farmers. Because of big technology and big economics, there is more abandoned land in the county than ever before. Many of the better farms are visibly deteriorating, for want of manpower and time and money to maintain them properly. The number of part-time farmers and ex-farmers increases every year. Our harvests depend more and more on the labor of old people and young children. The farm people live less and less from their own produce, more and more from what they buy. The best of them are more worried about money and more overworked than ever before. Among the people as a whole, the focus of interest has largely shifted from the household to the automobile; the ideals of workmanship and thrift have been replaced by the goals of leisure, comfort, and entertainment. For Henry County plays its full part in what Maurice Telleen calls "the world's first broad-based hedonism." The young people expect to leave as soon as they finish high school, and so they are without permanent interest; they are generally not interested in anything that cannot be reached by automobile on a good road. Few of the farmers' children will be able to afford to stay on the farm—perhaps even fewer will wish to do so, for it will cost too much, require too much work and worry, and it is hardly a fashionable ambition.

And nowhere now is there a market for minor produce: a bucket of cream, a hen, a few dozen eggs. One cannot sell milk from a few cows anymore; the law-required equipment is too expensive. Those markets were done away with in the name of sanitation—but, of course, to the enrichment of the large producers. We have always had to have "a good reason" for doing away with small operators, and in modern times the good reason has often been sanitation, for which there is apparently no small or cheap technology. Future historians will no doubt remark upon the inevitable association, with us, between sanitation and filthy lucre. And it is one of the miracles of science and hygiene that the germs that used to be in our food have been replaced by poisons.

In all this, few people whose testimony would have mattered have seen the connection between the "modernization" of agricultural techniques and the disintegration of the culture and the communities of farming—and the consequent disintegration of the structures of urban life. What we have called agricultural progress has, in fact, involved the forcible displacement of millions of people.

I remember, during the fifties, the outrage with which our political leaders spoke of the forced removal of the populations of villages in communist countries. I also remember that at the same time, in Washington, the word on farming was "Get big or get out"—a policy which is still in effect and which has taken an enormous toll. The only difference is that of method: the force used by the communists was military; with us, it has been economic—a "free market" in which the freest were the richest. The attitudes are equally cruel, and I believe that the results will prove equally damaging, not just to the concerns and values of the human spirit, but to the practicalities of survival.

And so those who could not get big have got out—not just in my community, but in farm communities all over the country. But as a social or economic goal, bigness is totalitarian; it establishes an inevitable tendency toward the *one* that will be the biggest of all. Many who got big to stay in are now being driven out by those who got bigger. The aim of bigness implies not one aim that is not socially and culturally destructive.

And this community-killing agriculture, with its monomania of bigness, is not primarily the work of farmers, though it has burgeoned on their weaknesses. It is the work of the institutions of agriculture: the university experts, the bureaucrats, and the "agribusinessmen," who have promoted so-called efficiency at the expense of community (and of real efficiency), and quantity at the expense of quality.

In 1973, 1000 Kentucky dairies went out of business. They were the victims of policies by which we imported dairy products to compete with our own and exported so much grain as to cause a drastic rise in the price of feed. And, typically, an agriculture expert at the University of Kentucky, Dr. John Nicolai, was optimistic about this failure of 1000 dairymen, whose cause he is supposedly being paid—partly with *their* tax money—to serve. They were inefficient producers, he said, and they needed to be eliminated....

And along with the rest of society, the established agriculture has shifted its emphasis, and its interest, from quality to quantity, having failed to see that in the long run the two ideas are inseparable. To pursue quantity alone is to destroy those disciplines in the producer that are the only assurance of quantity.

What is the effect on quantity of persuading a producer to produce an inferior product? What, in other words, is the relation of pride or craftsmanship to abundance? That is another question the "agribusinessmen" and their academic collaborators do not ask. They do not ask it because they are afraid of the answer: The preserver of abundance is excellence.

My point is that food is a cultural product; it cannot be produced by technology alone. Those agriculturists who think of the problems of food production solely in terms of technological innovation are oversimplifying both the practicalities of production and the network of meanings and values necessary to define, nurture, and preserve the practical motivations. That the discipline of agriculture should have been so divorced from other disciplines has its immediate cause in the compartmental structure of the universities, in which complementary, mutually sustaining and enriching disciplines are divided, according to "professions," into fragmented, one-eyed specialties. It is suggested, both by the organization of the universities and by the kind of thinking they foster, that farming shall be the responsibility only of the college of agriculture, that law shall be in the sole charge of the professors of law, that morality shall be taken care of by the philosophy department, reading by the English department, and so on. The same, of course, is true of government, which has become another way of institutionalizing the same fragmentation.

However, if we conceive of a culture as one body, which it is, we see that all of its disciplines are everybody's business, and that the proper university product is therefore not the whittled-down, isolated mentality of expertise, but a mind competent in all its concerns. To such a mind it would be clear that there are agricultural disciplines that have nothing to do with crop production, just as there are agricultural obligations that belong to people who are not farmers.

A culture is not a collection of relics or ornaments, but a practical necessity, and its corruption invokes calamity. A healthy culture is a communal order of memory, insight, value, work, conviviality, reverence, aspiration. It reveals the human necessities and the human limits. It clarifies our inescapable bonds to the earth and to each other. It assures that the necessary restraints are observed, that the necessary work is done, and that it is done well. A healthy *farm* culture can be based only upon familiarity and can grow only among a people soundly established upon the land; it nourishes and safeguards a human intelligence of the earth that no amount of technology can satisfactorily replace. The growth of such a culture was once a strong possibility in the farm communities of this country. We now have only the sad remnants of those communities. If we allow another generation to pass without doing what is necessary to enhance and embolden the possibility now perishing with them, we will lose it altogether. And then we will not only invoke calamity—we will deserve it. . . .

It is by the measure of culture, rather than economics or technology, that we can begin to reckon the nature and the cost of the country-to-city migration that has left our farmland in the hands of only five percent of the people. From a cultural point of view, the movement from the farm to the city involves a radical simplification of mind and of character.

A competent farmer is his own boss. He has learned the disciplines necessary to go ahead on his own, as required by economic obligation, loyalty to his place, pride in his work. His workdays require the use of long experience and practiced judgment, for the failures of which he knows that he will suffer. His days do not begin and end by rule, but in response to necessity, interest, and obligation. They are not measured by the clock, but by the task and his endurance; they last as long as necessary or as long as he can work. He has mastered intricate formal patterns in ordering his work within the overlapping cycles—human and natural, controllable and uncontrollable—of the life of a farm.

Such a man, upon moving to the city and taking a job in industry, becomes a specialized subordinate, dependent upon the authority and judgment of other people. His disciplines are no longer implicit in his own experience, assumptions, and values, but are imposed on him from the outside. For a complex responsibility he has substituted a simple dutifulness. The strict competences of independence, the formal mastery, the complexities of attitude and know-how necessary to life on the farm, which have been in the making in the race of farmers since before history, all are replaced by the knowledge of some fragmentary task that may be learned by rote in a little while.

Such a simplification of mind is easy. Given the pressure of economics and social fashion that has been behind it and the decline of values that has accompanied it, it may be said to have been gravity-powered. The reverse movement—a reverse movement *is* necessary, and some have undertaken it—is uphill, and it is difficult. It cannot be fully accomplished in a generation. It will probably require several generations—enough to establish complex local cultures with strong communal memories and traditions of care.

There seems to be a rule that we can simplify our minds and our culture only at the cost of an oppressive social and mechanical complexity. We can simplify our society—that is, make ourselves free—only by undertaking tasks of great mental and cultural complexity. Farming, the *best* farming, is a task that calls for this sort of complexity, both in the character of the farmer and in his culture. To simplify either one is to destroy it.

That is because the best farming requires a farmer—a husbandman, a nurturer—not a technician or businessman. A technician or a businessman—given the necessary abilities and ambitions—can be made in a little while, by training. A good farmer, on the other hand, is a cultural product; he is made by a sort of training, certainly, in what his time imposes or demands, but he is also made by generations of experience. This essential experience can only be accumulated, tested, preserved, handed down in settled households, friendships, and communities that are deliberately and carefully native to their own ground, in which the past has prepared the present and the present safeguards the future.

The concentration of the farmland into larger and larger holdings and fewer and fewer hands—with the consequent increase of overhead, debt, and dependence on machines—is thus a matter of complex significance, and its agricultural significance cannot be disentangled from its cultural significance. It *forces* a profound revolution in the farmer's mind: once his investment in land and machines is large enough, he must forsake the values of husbandry and assume those of finance and technology. Thenceforth his thinking is not determined by agricultural responsibility, but by financial accountability and the capacities of his machines. Where his money comes from becomes less important to him than where it is going. He is caught up in the drift of energy and interest away from the land. Production begins to override maintenance. The economy of money has infiltrated and subverted the economies of nature, energy, and the human spirit. The man himself has become a consumptive machine.

For some time now ecologists have been documenting the principle that "you can't do one thing"—which means that in a natural system whatever affects one thing ultimately affects everything. Everything in the Creation is related to everything else and dependent on everything else. The Creation is one; it is a uni-verse, a whole, the parts of which are all "turned into one."

A good agricultural system, which is to say a durable one, is similarly unified....

It remains only to say what has often been said before—that the best human cultures also have this unity. Their concerns and enterprises are not fragmented, scattered out, at variance or in contention with one another. The people and their work and their country are members of each other and of the culture. If a culture is to hope for any considerable longevity, then the relationships within it must, in recognition of their interdependence, be predominantly cooperative rather than competitive. A people cannot live long at each other's expense or at the expense of their cultural birthright—just as an agriculture cannot live long at the expense of its soil or its work force, and just as in a natural system the competitions among species must be limited if all are to survive.

In any of these systems, cultural or agricultural or natural, when a species or group exceeds the principle of usufruct (literally, the "use of the fruit"), it puts itself in danger. Then, to use an economic metaphor, it is living off the principal rather than the interest. It has broken out of the system of nurture and has become exploitive; it is destroying what gave it life and what it depends upon to live. In all of these systems a fundamental principle must be the protection of the source: the seed, the food species, the soil, the breeding stock, the old and the wise, the keepers of memories, the records.

And just as competition must be strictly curbed within these systems, it must be strictly curbed *among* them. An agriculture cannot survive long at the expense of the natural systems that support it and that provide it with models. A culture cannot survive long at the expense of either its agricultural or its natural sources. To live at the expense of the source of life is obviously suicidal. Though we have no choice but to live at the expense of other life, it is necessary to recognize the limits and dangers involved: past a certain point in a unified system, "other life" is our own.

The definitive relationships in the universe are thus not competitive but interdependent. And from a human point of view they are analogical. We can build one system only within another. We can have agriculture only within nature, and culture only within agriculture. At certain critical points these systems have to conform with one another or destroy one another.

Under the discipline of unity, knowledge and morality come together. No longer can we have that paltry "objective" knowledge so prized by the academic specialists. To know anything at all becomes a moral predicament. Aware that there is no such thing as a specialized—or even an entirely limitable or controllable—effect, one becomes responsible for judgments as well as facts. Aware that as an agricultural scientist he had "one great subject," Sir Albert Howard could no longer ask, What can I do with what I know? without at the same time asking, How can I be responsible for what I know?

And it is within unity that we see the hideousness and destructiveness of the fragmentary—the kind of mind, for example, that can introduce a production machine to increase "efficiency" without troubling about its effect on workers, on the product, and on consumers; that can accept and even applaud the "obsolescence" of the small farm and not hesitate over the possible political and cultural effects; that can recommend continuous tillage of huge monocultures, with massive use of chemicals and no animal manure or humus, and worry not at all about the deterioration or loss of soil. For cultural patterns of responsible cooperation we have substituted this moral ignorance, which is the etiquette of agricultural "progress."

12.2 FRANCES MOORE LAPPÉ AND JOSEPH COLLINS

Hasn't the Green Revolution "Bought Us Time"?

The development of new varieties of rice and other grains, which when accompanied by high-technology agriculture using fertilizers and chemical pesticides produces vastly improved yields, has been referred to as the Green Revolution. The tripling of world grain production and the doubling of per capita agricultural production between 1950 and 1990 is primarily attributable to the proliferation of Green Revolution agricultural crops and methods. While this remarkable increase in global food production outstripped population growth and resulted in the harvesting of sufficient food to theoretically provide all the world's people with an adequate diet, it by no means resulted in an end to starvation. Indeed, malnutrition in less developed countries has continued to increase.

Agricultural policy analysts, Frances Moore Lappé, author of the best-seller *Diet for a Small Planet,* 20th ed. (Ballantine, 1991), and Joseph Collins are co-founders of the Institute for Food and Development Policy. In their book *Food First: Beyond the Myth of Scarcity* (Houghton Mifflin, 1977), coauthored with Cary Fowler, senior officer of the United Nations Food and Agriculture Organization, they present a well-documented analysis and explanation of the apparent paradox of more hunger in the face of more food. Their general conclusion is that the root cause of this paradox is that agricultural resources are organized with the goal of producing private wealth rather than feeding people.

In the following excerpt from *Food First,* the authors explain how the introduction of Green Revolution technology enhanced the concentration of wealth in the hands of the few at the expense of the many. In other sections of the book not reprinted here, Lappé and Collins discuss the environmental problems that result from the nonsustainable technologies associated with the Green Revolution.

Key Concept: why world hunger has grown despite increased food production

Question: You say that the narrow focus on increased production has benefited the prosperous farmer. But why is that so terrible? More food at least has been produced. How could this hurt the poor? They need food more than anything else. You seem to expect the strategies to increase production, such as the Green Revolution, also to solve *social* problems. How could they? Increasing production is a scientific and a technological problem. Norman Borlaug, father of the Green Revolution, has said that the most the Green Revolution could do is to "buy us time" while we slow population growth and work on economic problems. Hasn't the Green Revolution accomplished all one could reasonably expect? Isn't progress with disruption better than no progress at all?

Our Response: To answer this question we had to look systematically at just what happens once the solution to hunger is sought in a big technological push to increase production. This search has revealed a powerful dynamic by which the better off have "progressed" *at the expense* of the majority. Around the world we find a strikingly consistent pattern. Although the question implies that the impact of the Green Revolution is limited to underdeveloped countries, it is not. . . .

MORE GRAIN: WHO GAINS?

For many outsiders looking at hunger in underdeveloped countries, the fact that greater production can bring cheaper grain appears as part of the solution. The mistake is in forgetting two points: First, many of the poor are also producers whose livelihood depends on selling their grain. Second, for those unable to take part in the new technology, yields have often *not* increased. With overall greater availability, however, and the failure of government policies to maintain prices, the poor farmers with the unimproved yields are in a worse plight than ever.

In Greece the agricultural credit corporations pressure farmers to sow foreign-bred HRV wheats. In the lowland areas occupied by large farms the result has been higher yields, thus increasing total Greek output. But in the mountains the HRV seeds yielded less than the varieties that had been grown for generations by the mountain people. As the national (and world) yields increased, wheat prices fell. The large commercial farms in the plains could withstand the price drop because their volume of production was large and increasing. But for the poorer farm on the mountain slopes the fall in income resulting from lower yields was often the final blow leading to the desertion of many mountain villages, as well as the loss to the world of wheat varieties unconsciously selected, over centuries, to thrive in more difficult conditions.

RENTS GO UP

Landlords in many countries have found they can transfer part of the burden of increased production costs onto the tenants or sharecroppers, in effect, forcing the tenants to pay for the new technology. For instance, with the introduction of the new technology, the cash rents tenants must pay have gone up by about one-third to one-half. Crop share rents are changing from the traditional 50–50 division between the landlord and the tenant to 70–30 in favor of the landlord, effectively cutting the tenant out of the production gains. In one area of India where the sharecropper used to get half the harvest, he now gets only one-third; another third goes to the landlord and the remaining third goes to pay off the debt for the tubewell the landlord purchased (it will, of course, go to the landlord once the tubewell is paid for). In one area of Malaysia tenants now have to provide 100 percent of the fertilizer costs.

Traditional landlords once had very clearcut, reciprocal obligations to tenants or sharecroppers. The landlord would have never considered passing on his obligations to the tenants. But now that more and more landlords are absentee city dwellers, traditional face-to-face dealings are being replaced by impersonal, money-based relationships. Landlords increasingly demand cash payment of rent instead of payment in kind. In the northern states of Malaysia cash payments are required at the *beginning* of the season. The tenant has to come up with rent at just the time he is least likely to have it. He therefore has to borrow at high rates of interest—thus reducing his total income. Moreover, if the crops fail, the tenant still has to come up with the rent.

By the same token, many landlords now prefer to pay in money wages rather than in farm produce. In times of inflating food prices, however, it is much better for the tenant-farmer to have part of the harvest than money. We learned of one district in India where landlords now pay only money wages, preferring to hoard and sell the rice later for enormous profit. In 1974, India's *Economic and Political Weekly* reported that in Thanjavur, Tamil Nadu, "hordes of the police were stationed in the paddy fields to quell disturbances arising out of the landlords' refusal to pay even a part of the wages in kind."

LAND VALUES SOAR

In countries where food-resources are still allowed to be held for private gain, the inpouring of government funds in the form of irrigation works and subsidies for fertilizers and machinery have combined with the higher potential yields of the new seeds to turn farming into the world's hottest growth industry. Instead of a way of livelihood for millions of small, self-provisioning farmers, agriculture is increasingly seen as a lucrative opportunity by a new class of "farmers" with the money or influence to get in on the action—moneylenders, military officers, bureaucrats, city-based speculators, and foreign corporations. In those areas targeted by the "production strategy" land values have gone up three-, four-, or even fivefold as these so-called farmers compete for the land that they believe, often rightly, will make them a fortune.

Here is development economist Wolf Ladejinsky's well-known account of how nonfarmers in India buy up land for speculation:

> The buyers are a motley group: some connected with land through family ties, some altogether new to agriculture. A few have unemployed rupees acquired through undeclared earnings, and most of them look upon farming as a tax-haven, which it is, and as a source of earning tax-free supplementary income. The medical doctor from Jullundur who turned part-time farmer is sitting pretty. The 15 acres purchased four years ago have tripled in value. To listen to him, he is in farming "for the good of the country." . . . His only vexation is whether or not he will succeed in buying another 10 acres he has his eyes on—and what a disappointed man he will be if they escape him! As we watched him supervise the threshing, he was anything but a "gentleman farmer."

Nonfarmers are taking over agriculture not only because governments have made investments attractive, but because increasingly only the wealthier urban dwellers can obtain credit or afford to buy the higher-priced land and the necessary inputs. As the price of land rises, purchase by the smallholder or tenant, if unlikely before, becomes completely out of the question. In countries where security of tenure is legally guaranteed after the tenant has continuously cultivated a given plot for a certain number of years, some landlords maneuver to ensure that their tenants are never given legal title to the land, now that land is more valuable. In Tanjore, India, landlords shift sharecroppers from one plot to another each year to successfully dodge such tenure regulations.

Moreover, as the market value of land increases, taxes increase. In Colombia wealthy potential buyers of small plots persuade tax authorities to revalue the land in order to put pressure on the small farmers. Peasants who cannot afford to plant the new varieties of coffee find that they cannot pay the higher tax bill and are forced to sell to larger landholders who usually can evade the tax by paying a bribe.

FEWER PEOPLE CONTROL MORE LAND

Fewer and fewer people control more and more of farm production. A pattern of increasing monopolization of agricultural land moves ahead in India, Bangladesh, Mexico, the Philippines, Colombia—in virtually all countries where officially subsidized "modernization" now means that high returns stem from the sheer amount of land one can control, not from how well one farms.

In the area of Tamesis, Colombia, the better-off coffee growers able to adopt the new seed varieties increased the average size of their holdings by 76 percent between 1963 and 1970. Similarly, in the government-subsidized irrigated zones of Morocco land concentration is increasing. In just five years, from 1965 to 1970 the average size of modern, Moroccan-owned farms in one irrigated area increased 30 percent. In the Indian Punjab, between the fifties and the mid-sixties, the amount of land owned by the largest farmers (those with 100–150 acres) increased at a rate four to ten times greater than that of the smaller-sized farms.

Another sign of increasing concentration of land ownership is that the smallest farmers are selling their land. In both 1969–1970 and 1972–1973 farmers in Bangladesh who owned less than one acre (accounting for about a quarter of all farmers) sold well over half of their land.

To some, the decline of the small farmer appears unfortunate, but, alas, inevitable. But the tightening of control over agricultural production is not inevitable. It results from the actions and even the planning of people. In the early 1950s, large farmers in the Mexican state of Sonora saw that land values were about to go up because of massive government irrigation plans for the area. They began to contrive to take over cheaply the land owned by thousands of smallholders. They turned to their friends within the National Agricultural Credit Bank—the government agency on which smallholders in the area depended for survival. The bank began to delay crop credit for smallholders. In some cases they received credit so late that their wheat, to take but one example, had to be planted out of season and thus failed during several years. The bank also began to provide sacks of wheat seed that some say turned to dust between their fingers and fertilizer which they are sure was nothing more than white powder. The smallholders' expenses soared. They had several disastrous years. Then came the final blow: The government foreclosed on all properties with outstanding debts to federal agencies. The large farmers had succeeded. The majority of the smallholders in one devastated settlement ended up selling their land for about one ninth of the market price to two of the largest and most politically influential landowners in the state.

Where much land has traditionally been communally worked, as in Africa, the new agricultural entrepreneurs might have an even easier time expanding commercial operations. Without private property rights and a small farm tradition that exists in Asia, little if anything stands in the way of tribal chieftains and foreign corporations who want to appropriate communal land for their private gain.

THE MAKING OF THE LANDLESS

In certain areas landlords are moving to push their tenants off the land. The landlords see several advantages. For instance, they are freed from tenants who might conceivably claim land under a land-to-the-tiller reform movement. Moreover, the large landowner finds it more profitable to mechanize production or take advantage of part-time laborers who have no claim on the land or on the harvest. A study for the World Bank on the size of farms in the Indian Punjab during the 1960s concluded that farms that had been mechanized grew by an average of 240 percent over a three-year period, primarily because the landlords decided to cultivate land they had previously rented out. The landlord's gain—higher cash income—was society's loss, as a substantial number of tenants could no longer rent the land they needed to support themselves. In

India, in 1969, there were 40,000 eviction suits against sharecroppers in Bihar alone and 80,000 in Karnatika (Mysore).

As the control of land tightens and more tenants are evicted, the number of landless laborers mounts. In all nonsocialist underdeveloped countries 30 to 60 percent of rural adult males are now landless....

So the number of landless mounts while the number of rural jobs shrinks. Traditionally in many countries even the poorest landless peasant had access to part of the harvest. In India, Bangladesh, Pakistan, and Indonesia the large landowner once felt obligated to permit all who wished to participate in the harvest to retain one sixth of what they harvested. Thus even the most impoverished were assured of work for a few bags of grain. Now, with the increased likelihood of profitable sales, the new agricultural entrepreneurs are rejecting the traditional obligations of the landowner to the poor. It is now common for landowners to sell the standing crop to an outside contractor before harvest. The outsider, with no local obligations, can seek the cheapest labor, even bringing in workers from neighboring areas.

In Java landless laborers were once permitted to squat on dry land in the off season to grow cassava and vegetables. With the new rice seeds, landlords are now interested in irrigating the land for year-round production for commercial markets. Squatters are no longer welcome.

In addition, the introduction of large-scale mechanization is a double-edged sword for the rural poor. Large landowners say that the only way they can make their new machines pay off is to reduce per acre cost by expanding their acreage. As we have already seen, expansion by the largeholders forces more and more tenants and small farmers off the land, thereby creating greater numbers of landless in search of farm work. Simultaneously, however, the machines drastically decrease the number and length of jobs available. A tractor cuts to a fifth the number of workers needed to prepare the same field with a bullock-drawn plow. The same is true of a mechanical reaper compared to a hand scythe....

The production strategy we have been describing often offers the poor the illusion of rural employment. One of the best documented cases is the Mexican state of Sonora. There, the clearing of new land and vast irrigation projects executed with public funds attracted workers to Sonora during the fifties and sixties. By 1971, when these projects had been completed, the laborers still needed jobs but, with the land now run as large highly mechanized operations, they could at best hope for six months' work a year. Thus those who actually had labored to make it possible for Sonora to become Mexico's production showcase were largely cut out of its bounty.

We have seen that with the introduction of new technologies into societies where small groups can monopolize agricultural resources, the price of land and therefore land rents go up and tenants are displaced from the land and laborers from their jobs. It is not surprising, therefore, to find peasants in these countries invariably sinking deeper into debt to stay alive. Cheap credit schemes often only worsen the predicament of the poor. In Malaysia, for example, landlords commonly obtain loans from rural banks by using their land as collateral. (The banks are reluctant to issue low-interest loans to tenants because they have no collateral.) The money obtained by the landlord is then re-lent to tenants at interest rates left to the landlord's own discretion. The result is the reinforcement of debt bondage, making the peasant permanently so indebted that he is obliged to accept wages for his labor at a rate 30 to 50 percent below the going market rate.

WHEN MORE FOOD MEANS MORE HUNGER

We have now arrived at the endpoint of the tragic decline of the rural majority. In country after country, where agricultural resources are allowed to be sources of private wealth, the drive to increase food production has made even worse the lives of the poor majority, despite per capita production increases. We have seen how:

- Land values go up, forcing tenants and small farmers off the land.
- Rents increase.
- Payments in money become the rule, yet money buys less food.
- The control of farmland becomes concentrated in fewer hands, many of whom are speculative entrepreneurs, not farmers.
- Even communal lands (as in many African villages or land-reform areas) are appropriated by powerful individuals such as chieftains or caciques for their private gain at the expense of the welfare of the community.
- Corporate control, often foreign, extends further into production.
- Peasants are trapped into debt bondage.
- Poverty and inequality deepen.
- Production totals, not the participation of the rural population in the production process (livelihood and nourishment), become the measure of success for agricultural planners.
- Quantity and market value, not nutritive value, become the goal of agricultural planning.

The net result? Hunger tightens its hold on the rural majority.

A series of major studies now being completed by the International Labor Organization (ILO) documents that in the seven South Asian countries comprising 70 percent of the rural population of the nonsocialist underdeveloped world, the rural poor have become worse off than they were ten or twenty years ago. The summary study notes that ironically *"the increase in poverty has been associated not with a fall but with a rise in cereal production per head, the main component of the diet of the poor."* Here are typical examples:

- The Philippines: Despite the fact that agricultural production increased by 3 to 4 percent per year during the last fifteen to twenty years, about one fifth of the rural households experienced a dramatic and *absolute* decline in living standards, which accelerated during the early seventies. By 1974 daily real wages in agriculture fell to almost one third of what they were in 1965.

- Bangladesh: Between 1963 and 1975, the proportion of rural households classified as absolutely poor increased by more than a third and that of those classified as extremely poor increased five times. Yet about 15 percent of the rural households in Bangladesh had significantly higher real incomes in 1975.
- West Malaysia: By 1970 the bottom 20 percent of rural households had experienced a fall of over 40 percent in their average income since 1957, while the average income of the next 20 percent fell 16 percent. By contrast, the top 20 percent of rural households increased their mean income 21 percent.
- Sri Lanka: Despite a rise in per capita income between 1963 and 1973, actual rice consumption *fell* for all except the highest income class. All workers experienced a fall in real wages, except for those in industry and commerce whose real wages remained static.

Part of the reason that most people have not been able to perceive this tragic retrogression is what we have come to call the "language of deception"—terms that obfuscate reality. One such term is "per capita." As these examples show, per capita production and income have been going up in the very countries where often the majority of the people have become worse off with each succeeding year. It is, of course, precisely the kind of development policy that measures itself in per capita terms that results in the absolute decline of the majority.

BOUGHT US TIME?

Many view the Green Revolution as a technical innovation and feel that, as such, it should not be expected to solve social problems. But what we have found is that there can be no separation between technical innovation and social change. Whether promotion of the wealthier class of farmers is deliberate government policy or not, inserting any profitable technology into a society shot through with power inequalities (money, landownership, privilege, access to credit) sets off the disastrous retrogression of the less powerful majority. The better off and powerful in a society further enrich themselves at the expense of the national treasury and the rural poor. As those initially better off gain even greater control over the production process, the majority of people are made marginal, in fact, totally irrelevant, to the process of agricultural production. In such societies the reserves of landless and jobless function only to keep wages down for those who do find jobs. Excluded from contributing to the agricultural economy, the poor majority are no longer its beneficiaries, for being excluded from production means being excluded from consumption. A thirty-six-cents-a-day laborer in Bihar, India, knows this truth well: "If you don't own any land, you never get enough to eat," he says, "even if the land is producing well."

The Green Revolution has *not* "bought us time" as the question suggests. "Modernization" overlaid on oppressive social structures entrenches the ownership classes who are now even better positioned and less willing to part with their new-found wealth. Thus, defining the problem of hunger as one of production is not merely incorrect. To focus only on raising production, without first confronting the issue of who controls and who participates in the production process, actually compounds the problem. It leaves the majority of people worse off than before. In a very real sense the idea that we are progressing is our greatest handicap. We cannot move forward—we cannot take the first step toward helping improve the welfare of the vast majority of the world's people—until we can see clearly that we are now moving backward.

12.3 LESTER BROWN

Food Scarcity: An Environmental Wakeup Call

Lester Brown started his career as a tomato farmer in southern New Jersey. Currently, he is the founder, president, and a senior researcher at the World Watch Institute in Washington, D.C. The *Washington Post* has referred to Brown as "one of the world's most influential thinkers," and he has been called "the guru of the global environmental movement" by the *Telegraph* of Calcutta, India. Brown's academic credentials include an M.S. in agricultural economics from the University of Maryland and an M.P.A. from Harvard University. While serving as an adviser to Secretary of Agriculture Orville Freeman in the mid-1960s, he was appointed administrator of the department's International Agricultural Development Service. In addition to being the senior editor of the influential World Watch Institute's yearly *State of the World* reports, and writing numerous other scholarly articles, he has authored or coauthored a series of highly regarded books, the most recent of which is *Beyond Malthus: Nineteen Dimensions of the Population Challenge* (W. W. Norton, 1999).

In "Food Scarcity: An Environmental Wakeup Call," *The Futurist* (January/February 1998), from which the following selection is taken, Brown warns of economic and social disruptions due to food scarcity. Brown states that continuing reductions in per capita cropland and irrigation water will cause the level of food security to drop below an acceptable level. He predicts that rampant environmental degradation in the form of deforestation and soil erosion, coupled with increasing greenhouse gas emissions, will lead to an economic decline of the global food system. In Brown's opinion, the most difficult challenge in building a sustainable economy that can assure adequate food supplies in the future is in stabilizing population and climate.

Key Concept: environmental degradation coupled with population growth as causes of food scarcity

The environmental deterioration of the last few decades cannot continue indefinitely without eventually affecting the world economy. Until now, most of the economic effects of environmental damage have been local: the collapse of a fishery here or there from overfishing, the loss of timber exports by a tropical country because of deforestation, or the abandonment of cropland because of soil erosion. But as the scale of environmental damage expands, it threatens to affect the global economy as well.

The consequences of environmental degradation are becoming more clear. We cannot continue to deforest the earth without experiencing more rainfall runoff, accelerated soil erosion, and more destructive flooding. If we continue to discharge excessive amounts of carbon into the atmosphere, we will eventually face economically disruptive climate change. If we continue to overpump the earth's aquifers, we will one day face acute water scarcity.

If we continue to overfish, still more fisheries will collapse. If overgrazing continues, so, too, will the conversion of rangeland into desert. Continuing soil erosion at the current rate will slowly drain the earth of its productivity. If the loss of plant and animal species continues at the rate of recent decades, we will one day face ecosystem collapse.

Everyone agrees that these trends cannot continue indefinitely, but will they stop because we finally do what we know we should do, or because the economic expansion that is causing environmental decline begins to be disrupted?

AGRICULTURE: THE MISSING LINK

The food system is likely to be the sector through which environmental deterioration eventually translates into economic decline. This should not come as a surprise. Archaeological evidence indicates that agriculture has often been the link between environmental deterioration and economic decline. The decline of the early Mesopotamian civilization was tied to the waterlogging and salting of its irrigated land. Soil erosion converted into desert the fertile wheatlands of North Africa that once supplied the Roman Empire with grain.

Rising grain prices will be the first global economic indicator to tell us that we are on an economic and demographic path that is environmentally unsustainable. Unimpeded environmental damage will seriously impair the capacity of fishers and farmers to keep up with the growth in demand, leading to rising food prices. The social consequences of rising grain prices will become unacceptable to more and more people, leading to political instability. What begins as environmental degradation eventually translates into political instability.

A doubling of grain prices, such as occurred briefly for wheat and corn in early 1996, would not have a major immediate effect on the world's affluent, both because they spend only a small share of their income for food and because their food expenditures are dominated more by processing costs than by commodity prices. But for the 1.3 billion in the world who live on a dollar a day or less, a prolonged period of higher grain prices would quickly become life-threatening.

Heads of households unable to buy enough food to keep their families alive would hold their governments responsible and take to the streets. The resulting bread or rice riots could disrupt economic activity in many countries. If the world could not get inflated food prices back down to traditional levels, this could negatively affect the earnings of multinational corporations, the performance of stock markets, and the stability of the international monetary system. In a world economy more integrated than ever before, the problems of the poor would then become the problems of the rich.

The consequences of environmental abuse that scientists have warned about can be seen everywhere:

- In the European Union, the allowable fish catch has had to be reduced by 20% or more in an effort to avert the collapse of the region's fisheries.
- In Saudi Arabia, overreliance on a fossil aquifer to expand grain production contributed to an abrupt 62% drop in the grain harvest between 1994 and 1996.
- The soil degradation and resulting cropland abandonment that invariably follows the burning off of the Amazon rain forest for agriculture has helped make Brazil the largest grain importer in the Western Hemisphere.

As the number of such situations multiplies, it becomes more and more difficult to feed a world population that is expanding by 80 million people per year. Even without further environmental degradation, we approach the new millennium with 800 million hungry and malnourished people.

These 800 million are hungry because they are too poor to buy enough food to satisfy their basic nutritional needs. If the price of grain were to double, as it already has for some types of seafood, it could impoverish hundreds of millions more almost overnight. In short, a steep rise in grain prices could impoverish more people than any event in history, including the ill-fated Great Leap Forward in China that starved 30 million people to death between 1959 and 1961.

IN SEARCH OF LAND

As the world's population, now approaching 5.8 billion, continues to expand, both the area of cropland and the amount of irrigation water per person are shrinking, threatening to drop below the amount needed to provide minimal levels of food security.

Over time, farmers have used ingenious methods to expand the area used to produce crops. These included irrigation, terracing, drainage, fallowing, and even, for the Dutch, reclaiming land from the sea. Terracing let farmers cultivate steeply sloping land on a sustainable basis, quite literally enabling them to farm the mountains as well as the plains. Drainage of wetlands opened fertile bottomlands for cultivation. Alternate-year fallowing to accumulate moisture helped farmers extend cropping into semiarid regions.

By the middle of this century, the frontiers of agricultural settlement had largely disappeared, contributing to a dramatic slowdown in the growth in area planted to grain. Between 1950 and 1981, the area in grain increased from 587 million to 732 million hectares, a gain of nearly 25%. After reaching a record high in 1981, the area in grain declined, dropping to 683 million hectares in 1993. It has turned upward since then, increasing to 696 million hectares in 1996 as idled cropland was returned to production and as record grain prices in the spring of 1996 led farmers to shift land out of soybeans and other oilseeds.

While the world grain harvested area expanded from 1950 until it peaked in 1981, the growth was quite slow compared with that of population. As a result, the grainland area per person has been declining steadily since mid-century, shrinking from 0.23 hectares in 1950 to 0.12 hectares in 1996. If grain-land gains and losses continue to offset each other in the decades ahead, the area will remain stable at roughly 700 million hectares. But with population projected to grow at some 80 million a year over the next few decades, the amount of cropland available to produce grain will continue to decline, shrinking to 0.08 hectares per person in 2030.

IN SEARCH OF WATER

The world's farmers are also facing water scarcity. The expanding demand for water is pushing beyond the sustainable yield of aquifers in many countries and is draining some of the world's major rivers dry before they reach the sea. As the demand for water for irrigation and for industrial and residential uses continues to expand, the competition between countryside and city for available water supplies intensifies. In some parts of the world, meeting growing urban needs is possible only by diverting water from irrigation.

One of the keys to the near tripling of the world grain harvest from 1950 to 1990 was a 2.5-fold expansion of irrigation, a development that extended agriculture into arid regions with little rainfall, intensified production in low-rainfall areas, and increased dry-season cropping in countries with monsoonal climates. Most of the world's rice and much of its wheat is produced on irrigated land.

A critical irrigation threshold was crossed in 1979. From 1950 until then, irrigation expanded faster than population, increasing the irrigated area per person by nearly one-third. This was closely associated with the worldwide rise in grain production per person of one-third. But since 1979, the growth in irrigation has fallen behind that of population, shrinking the irrigated area per person by some 7%. This trend, now well established, will undoubtedly continue as the demand for water presses ever more tightly against available supplies.

As countries and regions begin to press against the limits of water supplies, the competition between cities and the countryside intensifies. And the cities almost always win. As water is pulled away from agriculture, production often drops, forcing the country to import grain. Importing a ton of grain is, in effect, importing thousands of tons of water. For countries with water shortages, importing grain is the most efficient way to import water. Just as land scarcity has shaped international grain trade patterns historically, water scarcity is now beginning to do the same.

The bottom line is that the world's farmers face a steady shrinkage in both grainland and irrigation water per person. As cropland and irrigation water become ever more scarce, prices of both are likely to rise, pushing grain prices upward.

Aquifer depletion and the future cutbacks in water supplies that will eventually follow may pose a far greater threat to economic progress than most people realize. If aquifer depletion were simply a matter of a few isolated instances, it would be one thing, but it is now in evidence in scores of countries. Among those suffering from extensive aquifer depletion are China, India, and the United States—the three countries that collectively account for about half of the world grain harvest.

THE ONSET OF FOOD SCARCITY

Evidence that the degradation of the earth is leading to food scarcity has been accumulating for many years. The oceanic fish catch, for example, plagued by overfishing and pollution, has grown little after increasing from 19 million tons in 1950 to 89 million tons in 1989. Grainland productivity increased by more than 2% a year from 1950 to 1990, but dropped to scarcely 1% a year from 1990 to 1995—well below the growth in demand.

All the key food-security indicators signal a shift from surplus to scarcity. During the mid-1990s, the United States began using again all the cropland that had been idled under commodity programs in an effort to offset the slower rise in land productivity. Even so, in 1996 world carryover stocks of grain, perhaps the most sensitive indicator of food security, dropped to the lowest level on record—a mere 52 days of consumption. Even with the exceptional harvest of 1996, stocks were rebuilt to only 57 days of consumption, far below the 70 days needed to provide a minimal buffer against a poor harvest. If grain stocks cannot be rebuilt with an outstanding harvest, when can they be?

During the late spring and early summer of 1996, world wheat and corn prices set record highs under pressure from a 1995 harvest reduced by heat waves in the U.S. Corn Belt and from China's emergence as the world's second-largest grain importer. Wheat traded at over $7 a bushel, more than double the price in early 1995. in mid-July, corn traded at an all-time high of $5.54 a bushel, also double the level of a year earlier.

In the summer of 1996, the government of Jordan, suffering from higher prices for imported wheat and a growing fiscal deficit, was forced to eliminate the bread subsidy. The resulting bread riots lasted several days and threatened to bring down the government.

Food scarcity may provide the environmental wakeup call the world has long needed. Rising food prices may indicate the urgency of reversing the trends of environmental degradation before resulting political instability reaches the point where economic progress is no longer possible.

AN UNPRECEDENTED CHALLENGE

Making sure that the next generation has enough food is no longer merely an agricultural matter. Achieving an acceptable balance between food and people depends as much on family planners as on farmers. Decisions made in the ministries of energy that will affect future climate stability may have as much effect on the food security of the next generation as those made in agricultural ministries.

The two most difficult components of the effort to secure future food supplies and build an environmentally sustainable economy are stabilizing population and climate. The former depends on a revolution in human reproductive behavior; the latter, on a restructuring of the global energy economy. Either would thoroughly challenge a single generation, but our generation must attempt both simultaneously. In addition, building an environmentally sustainable economy depends on reversing deforestation, arresting the loss of plant and animal species, and stabilizing fisheries, aquifers, and soils.

In a world where both the seafood catch and the grain harvest per person are declining, it may be time to reassess population policy. For example, some governments, facing a deterioration in their food situation, may have to ask if couples are morally justified in having more than two children, the number needed to replace themselves.

The world has taken one small step in the right direction with the stabilization of population in some 32 countries—all of which, except Japan, are in Europe. These countries, home to some 14% of the world's people, clearly demonstrate that population stabilization is possible.

Stabilizing climate means reducing carbon emissions and, hence, fossil-fuel burning—not an easy undertaking given that 85% of all commercial energy comes from fossil fuels. The outline of a solar/hydrogen economy that is likely to replace the fossil-fuel-based economy of today is beginning to emerge. Both the technology and the economics of harnessing solar and wind energy on a massive scale are beginning to fall into place. Although still small compared with fossil-fuel use, wind-generated electricity is expanding by more than 20% a year, and the use of solar cells is growing almost as fast.

The second major opportunity for reducing carbon emissions is raising the efficiency of energy use. The impressive gains in boosting energy efficiency following the oil price shocks of the 1970s have waned in recent years. Adoption of a carbon tax (offset by a reduction in income taxes) that even partly reflected the costs of air pollution, acid rain, and climate disruption from burning fossil fuels would accelerate investment in solar and wind energy as well as in energy efficiency.

The shift from surplus to scarcity will affect land-use policy. During the last half century, when the world was plagued with farm surpluses and farmers were paid to idle cropland, there seemed little need to worry about the conversion of cropland to nonfarm uses. Cropland was a surplus commodity. But in a world of food scarcity, land use suddenly emerges as a central issue. Already, a group of leading scientists in China has issued a white paper challenging the decision by the Ministry of Heavy Industry to develop an auto-centered transport system, arguing that the country does not have enough land both to provide roads, highways, and parking lots and to feed its people. They argue instead for a state-of-the-art rail passenger system augmented by bicycles.

Perhaps the best model of successful cropland protection is Japan. The determination to protect its riceland with land-use zoning can be seen in the hundreds of small rice fields within the city boundaries of Tokyo. By tenaciously protecting the land needed for rice, Japan remains self-sufficient in staple food.

In addition to protecting cropland from conversion to nonfarm uses, either through zoning or through a stiff tax on conversion, future food security depends on reducing the loss of topsoil from wind and water erosion. In a world facing food scarcity, every ton of topsoil lost from erosion today threatens the food security of the next generation. Here, the United States has emerged as a leader, with its Conservation Reserve Program. Among other things, it promotes the conversion of highly erodible cropland into grass, transforming it to grazing land before it becomes wasteland. This program also denies the benefits of any government programs to farmers with excessive soil erosion on their land if they do not adopt an approved soil conservation management program.

Like land, water is also being diverted to nonfarm uses. With water scarcity now constraining efforts to expand food production in many countries, raising the efficiency of water use is emerging as a key to expanding food production. A shift to water markets, requiring users to pay the full cost of water, would lead to substantial investments in efficiency. The common practice of supplying water either free of charge or at a nominal cost to farmers, industries, and urban dwellers leads to water waste.

Stretching water supplies enough to satisfy future food needs means boosting the efficiency of water use, emulating the achievements of Israel—the pacesetter in this field. *Land productivity* has long been part of our vocabulary, an indicator that we measure in yield per hectare. But the term *water productivity* is rarely heard. Until it, too, becomes part of our everyday lexicon, water scarcity will cloud our future.

FEEDING THE FUTURE

Securing future food supplies will affect every facet of human existence—from land-use policy to water-use policy to how we use leisure time. If food security is the goal, then the dream of some of having a car in every garage, a swimming pool in every backyard, and a golf course in every community may remain simply a dream.

Until recently, the world had three reserves it could call on in the event of a poor harvest—cropland idled under farm programs, surplus stocks of grain in storage, and the one-third of the world grain harvest that is fed to livestock, poultry, and fish. By 1997, the first two of these reserves had largely disappeared. The only one remaining that can be tapped in a world food emergency is the grain used as feed. This is much more difficult to draw on. Higher prices, of course, will encourage the world's affluent to eat less grain-intensive livestock products, but prices high enough to have this effect would also threaten the survival of the world's low-income consumers.

In the event of a world food emergency, one way to restrict the rise in grain prices and restore market stability would be to level a tax on the consumption of livestock products, offsetting it with a reduction in income taxes. Lowering the demand for grain would also lower its price. Unpopular though it would be, such a tax might be acceptable if it were the key to maintaining political stability and sustaining economic progress in low-income countries. Such a step would not solve the food problem, but as a temporary measure it would buy some additional time to stabilize population.

It appears that future food security depends on creating an environmentally sustainable economy. Simply put, if political leaders do manage to secure food supplies for the next generation, it will be because they have moved the world economy off the current path of environmental deterioration and eventual economic disruption and onto an economic and demographic path that is environmentally sustainable.

CHAPTER 13 Pest Control

13.1 RACHEL CARSON

Silent Spring

"The beauty of the living world I was trying to save has always been uppermost in my mind—that, and anger at the senseless, brutish things that were being done. . . . Now I can believe I have at least helped a little." This was Rachel Carson's modest appraisal of the impact of her book *Silent Spring* (Houghton Mifflin, 1962), expressed in a letter to a friend shortly before her death. In fact, the book was an enormous success, remaining on the best seller list for almost a year. It is widely attributed with raising the public consciousness and stimulating the activism that gave rise to the ongoing environmental movement.

Carson (1907–1964) developed her professional interest in the sea, the subject of her earlier literary works, when, at the age of 22, she went to work as a researcher at the Marine Biological Laboratory at Woods Hole in Falmouth, Massachusetts. She spent much of her career working for the U.S. Fish and Wildlife Service, where she became increasingly concerned about the destructive ecological impacts of the growing use of DDT and other chlorinated pesticides. When *Silent Spring* first appeared in serial form in *The New Yorker,* it was greeted by a vociferous attack orchestrated by the chemical and agribusiness industries. Carson was charged with being an extremist, and her indictment of pesticides was branded a gross exaggeration. But her careful documentation of the facts and calm defense of her arguments was more than a match for the industrial mudslingers. A growing chorus of environmentally concerned scientists joined the fray on Carson's side. President John F. Kennedy, who had read and been moved by the book, discussed it at a press conference and appointed a panel of experts to examine its findings. The report of the panel vindicated Carson and castigated the irresponsible indifference of her attackers.

The following selection excerpts from *Silent Spring* illustrates the eloquence and passion of Carson's prose as well as the ecological ethic that underlies her analysis.

Key Concept: the folly of trying to control nature with chemical poisons

A FABLE FOR TOMORROW

There was once a town in the heart of America where all life seemed to live in harmony with its surroundings. The town lay in the midst of a checkerboard of prosperous farms, with fields of grain and hillsides of orchards where, in spring, white clouds of bloom drifted above the green fields. In autumn, oak and maple and birch set up a blaze of color that flamed and flickered across a backdrop of pines. Then foxes barked in the hills and deer silently crossed the fields, half hidden in the mists of the fall mornings.

Along the roads, laurel, viburnum and alder, great ferns and wildflowers delighted the traveler's eye through much of the year. Even in winter the roadsides were places of beauty, where countless birds came to feed on the berries and on the seed heads of the dried weeds rising above the snow. The countryside was, in fact, famous for the abundance and variety of its bird life, and when the flood of migrants was pouring through in spring and fall people traveled from great distances to observe them. Others came to fish the streams, which flowed clear and cold out of the hills and contained shady pools where trout lay. So it had been from the days many years ago when the first settlers raised their houses, sank their wells, and built their barns.

Then a strange blight crept over the area and everything began to change. Some evil spell had settled on the community: mysterious maladies swept the flocks of chickens; the cattle and sheep sickened and died. Everywhere was a shadow of death. The farmers spoke of much illness among their families. In the town the doctors had become more and more puzzled by new kinds of sickness appearing among their patients. There had been several sudden and unexplained deaths, not only among adults but even among children, who would be stricken suddenly while at play and die within a few hours.

There was a strange stillness. The birds, for example—where had they gone? Many people spoke of them, puzzled and disturbed. The feeding stations in the backyards were deserted. The few birds seen anywhere were moribund; they trembled violently and could not fly. It was a spring without voices. On the mornings that had once throbbed with the dawn chorus of robins, catbirds, doves, jays, wrens, and scores of other bird voices there was now no sound; only silence lay over the fields and woods and marsh.

On the farms the hens brooded, but no chicks hatched. The farmers complained that they were unable to raise any pigs—the litters were small and the young survived only a few days. The apple trees were coming into bloom but no bees droned among the blossoms, so there was no pollination and there would be no fruit.

The roadsides, once so attractive, were now lined with browned and withered vegetation as though swept by fire. These, too, were silent, deserted by all living things. Even the streams were now lifeless. Anglers no longer visited them, for all the fish had died.

In the gutters under the eaves and between the shingles of the roofs, a white granular powder still showed a few patches; some weeks before it had fallen like snow upon the roofs and the lawns, the fields and streams.

No witchcraft, no enemy action had silenced the rebirth of new life in this stricken world. The people had done it themselves.

This town does not actually exist, but it might easily have a thousand counterparts in America or elsewhere in the world. I know of no community that has experienced all the misfortunes I describe. Yet every one of these disasters has actually happened somewhere, and many real communities have already suffered a substantial number of them. A grim specter has crept upon us almost unnoticed, and this imagined tragedy may easily become a stark reality we all shall know....

THE OBLIGATION TO ENDURE

The history of life on earth has been a history of interaction between living things and their surroundings. To a large extent, the physical form and the habits of the earth's vegetation and its animal life have been molded by the environment. Considering the whole span of earthly time, the opposite effect, in which life actually modifies its surroundings, has been relatively slight. Only within the moment of time represented by the present century has one species —man—acquired significant power to alter the nature of his world.

During the past quarter century this power has not only increased to one of disturbing magnitude but it has changed in character. The most alarming of all man's assaults upon the environment is the contamination of air, earth, rivers, and sea with dangerous and even lethal materials. This pollution is for the most part irrecoverable; the chain of evil it initiates not only in the world that must support life but in living tissues is for the most part irreversible. In this now universal contamination of the environment, chemicals are the sinister and little-recognized partners of radiation in changing the very nature of the world—the very nature of its life. Strontium 90, released through nuclear explosions into the air, comes to earth in rain or drifts down as fallout, lodges in soil, enters into the grass or corn or wheat grown there, and in time takes up its abode in the bones of a human being, there to remain until his death. Similarly, chemicals sprayed on croplands or forests or gardens lie long in soil, entering into living organisms, passing from one to another in a chain of poisoning and death. Or they pass mysteriously by underground streams until they emerge and, through the alchemy of air and sunlight, combine into new forms that kill vegetation, sicken cattle, and work unknown harm on those who drink from once pure wells. As Albert Schweitzer has said, "Man can hardly even recognize the devils of his own creation."

It took hundreds of millions of years to produce the life that now inhabits the earth—eons of time in which that developing and evolving and diversifying life reached a state of adjustment and balance with its surroundings. The environment, rigorously shaping and directing the life it supported, contained elements that were hostile as well as supporting. Certain rocks gave out dangerous radiation; even within the light of the sun, from which all life draws its energy, there were short-wave radiations with power to injure. Given time—time not in years but in millennia—life adjusts, and a balance has been reached. For time is the essential ingredient; but in the modern world there is no time.

The rapidity of change and the speed with which new situations are created follow the impetuous and heedless pace of man rather than the deliberate pace of nature. Radiation is no longer merely the background radiation of rocks, the bombardment of cosmic rays, the ultraviolet of the sun that have existed before there was any life on earth; radiation is now the unnatural creation of man's tampering with the atom. The chemicals to which life is asked to make its adjustment are no longer merely the calcium and silica and copper and all the rest of the minerals washed out of the rocks and carried in rivers to the sea; they are the synthetic creations of man's inventive mind, brewed in his laboratories, and having no counterparts in nature.

To adjust to these chemicals would require time on the scale that is nature's; it would require not merely the years of a man's life but the life of generations. And even this, were it by some miracle possible, would be futile, for the new chemicals come from our laboratories in an endless stream; almost five hundred annually find their way into actual use in the United States alone. The figure is staggering and its implications are not easily grasped—500 new chemicals to which the bodies of men and animals are required somehow to adapt each year, chemicals totally outside the limits of biologic experience.

Among them are many that are used in man's war against nature. Since the mid-1940's over 200 basic chemicals have been created for use in killing insects, weeds, rodents, and other organisms described in the modern vernacular as "pests"; and they are sold under several thousand different brand names.

These sprays, dusts, and aerosols are now applied almost universally to farms, gardens, forests, and homes—nonselective chemicals that have the power to kill every insect, the "good" and the "bad," to still the song of birds and the leaping of fish in the streams, to coat the leaves with a deadly film, and to linger on in soil—all this though the intended target may be only a few weeds or insects. Can anyone believe it is possible to lay down such a barrage of poisons on the surface of the earth without making it unfit for all life? They should not be called "insecticides," but "biocides."

The whole process of spraying seems caught up in an endless spiral. Since DDT was released for civilian use, a process of escalation has been going on in which ever more toxic materials must be found. This has happened because insects, in a triumphant vindication of Darwin's principle of the survival of the fittest, have evolved super races immune to the particular insecticide used, hence a deadlier one has always to be developed—and then a deadlier one than that. It has happened also because, for reasons to be described later, destructive insects often undergo a "flareback," or resurgence, after spraying, in numbers greater than before. Thus the chemical war is never won, and all life is caught in its violent crossfire.

Along with the possibility of the extinction of mankind by nuclear war, the central problem of our age has therefore become the contamination of man's total environment with such substances of incredible potential for harm—substances that accumulate in the tissues of plants and animals and even penetrate the germ cells to shatter or alter the very material of heredity upon which the shape of the future depends. . . .

THE OTHER ROAD

We stand now where two roads diverge. But unlike the roads in Robert Frost's familiar poem, they are not equally fair. The road we have long been traveling is deceptively easy, a smooth superhighway on which we progress with great speed, but at its end lies disaster. The other fork of the road—the one "less traveled by"—offers our last, our only chance to reach a destination that assures the preservation of our earth.

The choice, after all, is ours to make. If, having endured much, we have at last asserted our "right to know," and if, knowing, we have concluded that we are being asked to take senseless and frightening risks, then we should no longer accept the counsel of those who tell us that we must fill our world with poisonous chemicals; we should look about and see what other course is open to us.

A truly extraordinary variety of alternatives to the chemical control of insects is available. Some are already in use and have achieved brilliant success. Others are in the stage of laboratory testing. Still others are little more than ideas in the minds of imaginative scientists, waiting for the opportunity to put them to the test. All have this in common: they are *biological* solutions, based on understanding of the living organisms they seek to control, and of the whole fabric of life to which these organisms belong. Specialists representing various areas of the vast field of biology are contributing—entomologists, pathologists, geneticists, physiologists, biochemists, ecologists—all pouring their knowledge and their creative inspirations into the formation of a new science of biotic controls. . . .

Through all these new, imaginative, and creative approaches to the problem of sharing our earth with other creatures there runs a constant theme, the awareness that we are dealing with life—with living populations and all their pressures and counterpressures, their surges and recessions. Only by taking account of such life forces and by cautiously seeking to guide them into channels favorable to ourselves can we hope to achieve a reasonable accommodation between the insect hordes and ourselves.

The current vogue for poisons has failed utterly to take into account these most fundamental considerations. As crude a weapon as the cave man's club, the chemical barrage has been hurled against the fabric of life—a fabric on the one hand delicate and destructible, on the other miraculously tough and resilient, and capable of striking back in unexpected ways. These extraordinary capacities of life have been ignored by the practitioners of chemical control who have brought to their task no "high-minded orientation," no humility before the vast forces with which they tamper.

The "control of nature" is a phrase conceived in arrogance, born of the Neanderthal age of biology and philosophy, when it was supposed that nature exists for the convenience of man. The concepts and practices of applied entomology for the most part date from that Stone Age of science. It is our alarming misfortune that so primitive a science has armed itself with the most modern and terrible weapons, and that in turning them against the insects it has also turned them against the earth.

13.2 ROBERT VAN DEN BOSCH

The Pesticide Conspiracy

The attempts by the supporters of uncontrolled pesticide use to discredit ecologist Rachel Carson, author of *Silent Spring* (Houghton Mifflin, 1962), in which she warns of the dangers of indiscriminate pesticide use, emphasize the fact that although she was a scientist, her training and expertise were not specifically in the field of pest management. No such claim could be made about Robert van den Bosch, a highly accomplished scientist whose research career was devoted to the problem of controlling pest populations. Although van den Bosch's prose lacks the grace and literary merit that helped to make *Silent Spring* a best-seller, his well-documented book *The Pesticide Conspiracy* (Doubleday, 1978) is an even more potent indictment of the pesticide industry. In particular, he exposes the industry's influence in turning the U.S. Department of Agriculture into a willing accomplice in promoting the irresponsible, ecologically disastrous, and ineffective overuse of chemical poisons.

In the following excerpt from *The Pesticide Conspiracy,* van den Bosch not only details the reasons why chemical poisons alone have not, and cannot, control agricultural pests, but he emphasizes that a much more effective alternative exists. This alternative is integrated pest management (IPM), which relies on a wide variety of technologies (including judicious use of appropriate chemicals), based on detailed knowledge of the pest in question. Van den Bosch was one of the developers and early advocates of IPM. The fact that it has taken more than three decades since *Silent Spring* for government agricultural policymakers to embrace IPM can in some part be attributed to the untimely death of van den Bosch shortly after the publication of *The Pesticide Conspiracy.*

Key Concept: integrated pest management

INTRODUCTION

A Can of Worms

In the early summer of 1976, a popular California radio station broadcast to growers an insecticide advertisement prepared for a major chemical company by a New York ad agency. The broadcast warned the growers of the imminent appearance of a "menacing" pest in one of their major crops and advised that as soon as the bugs "first appear" in the fields the growers should start a regular spray program, using, of course, the advertised insecticide. The broadcast also claimed that the material was *the one* insecticide the growers in the area could depend on for effective and economical control of the threatening pest, and further told the growers that through its use they would get a cleaner crop and more profit at harvest time.

The advertisement epitomizes what is wrong with the American way of killing bugs, a practice more often concerned with merchandising gimmickry than it is with applied science. In connection with this gimmickry, much of modern chemical pest control is dishonest, irresponsible, and dangerous. This was true of the radio advertisement just described. It was *dishonest* in its claim that the touted insecticide was *the one* material that growers could depend upon, for in actuality there are several equally effective insecticides and none will assure a cleaner crop and more profit. The advertisement was *irresponsible* in advising growers to initiate a regular spray program upon "first appearance" of the pest. Intensive research has shown that spraying of the crop should be undertaken only when the pest population reaches and maintains a prescribed level during the budding season and that sprays should never be applied on a regular schedule. Finally, the advertisement was *dangerous,* because the advised spraying, if widely adopted by the growers, would have resulted in the senseless dumping of huge amounts of a highly hazardous poison into the environment.

As a veteran researcher in insect control, I have long been disturbed by the dishonest, irresponsible, and dangerous nature of our prevailing chemical control strategy, but I am even more distressed by the knowledge that this simplistic strategy cannot possibly contain the versatile, prolific, and adaptable insects. For a third of a century following the emergence of DDT, we have been locked onto this costly and hazardous insect control strategy, which for biological and ecological reasons, never had a chance to succeed.

What is most disturbing of all is our inability to clean up the mess by shifting to the workable, ecologically based, alternative strategy that modern pest-control specialists term *integrated control* (also termed integrated pest management). Integrated control, as the name implies, is a holistic strategy that utilizes technical information, continuous pest-population monitoring, resource (crop) assessment, control-action criteria, materials, and methods, in concert with natural mortality factors, to manage pest populations in a safe, economical, and effective way. Integrated control is the only strategy that will work effectively against the insects, because it systematically utilizes all possible tactics in such a way that they attain full individual impact, function collectively for maximum mutual effect, and cause minimum detriment to the surrounding environment. In other words, unlike the prevailing chemical control strategy, with its emphasis on product merchandising, integrated control is a technology. It is scientific pest control and, as such, the only way we can hope to gain the upper hand in our battle with the insects. In every respect, integrated control makes sense, and it works. Despite this, our swing to this better pest-management strategy has been painfully slow, and for a clear reason. The impediment has been a powerful coalition of individuals, corporations, and agencies that profit from the prevailing chemical control strategy and brook no interference with the status quo. This power consortium has been unrelenting in its efforts to keep things as they are and as so frequently happens in our society, the games it plays to maintain the status quo are often corruptive, coercive, and sinister.

This book, then, is a tale of a contemporary technology gone sour under the pressures generated by a powerful vested interest. Bugs provide the theme, but politics, deceit, corruption, and treachery are its substance. I feel that the story is a most timely one, for it describes an ecological rip-off and how this atrocity is being perpetuated by tacticians of pure Watergate stripe. The book is largely based on personal experiences and insights gained from more than a quarter century of battling the bugs and their human allies who devised and maintain the inadequate chemical control strategy. It is a tale of personal outrage that I hope proves highly infectious. . . .

INTEGRATED CONTROL— A BETTER WAY TO BATTLE THE BUGS

The 1975 meeting of the Entomological Society of America was the scene of an interesting comparison between the contrasting insect-control strategies of two of the world's great nations, the People's Republic of China and the United States of America. And from what transpired, it appears as though the Chinese pest-control system has more going for it than does ours. I would like to dwell on this matter a bit, for not only does it cast light on the right and wrong ways to combat insects but also because, if we are willing to read the signals honestly, it gives us considerable insight into what is going wrong with the American way of doing things. There may be something of value in such an exercise.

Insect control in China was described, to an audience of two thousand attending the opening plenary session of the Entomological Society, by a panel of America's leading entomologists who earlier in the year had visited China under the China-U.S. cultural exchange. I know most of the panelists, some intimately, and would characterize them largely as politically moderate Middle Americans. In other words, they had no ax to grind on behalf of China and its Marxist political ideology but reported things as they witnessed and recorded them. From what they had to say, it seems that China's entomologists constantly sift the world's literature and other information sources for relevant techniques, methods, and materials, and integrate them along with their own technical developments into a highly effective national integrated pest-management system. Under this system there is continuous monitoring of pest populations, use of action-precipitating pest-population thresholds (economic thresholds), and the implementation of a variety of tactics, including chemical, cultural, and biological controls, as circumstances dictate.

This program is serving China well. For example: using this pest-control system, China grows 39 per cent of the world's rice, which not only feeds her 900 million citizens but enables her to be a major rice exporter. China also utilizes her pest-management system against disease-transmitting and nuisance insects such as mosquitoes and flies. It is interesting that in mosquito control she employs virtually no DDT, apparently relying instead on reduction of mosquito breeding sources, mosquito exclusion tactics, natural controls, and the judicious use of "safe" insecticides. In this latter connection it is especially noteworthy that China, though producing about one hundred insecticides, relies heavily on seven organophosphates because of their limited hazard to warm-blooded animals. And under her insect-control system, she uses these materials judiciously.

Now let's see how we do things in the U.S.A. Two days after the China report, the Entomological Society heard Assistant Agriculture Secretary Robert Long tell us all about it. On this occasion we were a captive audience, since the convention registration fee included the price of a ticket to the Society's annual awards luncheon, before which industry's spokesman Long performed as "distinguished" guest speaker. In reading the fine print of the meeting program I had earlier discovered that Long's visit to New Orleans was arranged at the behest of the agri-chemical industry. And it didn't take long for him to burst into his expected song as he unleashed a vicious attack on industry's great tormenter, the Environmental Protection Agency [EPA]. In his speech, Long first chortled over the recently enacted, politically inspired amendments to the Federal Insecticide, Fungicide, and Rodenticide Act (FIFRA), which give USDA [the U.S. Department of Agriculture] considerable veto power over EPA pesticide decisions. But then he made it abundantly clear that this was not enough. Despite the FIFRA amendments, Long left little doubt that in his mind EPA still had too much control over the registration and regulation of pesticides, particularly as regards EPA's intentions to seek re-registration of America's fourteen hundred pesticide species and their thirty thousand formulations. Here he ran up the alarm pennant by maintaining that EPA's protocols were so deeply mired in bureaucratic stickum that the agri-chemical industry simply would not make the effort to re-register their materials. In other words, he flatly told us that we were about to lose our thirty thousand pesticides, and he painted a terrifying picture of impending starvation, pestilence, and disease in the wake of this loss.

This rhetoric, as it was intended to do, quite probably frightened the naïve in the crowd while bringing joy to the hearts of Long's chemical-company sponsors. Robert Long, a glib spellbinder, well knew that his prediction of an imminent pesticide wipe-out was complete nonsense. Legal road-blocks and political gamesmanship make this a virtual impossibility. What Long was actually telling us was that the U.S. Department of Agriculture, with powerful political backing, intended to hound EPA into loosening its control over pesticide registration and regulation, to the point where the agri-chemical industry would have things just about as they were in the days before passage of the National Environmental Policy Act. The speech was simply a trial run, with Long using the entomologists to perfect the pitch with which he and other USDA brass planned to bushwhack EPA in forthcoming political jousting.

What he and his sponsors hoped to accomplish, then, was an easing of the way for the American agri-chemical industry to unload its fourteen hundred pesticides in their thirty thousand varieties onto the environment, with USDA bulldozing the path. Fortunately, the 1976 presidential election aborted this plan, which, if it had unfolded, would have permitted the interests of the American chemical industry to transcend environmental quality, public health, and the economic well-being of the farmer and consumer. Madison Avenue would have predominated, while scientific pest control would have remained a fuzzy dream in the minds of a few radical researchers.

But let's return to China. How can she feed, and protect from pestilence, 900 million people, with just a handful of insecticides, while we are led to believe that we must have thousands of poisons or otherwise be overwhelmed by an insect avalanche? Is it that we have a vastly more severe pest problem? I hardly think so. Malaria is nowhere endemic in the United States, but it is in China, as are other horrible, insect-borne diseases. Nor do we have 900 million mouths to feed. What, indeed, has happened is that China has used her intelligence to invoke a national *integrated pest-management strategy,* while our strategy is chemical control dominated by the marketing thrust of the agri-chemical industry. Result: pest-control chaos, and if we care to look about us, we will find that similar chaos characterizes many of the other things that we do.

But it isn't too late to change our ways in pest control or, for that matter, in other aspects of applied technology. As I have mentioned several times, it was a mistake to challenge the insects head on with crude chemical weapons. The bugs are too diverse, adaptable, and prolific to be beaten by such a simple strategy. But we were so dazzled by DDT's great killing efficiency and, perhaps, our cleverness in concocting the stuff, that we ignored the possibility of a bug backlash and plunged full blast into the chemical "extermination" campaign. And once we had made our move, we were hooked onto an insecticide treadmill just like an addict on junk.

Now, suddenly, in the midst of the nightmare, when our addiction demands heavier doses and more frequent fixes, the chemicals are hard to get and very expensive. Alarmingly, with famine an increasing global concern, many of the chemical eggs in our bug-control basket are no longer effective. The insects, our great rivals for the earthly bounty, are gearing up to march through our gardens, groves, forests, and fields largely immune to our chemical weapons and freed from natural controls. And in the disease area, too, the breakdown is having a disturbing effect, as malaria makes its dreadful resurgence largely because of mosquito resistance to DDT and other insecticides.

The situation would be much more frightening but for a handful of pest-control radicals who never tumbled to the chemical strategy. These are the renegades who quietly worked away on integrated control programs while most in the pest-control arena were on the chemical kick. Though integrated control is still limited in scope, there are enough programs in operation or under development to offer encouragement that there is indeed a better way to battle the bugs.

What Is Integrated Control?

Integrated control is simply rational pest control: the fitting together of information, decision-making criteria, methods, and materials with naturally occurring pest mortality into effective and redeeming pest-management systems.

Under integrated control, natural enemies, cultural practices, resistant crop and livestock varieties, microbial agents, genetic manipulation, messenger chemicals, and yes, even pesticides become mutually augmentative instead of individually operative or even antagonistic, as is often the case under prevailing practice (e.g., insecticides versus natural enemies). An integrated control program entails six basic elements: (1) man, (2) knowledge/information, (3) monitoring, (4) the setting of action levels, (5) methods, and (6) materials.

Man conceives the program and makes it work. *Knowledge* and *information* are used to develop a system and are vital in its day-to-day operation. *Monitoring* is the continuous assessment of the pest-resource system. *Action levels* are the pest densities at which control methods are invoked. *Methods* are the pathways of action taken to manipulate pest populations. *Materials* are the tools of manipulation.

Sounds like what's going on in China, doesn't it!

Integrated control systems are dynamic, involving continuous information gathering and evaluation, which in turn permit flexibility in decision-making, alteration of the pathways of action, and variation in the agents used. It is the pest-control adviser who gives integrated control its dynamism. By constantly "reading" the situation and invoking tactics and materials as conditions dictate, he acts as a surrogate insecticide, "killing" insects with knowledge and information as well as pesticides, pathogens, parasites, and predators. Integrated control's dynamism is a major factor that sets it off from conventional pest control. Thus, though the latter involves some of the same elements, it lacks dynamism in that it is essentially preprogrammed to the prophylactic or therapeutic use of pesticides. In other words, pesticides dominate the system and constitute its rigid backbone. Where a crop is involved, there is little or no on-going assessment of the crop ecosystem and the dynamic interplay of plant, pests, climate, and natural enemies. This pest-control pattern prevails even in California, our most advanced agro-technology, where over one hundred research entomologists busily at work killing bugs for more than a quarter century have developed fewer than half a dozen valid economic thresholds for the hundreds of pest species. A perusal of the stack of official University of California pest-control recommendations reveals the following kinds of pest-control action criteria:

- when damaging plants
- when present
- when damage occurs
- when they first appear
- when colonies easily found
- when abundant
- when needed
- early season
- when present in large numbers before damage occurs
- anytime when present
- early, mid, and late season
- on small plants as needed
- when present and injuring the plants
- when feeding on the pods
- throughout the season

- when infestation spotty
- when plants are three feet tall.

What this long menu of senseless gobbledygook implies is that in California the insecticide folks have a wide-open field in which to hustle their chemicals, and this they do with greater success than anywhere else in the world.

Under the prevailing chemical control strategy, there is virtually no flexibility in decision-making, particularly as regards alternative pathways of action. The game plan is set at the start and it is stubbornly followed. Result, the familiar case of the fruit grower who year after year automatically sprays his orchard a dozen times or more with the calendar as his main decision-making guide. Or the cotton grower who typically sprays when a chemical-company fieldman drops around and tells him that a few stinkbugs, bollworms, or army worms are showing up in the south forty.

In conventional pest control, one turns on the chemical switch, sits back, and lets the insecticides do the job. It is the lazy man's approach, which characterizes so many aspects of modern life and for which society and the environment pay dearly. A measure of this cost can be gained from a brief analysis of pest control in California.

California's pest control is locked to chemical pesticides. The state is the country's greatest user of these materials, and as stated earlier, receives about 5 per cent of the world's pesticide load. It appears that along with its primacy in smog and earthquakes, California has another distinction: leadership in pesticide pollution. Little wonder! More than fourteen hundred chemical company fieldmen (salesmen) prowl the state, servicing the prevailing pest-control system. They assure a sustained chemical blizzard as well as a fat market for the agri-chemical industry. And at what a cost! Not only does this horde of hustling polluters dump hundreds of tons of unneeded pesticides into the environment, but in the bargain they annually cost California's economy about $50 million to support their huckstering. The chemical companies and many of the major pesticide users (growers, mosquito abaters, forest pest controllers, and pest-control operators) don't pay the bill, they simply pass it on to the consumer, who doubles as taxpayer. But the story doesn't end with money needlessly spent; there are also ecological and social impacts, which add immensely to the cost of the prevailing chemical control strategy.

What I have just described for California pretty much characterizes pest control for the United States in general, and for that matter, other of the world's modern agri-technologies. Chemical pest control, like so many of our modern practices, is a technology gone wild under the merchandising imperative. And as with our other excesses, this rampant technology must be brought under rein if irreparable damage is to be avoided. I am convinced that we pest-control researchers (particularly entomologists) have the capacity to turn things around through integrated control, and perhaps coincidentally establish a model of technological responsibility for other disciplines.

CHAPTER 14 Environmental Carcinogens and Hormone Mimics

14.1 SANDRA STEINGRABER

Living Downstream: An Ecologist Looks at Cancer and the Environment

Biologist Sandra Steingraber is the author of *Post-Diagnosis* (Firebrand Books, 1995), a volume of poetry based on personal experiences during her battle with bladder cancer. That struggle, the subsequent loss of a close, young friend to cancer, and the knowledge that cancer has affected many members of her immediate family, inspired Steingraber to carefully study the notes produced by Rachel Carson during her research for the writing of *Silent Spring* (Houghton Mifflin, 1962). Steingraber then explored the numerous sources of environmental pollution that she was personally exposed to while growing up in central Illinois. Fields were frequently doused with pesticides and industrial plants located on the banks of the Illinois River emitted various toxins into the air and water. This research resulted in the writing of *Living Downstream: An Ecologist Looks at Cancer and the Environment* (Addison-Wesley, 1997), from which the following selection is taken.

In this widely acclaimed book, Steingraber blends poetic anecdotes and vivid descriptions of reckless industrial and agricultural pollution with a wealth of data from scientific and medical literature. The result is a compelling analysis of what is known and unknown about the relationship between environmental factors and cancer. Steingraber bemoans the imbalance between funding devoted to studies of genetic predispositions to cancer and the relative paucity of funding devoted to studies of potential environmental contributions to cancer incidence. She argues persuasively that while we can do little to change our genetic inheritance, there is much that can be done to reduce human exposure to environmental carcinogens.

Living Downstream became the focus of a heated controversy in December 1997. It was revealed that the *New England Journal of Medicine* was guilty of an ethical lapse by publishing a scathing review of the book without identifying the reviewer as a senior official at W. R. Grace & Company. The book contains a detailed discussion of W. R. Grace's role in the pollution of the water in Woburn, Massachusetts; a town where a cluster of fatal childhood leukemia cases has occurred. Steingraber was recently vindicated when she was appointed to the National Action Plan on Breast Cancer by the U.S. Department of Health and Human Services.

Key Concept: the need for further research on environmental factors in cancer incidence

I had bladder cancer as a young adult. If I tell people this fact, they usually shake their heads. If I go on to mention that cancer runs in my family, they usually start to nod. *She is from one of those cancer families,* I can almost hear them thinking. Sometimes, I just leave it at that. But, if I am up for blank stares, I add that I am adopted and go on to describe a study of cancer among adoptees that found correlations within their adoptive families but not within their biological ones. ("Deaths of adoptive parents from cancer before the age of 50 increased the rate of mortality from cancer fivefold among the adoptees.... Deaths of biological parents from cancer had no detectable effect on the rate of mortality from cancer among the adoptees.") At this point, most people become very quiet.

These silences remind me how unfamiliar many of us are with the notion that families share environments as well as chromosomes or with the concept that our genes work in communion with substances streaming in from the larger, ecological world. What runs in families does not necessarily run in blood. And our genes are less an inherited set of teacups enclosed in a cellular china cabinet than they are plates used in a busy diner. Cracks, chips, and scrapes accumulate. Accidents Happen.

My Aunt Jean died of bladder cancer. Raymond and Violet both died of colon cancer. LeRoy is currently under treatment. These are my father's relatives. About Uncle Ray I remember very little, except that he, along with my dad, was one of the less loud of the concrete-pouring, brick-laying Steingraber brothers.

Aunt Jean laughed a lot and once asked me to draw a pig so she could tape it to her refrigerator door. Red-haired Aunt Vi cooked magnificent dinners, was partial to wearing pink, and was married to a man truly untempted by silence. Together, she once remarked, the two of them sure knew how to enjoy themselves. Her widowed husband, my Uncle Ed, is now being treated aggressively for prostate cancer. Nonetheless, at last report, he was busy building a shrine to his wife out in the backyard. When it comes to expressions of grief, my father's side of the family tends toward large-scale construction projects.

The man who was to be my brother-in-law was stricken with intestinal cancer at the age of twenty-one. He cleaned out chemical drums for a living. Three years before Jeff's diagnosis, I was diagnosed with bladder cancer, and three years before my diagnosis, my mother learned she had metastatic breast cancer. That she is still alive today is a topic of considerable wonder among her doctors. Mom is matter-of-fact about this, although she will, if prompted, shyly point out that she has outlived her oncologist and three of her other doctors, two of whom died of cancer.

My mother was first diagnosed in 1974, a year that is considered an anomaly in the annals of breast cancer. Graphs displaying U.S. breast cancer incidence rates across the decades show a gently rising line that suddenly zooms skyward, falls back, then continues its slow ascent. The story behind the blip of '74 has been deemed a textbook lesson in statistical artifacts.

In this year, First Lady Betty Ford and Second Lady Happy Rockefeller both underwent mastectomies. The words *breast cancer* entered public conversation. Women who might otherwise have delayed routine checkups or who were hesitant to seek medical opinion about a lump were propelled into doctors' offices. The result was that a lot of women were diagnosed with breast cancer within a short period of time, my mother among them.

When I, at age fifteen, inquired why my mother was in the hospital, the answer was "Because she has what Mrs. Ford has." When my mother, at age forty-four, questioned whether a radical mastectomy was necessary, she was told, "If it's good enough for Happy, it's good enough for you." . . .

ECOLOGICAL ROOTS

In 1983, I took the train home to Illinois for the holidays—and an appointment at the hospital.

The scheduling of cancer checkups is always an elaborate decision. The calendar date must sound auspicious. Monday or Tuesday appointments are best; otherwise, one risks waiting through the weekend for the results of a laggard lab test or delayed radiology report. It's also best if these appointments fall within a hectic, deadline-filled month so that frenetic activity can preclude fretfulness. During the years I was a graduate student, this meant the ends of semesters, which explains why some half dozen Christmas carols now remind me of outpatient waiting rooms. This particular appointment was destined to turn out fine. What I remember most clearly is my journey there by train.

Something about the landscape changes abruptly between northern and central Illinois. I am not sure what it is exactly, but it happens right around the little towns of Wilmington and Dwight. The horizon recedes, and the sky becomes larger. Distances increase, as though all objects are moving slowly away from each other. Lines become more sharply drawn. These changes always make me restless and, when driving, drive faster. But since I am in a train, I close the book I am reading and begin impatiently straightening the pages of a newspaper strewn over the adjacent seat.

That is when my eye catches the headline of a back-page article: SCIENTISTS IDENTIFY GENE RESPONSIBLE FOR HUMAN BLADDER CANCER. Pulling the newspaper onto my lap, I stare out the window and become very still. It is only early evening, but the fields are already dark, a patchwork of lights quilted over and across them. They have always soothed me. I look for signs of snow. There are none. Finally, I read the article.

Researchers at the Massachusetts Institute of Technology, it seems, had extracted DNA from the cells of a human bladder tumor and used it to transform normal mouse cells into cancerous ones. Through this process, they located the segment of DNA responsible for the transformation. And by comparing this segment to its unmutated form in noncancerous human cells, they were able to pinpoint the exact alteration that had caused a respectable gene to go bad.

In this case, the mutation turned out to be a substitution of one unit of genetic material for another in a single rung of the DNA ladder. Namely, at some point during DNA replication, a double-ringed base called guanine was swapped for the single-ringed thymine. Like a typographical error in which one letter replaces another—*snow* instead of *show, block* instead of *black*—the message sent out by this gene was utterly changed. Instead of instructing the cell to manufacture the amino acid glycine, the altered gene now specified for valine. (Nine years later, other researchers would determine that this substitution alters the structure of proteins involved in signal transduction—the crucial line of communication between the cell membrane and the nucleus that helps coordinate cell division.)

Guanine instead of thymine. Valine instead of glycine. I look away again—this time at my face superimposed over the landscape by the window's mirror. If, in fact, this mutation was involved in my cancer, when did it happen? Where was I? Why had it escaped repair? I had been betrayed. But by what?

Thirteen years later, I possess a bulging file of scientific articles documenting an array of genetic changes involved in bladder cancer. Besides the oncogene just described, two tumor suppressor genes, p15 and p16, have also been discovered to play a role. Their deletion is a common event in transitional cell carcinoma, the kind of cancer I had. Mutations of the famous p53 tumor suppressor gene, with guest-star appearances in so many different cancers, have been detected in more than half of invasive bladder tumors. Also associated with transitional cell carcinomas are surplus numbers of growth factor receptors. Their overexpression has been linked to the kinds of gross genetic injuries that appear near the end of the malignant process.

The nature of the transaction between these various genes and certain bladder carcinogens has likewise been worked out in the years since a newspaper article introduced me to the then new concept of oncogenes. Consider, for example, that redoubtable class of bladder carcinogens called aromatic amines —present as contaminants in cigarette smoke; added to rubber during vulcanization; formulated as dyes for cloth, leather, and paper; used in printing and color photography; and featured in the manufacture of certain pharmaceuticals and pesticides. Aniline, benzidine, naphthylamine, and *o*-toluidine are all members of this group. The first reports of excessive bladder cancers among workers in the aniline dye industry were published in 1895. . . . More than a century later, we now know that anilines and other aromatic amines ply their wickedness by forming DNA adducts in the cells of the tissues lining the bladder, where they arrive as contaminants of urine.

We also now know that aromatic amines are gradually detoxified by the body through a process called acetylation. Like all such processes, it is carried out by a special group of detoxifying enzymes whose actions are controlled and modified by a number of genes. People who are slow acetylators have low levels of these enzymes and are at greater risk of bladder cancer from exposure to aromatic amines. Members of this population can be readily identified because they bear significantly higher burdens of adducts than fast acetylators at the same exposure levels. These genetically suspectible individuals hardly constitute a tiny minority: more than half of Americans and Europeans are estimated to be slow acetylators.

Very likely, I am one. You may be one, too.

We know a lot about bladder cancer. Bladder carcinogens were among the earliest human carcinogens ever identified, and one of the first human oncogenes ever decoded was isolated from some unlucky fellow's bladder tumor. More than most malignancies, bladder cancer has provided researchers with a picture of the sequential genetic changes that unfold from initiation through promotion to progression, from precursor lesions to increasingly more aggressive tumors.

Sadly, all this knowledge about genetic mutations, inherited risk factors, and enzymatic mechanisms has not translated into an effective campaign to prevent the disease. The fact remains that the overall incidence rate of bladder cancer increased 10 percent between 1973 and 1991. Increases are especially dramatic among African Americans: among black men, bladder cancer incidence has risen 28 percent since 1973, and among black women, 34 percent.

Somewhat less than half of all bladder cancers among men and one-third of all cases among women are thought to be attributable to cigarette smoking, which is the single largest known risk factor for this disease. . . . [T]he question still remains: What is causing bladder cancer in the rest of us, the majority of bladder cancer patients for whom tobacco is not a factor?

I also possess another bulging file of scientific articles. These concern the ongoing presence of known and suspected bladder carcinogens in rivers, groundwater, dump sites, and indoor air. For example, industries reporting to the Toxics Release Inventory disclosed environmental releases of the aromatic amine *o*-toluidine that totaled 14,625 pounds in 1992 alone. Detected also in effluent from refineries and other manufacturing plants, *o*-toluidine exists as residues in the dyes of commercial textiles, which may, according to the *Seventh Annual Report on Carcinogens,* expose members of the general public who are consumers of these goods: "The presence of *o*-toluidine, even as a trace contaminant, would be a cause for concern." A 1996 study investigated a sixfold excess of bladder cancer among workers exposed years before to *o*-toluidine and aniline in the rubber chemicals department of a manufacturing plant in upstate New York. Levels of these contaminants are now well within their legal workplace limits, and yet blood and urine collected from current employees were found to contain substantial numbers of DNA adducts and detectable levels of *o*-toluidine and aniline. Another recent investigation revealed an eightfold excess of bladder cancer among workers employed in a Connecticut pharmaceuticals plant that manufactured a variety of aromatic amines. This study was reported as having national implications because the main suspect, dichlorobenzidine, has been widely used throughout the United States.

What my various file folders do *not* contain is a considered evaluation of all known and suspected bladder carcinogens—their sources, their possible interactions with each other, and our various routes of exposure to them. As we have seen, trihalomethanes—those unwanted by-products of water chlorination—have been linked to bladder cancer, as has the dry-cleaning solvent and sometime-contaminant of drinking-water pipes, tetrachloroethylene. I possess individual reports on each of these topics. What I do not have is a comprehensive description of how all these substances behave in combination. What are the risks of multiple trace exposures? What happens when we drink trihalomethanes, absorb aromatic amines, and inhale tetrachloroethylene? Furthermore, what is the ecological fate of these substances once they are released into the environment? What happens when dyed cloth, colored paper, and leather goods are laundered, landfilled, or incinerated? And why—almost a century after some of them were so identified—do powerful bladder carcinogens such as amine dyes continue to be manufactured, imported, used, and released into the environment in the first place? However improved the record of effort to regulate them, why have safer substitutes not replaced them all? These questions remain, to my knowledge, largely unaddressed by the cancer research community.

Several obstacles, I believe, prevent us from addressing cancer's environmental roots. An obsession with genes and heredity is one.

Cancer research currently directs considerable attention to the study of inherited cancers. Most immediately, this approach facilitates the development of genetic testing, which attempts to predict an individual's risk of succumbing to cancer, based on the presence or absence of certain genetic alterations. These efforts may also reveal which genes are common targets of acquired mutation in the general population. (Hereditary mutations are present at the time of conception, and they are carried in the DNA of all body cells; acquired mutations, which accumulate over an individual's lifetime, are passed only to the direct descendents of the cells in which they arise.)

Hereditary cancers, however, are the rare exception. Collectively, fewer than 10 percent of all malignancies are thought to involve inherited mutations. Between 1 and 5 percent of colon cancers, for example, are of the hereditary variety, and only about 15 percent exhibit any sort of familial component. The remaining 85 percent of colon cancers are officially classified as "sporadic," which, confesses one prominent researcher, "is a fancy medical term for 'we don't know what the hell causes it.'" Breast cancer also shows little connection to heredity (probably between 5 and 10 percent). Finding "cancer genes" is not going to prevent the vast majority of cancers that develop.

Moreover, even when rare, inherited mutations play a role in the development of a particular cancer, environmental influences are inescapably involved as well. Genetic risks are not exclusive of environmental risks. Indeed, the direct consequence of some of these damaging mutations is that people become even more sensitive to environmental carcinogens. In the case of hereditary colon cancer, for example, what is passed down the generations is a faulty DNA repair gene. Its human heirs are thereby rendered less capable of coping with environmental assaults on their genes or repairing the spontaneous mistakes that occur during normal cell division. These individuals thus become more likely to accumulate the series of *acquired* mutations needed for the formation of a colon tumor.

Cancer incidence rates are not rising because we are suddenly sprouting new cancer genes. Rare, heritable genes that predispose their hosts to cancer by creating special susceptibilities to the effects of carcinogens have undoubtedly been with us for a long time. The ill effects of some of these genes might well be diminished by lowering the burden of environmental carcinogens to which we are all exposed. In a world free of aromatic amines, for example, being born a slow acetylator would be a trivial issue, not a matter of grave consequence. The inheritance of a defective carcinogen-detoxifying gene would matter less in a culture that did not tolerate carcinogens in air, food, and water. By contrast, we cannot change our ancestors. Shining the spotlight on inheritance focuses us on the one piece of the puzzle we can do absolutely nothing about....

During the last year of her life, Rachel Carson discussed before a U.S. Senate subcommittee her emerging ideas about the relationship between environmental contamination and human rights. The problems addressed in *Silent Spring,* she asserted, were merely one piece of a larger story—namely, the threat to human health created by reckless pollution of the living world. Abetting this hidden menace was a failure to inform common citizens about the senseless and frightening dangers they were being asked, without their consent, to endure. In *Silent Spring,* Carson had predicted that full knowledge of this situation would lead us to reject the counsel of those who claim there is simply no choice but to go on filling the world with poisons. Now she urged recognition of an individual's right to know about poisons introduced into one's environment by others and the right to protection against them. These ideas are Carson's final legacy.

The process of exploration that results from asserting our right to know about carcinogens in our environment is a different journey for every person who undertakes it. For all of us, however, I believe it necessarily entails a three-part inquiry. Like the Dickens character Ebenezer Scrooge, we must first look back into our past, then reassess our present situation, and finally summon the courage to imagine an alternative future.

We begin retrospectively for two reasons. First, we carry in our bodies many carcinogens that are no longer produced and used domestically but which linger in the environment and in human tissue. Appreciating how, even today, we remain in contact with banned chemicals such as PCBs and DDT requires a historical understanding. Second, because cancer is a multicausal disease that unfolds over a period of decades, exposures during young adulthood, adolescence, childhood—and even prior to birth—are relevant to our present cancer risks. We need to find out what pesticides were sprayed in our neighborhoods and what sorts of household chemicals were stored under our parents' kitchen sink. Reminiscing with neighbors, family members, and elders in the community where one grew up can be an eye-opening first step.

This part of the journey is, in essence, a search for our ecological roots. Just as awareness of our genealogical roots offers us a sense of heritage and cultural identity, our ecological roots provide a particular appreciation of who we are biologically. It means asking questions about the physical environment we have grown up within and whose molecules are woven together with the strands of DNA inherited from our genetic ancestors. After all, except for the original blueprint of our chromosomes, all the material that is us—from bone to blood to breast tissue—has come to us from the environment....

In full possession of our ecological roots, we can begin to survey our present situation. This requires a human rights approach. Such an approach recognizes that the current system of regulating the use, release, and disposal of known and suspected carcinogens—rather than preventing their generation in the first place—is intolerable. So is the decision to allow untested chemicals free access to our bodies, until which time they are finally assessed for carcinogenic properties. Both practices show reckless disregard for human life.

A human rights approach would also recognize that we do not all bear equal risks when carcinogens are allowed to circulate within our environment. Workers who manufacture carcinogens are exposed to higher levels, as are those who live near the chemical graveyards that serve as their final resting place. Moreover, people are not uniformly vulnerable to effects of environmental carcinogens. Individuals with genetic predispositions, infants whose detoxifying mechanisms are not yet fully developed, and those with significant prior exposures may all be affected more profoundly. Cancer may be a lottery, but we do not each of us hold equal chances of "winning." When carcinogens are deliberately or accidentally introduced into the environment, some number of vulnerable persons are consigned to death. The impossibility of tabulating an exact body count does not alter this fact. A human rights approach to cancer strives, nonetheless, to make these deaths visible.

Suppose we assume for a moment that the most conservative estimate concerning the proportion of cancer deaths due to environmental causes is absolutely accurate. This estimate, put forth by those who dismiss environmental carcinogens as negligible, is 2 percent. Though others have placed this number far higher, let's assume for the sake of argument that this lowest value is absolutely correct. Two percent means that 10,940 people in the United States die each year from environmentally caused cancers. This is more than the number of women who die each year from hereditary breast cancer—an issue that has launched multi-million-dollar research initiatives. This is more than the number of children and teenagers killed each year by firearms—an issue that is considered a matter of national shame. It is more than three times the number of nonsmokers estimated to die each year of lung cancer caused by exposure to secondhand smoke—a problem so serious it warranted sweeping changes in laws governing air quality in public spaces. It is the annual equivalent of wiping out a small city. It is thirty funerals every day.

None of these 10,940 Americans will die quick, painless deaths. They will be amputated, irradiated, and dosed with chemotherapy. They will expire privately in hospitals and hospices and be buried quietly. Photographs of their bodies will not appear in newspapers. We will not know who most of them are. Their anonymity, however, does not moderate violence. These deaths are a form of homicide.

... [A]ll activities with potential public health consequences should be guided by the *principle of the least toxic alternative,* which presumes that toxic substances will not be used as long as there is another way of accomplishing the task. This means choosing the least harmful way of solving problems—whether it be ridding fields of weeds, school cafeterias of cockroaches, dogs of fleas, woolens of stains, or drinking water of pathogens. Biologist Mary O'Brien advocates a system of alternatives assessment in which facilities regularly evaluate the availability of alternatives to the use and release of toxic chemicals. Any departure from zero should be preceded by a finding of necessity. These efforts, in turn, should be coordinated with active attempts to develop and make available affordable, nontoxic alternatives for currently toxic processes and with systems of support for those making the transition—whether farmer, corner dry-cleaner, hospital, or machine shop. Receiving the highest priority for transformation should be all processes that generate dioxin or require the use or release of any known human carcinogen such as benzene and vinyl chloride.

The principle of the least toxic alternative would move us away from protracted, unwinnable debates over how to quantify the cancer risks from each individual carcinogen released into the environment and where to set legal maximum limits for their presence in air, food, water, workplace, and consumer goods. As O'Brien observed, "Our society proceeds on the assumption that toxic substances *will* be used and the only question is how much.

Under the current system, toxic chemicals are used, discharged, incinerated, and buried without ever requiring a finding that these activities are necessary." The principle of the least toxic alternative looks toward the day when the availability of safer choices makes the deliberate and routine release of chemical carcinogens into the environment as unthinkable as the practice of slavery.

14.2 THEO COLBORN, DIANNE DUMANOSKI, AND JOHN PETERSON MYERS

Our Stolen Future

Until recently most efforts to assess the potential toxicity of synthetic industrial chemicals focused almost exclusively on their role as carcinogens. This was because of legitimate public concern about rising cancer rates and the belief among scientists that cancer was most likely caused by exposure to low levels of synthetic chemicals. Although some scientists urged public health officials to give serious consideration to other possible health effects of environmental contaminants, they were generally ignored. This was due to limited funding for research and the common belief that the toxic effects of synthetic chemicals required larger exposure than what the public ordinarily experienced.

In the late 1980s Theo Colborn, a research scientist for the World Wildlife Fund and who was then working on a study of pollution in the Great Lakes, began the process of linking together the results of a growing series of isolated studies. Researchers in the Great Lakes region, as well as in Florida, on the West Coast of the United States, and in Northern Europe, had observed widespread evidence of serious and frequently lethal physiological problems. These problems included abnormal reproductive development, unusual sexual behavior, and neurological impairment, and were exhibited by a diverse group of animal species. As a result of Colborn's insights, communications among researchers, and further studies, a hypothesis was developed that all of these wildlife problems were manifestations of abnormal estrogenic activity. The causative agents were identified as more than 50 synthetic chemical compounds that have been shown in laboratory studies to either mimic the action, or disrupt the normal function, of the powerful hormones responsible for sexual development and many other biological functions.

The following selection is taken from *Our Stolen Future* (Dutton, 1996) in which Colborn and her coauthors, journalist Dianne Dumanoski and W. Alton Jones Foundation director John Peterson Myers, summarize their research on the effects of exposure to environmental hormone mimics. The title of the book reflects their warning that unless action is taken to reduce such exposure, the fertility, intelligence, and even the survival of future generations may be threatened. In fact, such action has begun. In response to the mounting evidence that environmental estrogens may be a serious health threat, the United States Congress passed legislation requiring that all pesticides be screened for estrogenic activity and that the Environmental Protection Agency develop procedures for detecting environmental estrogenic contaminants in drinking water supplies. Government sponsored studies of synthetic endocrine disruptors and other hormone mimics are being conducted in the United Kingdom and Germany. Recently, the European Chemical Industry Council allocated $7 million for research on the possible endocrine-disrupting effects of chemicals.

Key Concept: disruptive developmental effects of environmental hormone mimics

ALTERED DESTINIES

"Our fate is connected with the animals," Rachel Carson wrote than three decades ago in *Silent Spring,* a now classic indictment of synthetic pesticides and human hubris that helped launch the modern environmental movement. This has long been a guiding belief among environmentalists, wildlife biologists, and others who recognize two fundamental realities—our shared evolutionary inheritance and our shared environment. What is happening to the animals in Florida, English rivers, the Baltic, the high Arctic, the Great Lakes, and Lake Baikal in Siberia has immediate relevance to humans. The damage seen in lab animals and in wildlife has ominously foreshadowed symptoms that appear to be increasing in the human population.

... [B]asic physiological processes such as those governed by the endocrine system have persisted relatively unchanged through hundreds of millions of years of evolution. Evolutionary narratives tend to highlight the innovations of natural selection, ignoring the stubborn conservative streak that has marked the history of life on Earth. At the same time that evolution experimented greatly with form, shaping the vessels in various and wondrous ways, it has strayed surprisingly little from an ancient recipe for life's biochemical brew. In examining our place in the evolutionary lineage, humans tend to focus inordinately on those characteristics that make us unique. But these differences are small, indeed, when compared to how much we share not only with other primates such as chimpanzees and gorillas, but with mice, alligators, turtles, and other vertebrates. Though turtles and humans bear little physical similarity, our kinship is unmistakable. The estrogen circulating in the painted turtle seen basking on logs during lazy summer afternoons is exactly the same as the estrogen rushing through the human bloodstream.

Humans and animals share a common environment as well as a common evolutionary legacy. Living in a man-made landscape, we easily forget that our well-being is rooted in natural systems. Yet all human enterprise rests on the foundation of natural systems that provide a myriad of invisible life-support services. Our connections to these natural systems may be less direct and obvious than those of an eagle or an otter, but we are no less deeply implicated in life's web. No one has stated this fundamental ecological principle more simply than the early twentieth-century American environmental philosopher, John Muir. "When we try to pick out anything by itself, we find that it is bound by a thousand invisible cords to everything in the universe."

Our regrettable experience with persistent chemicals over the past half century has demonstrated the reality of this deep and complex interconnection.

Whether we live in Tokyo, New York, or a remote Inuit village in the Arctic thousands of miles from farm fields or sources of industrial pollution, all of us have accumulated a store of persistent synthetic chemicals in our body fat. Through this web of inescapable connection, these chemicals have found their way to each and every one of us just as they have found their way to the birds, seals, alligators, panthers, whales, and polar bears. With this shared biology and shared contamination, there is little reason to expect that humans will in the long term have a separate fate.

Yet, some skeptics question whether animal studies provide a useful tool for forecasting threats to humans. The refrain that "mice are not little people" has been heard frequently in the ongoing debate about whether animal testing accurately predicts whether a chemical poses a cancer risk to humans. Critics have also attacked testing procedures for using unrealistically high doses, arguing, for example, that mice tested to discover if DDT caused cancer were fed more than eight hundred times the average amount humans would take in eating a typical diet.

Whatever the merits of these criticisms regarding cancer testing, they have little relevance to the use of animals to predict the effects of hormone-disrupting chemicals. Because scientists have only an incomplete understanding of the basic mechanisms that induce cancer, extrapolating from one species to another has admitted uncertainties. In contrast, scientists have a good grasp of the mechanisms and actions of hormones. They understand how chemical messages are sent and received and how some synthetic chemicals disrupt this communication. They know that hormones guide development in basically the same way in all mammals, and if there was any doubt, the DES [the synthetic hormone drug diethylstilbestrol] experience has verified the similarity of disruption across many species, including humans. Time after time, the abnormalities first seen in laboratory experiments with DES later showed up in the children of women who had taken this drug during pregnancy.

The relevance of the DES experience to the threat from environmental endocrine disruptors has also been questioned because of the very high doses given to pregnant women and to laboratory animals in experiments. While most of the early experiments did indeed use high doses, recent studies using much lower doses have produced no less alarming results. In fact, in some cases a high dose may paradoxically cause *less* damage than a lower dose. In exploring the effects of much lower doses of DES, Fred vom Saal has found that the response increases with dose for a time and then, with even higher doses, begins to diminish.

Vom Saal's dose response curve looks like an inverted U. Its shape is profoundly important to the interaction between the endocrine system and synthetic contaminants. Neither linear nor always moving in the same direction, the inverted U seems characteristic of hormone systems and it means that they do not conform to the assumptions that underlie classical toxicology—that a biological response always increases with dose. It means that testing with very high doses will miss some effects that would show up if the animals were given lower doses. The inverted U is another example of how the action of endocrine disruptors challenges prevailing notions about toxic chemicals. Extrapolation from high-dose tests to lower doses may in some cases seriously underestimate risks rather than exaggerate them.

Because the endocrine disruption question has surfaced so recently, the scientific case on the extent of the threat is still far from complete. Nevertheless, if one looks broadly at a wide array of existing studies from various branches of science and medicine, the weight of the evidence indicates that humans are in jeopardy and are perhaps already affected in major ways. Taken together, the pieces of this scientific patchwork quilt have, despite admitted gaps, a cumulative power that is compelling and urgent.

This was the lesson from the historic meeting on endocrine disruption that took place in July 1991 at the Wingspread Conference Center in Racine, Wisconsin. Over the years, dozens of scientists have explored isolated pieces of the puzzle of hormone disruption, but the larger picture did not emerge until Theo Colborn and Pete Myers finally brought twenty-one of the key researchers together. At this unique gathering, specialists from diverse disciplines ranging from anthropology to zoology shared what they knew about the role of hormones in normal development and about the devastating impacts of hormone-disrupting chemicals on wildlife, laboratory animals, and humans. For the first time, Ana Soto, Frederick vom Saal, Michael Fry, Howard Bern, John McLachlan, Earl Gray, Richard Peterson, Peter Reijnders, Pat Whitten, Melissa Hines, and others explored the exciting connections between their work and the ominous implications that arose from this exercise. As the evidence was laid out, the parallels proved remarkable and deeply disturbing. The conclusion seemed inescapable: the hormone disruptors threatening the survival of animal populations are also jeopardizing the human future.

At the end of the session, the scientists issued the Wingspread Statement, an urgent warning that humans in many parts of the world are being exposed to chemicals that have disrupted development in wildlife and laboratory animals, and that unless these chemicals are controlled, we face the danger of widespread disruption in human embryonic development and the prospect of damage that will last a lifetime.

The pressing question is whether humans are already suffering damage from half a century of exposure to endocrine-disrupting synthetic chemicals. Have these man-made chemicals already altered individual destinies by scrambling the chemical messages that guide development?

Many of those familiar with the scientific case believe the answer is yes. Given human exposure to dioxinlike chemicals, for example, it is probable that some humans, especially the most sensitive individuals, are suffering some effects. But whether hormone-disrupting chemicals are now having a broad impact across the human population is difficult to assess and even harder to prove. This is inescapable in light of the nature of the contamination, the transgenerational effects, the often long lag time before damage becomes evident, and the invisible nature of much of this damage. Those trying to document whether perceived increases in specific problems reflect genuine trends in human health find themselves thwarted by a dearth of reliable medical data. Few disease registries exist for anything except cancers. A number of pediatricians from various parts of the United States have expressed their concern about an increasing frequency of genital abnormalities in children such as undescended testicles, extremely small penises, and hypospadias, a defect in which the urethra that carries urine does not extend to the end of the penis, but it is virtually impossible to document these anecdotal reports. Unfortunately, the problems caused by endocrine disruption may have to reach crisis proportion before we have a clear sign that something serious is happening.

In the face of these difficulties, the animal studies provide a touchstone for identifying and investigating what might be happening in humans. They can alert us to the probable kinds of disruptions and help focus research efforts. They can also provide early warnings about the hazards of current levels of contamination. Because of the diversity of life, some animals are bound to be more readily exposed to contaminants than humans. Transgenerational effects, such as changes in behavior and diminished fertility, are also likely to show up faster in wildlife because most animals mature and reproduce more quickly than humans. Experimental work with animals adds another equally invaluable dimension. As the history of DES demonstrates, laboratory experiments with rats and mice accurately forecasted damage that later showed up in humans. The tragedy is that we ignored the warnings.

PART SIX

Environment and Society

On the Internet . . .

Sites appropriate to Part Six

This is the home page of the EcoJustice Network, which is an organization devoted to promoting environmental equity and justice.

`http://www.igc.apc.org/envjustice/`

National Councils for Sustainable Development is a program of the United Nations Earth Council and is devoted to exploring the implementation of sustainable development initiatives. This is the home page of the council.

`http://www.ncsdnetwork.org`

The Union of Concerned Scientists (UCS) describes itself as "citizens and scientists working together for a common goal: a healthy environment and a safe world, for today and for the next century." This Web site contains information about UCS and its goals.

`http://www.ucsusa.org`

The Worldwatch Institute is a nonprofit public policy research organization dedicated to informing policymakers and the public about emerging global problems and the complex links between the world economy and its environmental support systems. This is the home page of the World Watch Institute.

`http://www.worldwatch.org`

CHAPTER 15 Political and Economic Issues

15.1 MARK SAGOFF

At the Shrine of Our Lady of Fatima or *Why Political Questions Are Not All Economic*

The fact that at a local, national, or global level, insufficient public and private resources exist to address every conceivable environmental problem is not seriously contested by anyone. For this reason alone it is necessary to establish a system of priorities to guide those responsible for setting ecological and environmental protection policies. How to achieve this task is not obvious. Who should decide? On what basis should the decisions be made? Numerous surveys have shown that the ranking of the significance of environmental threats—such as hazardous wastes, air pollutants, pesticides, and ozone-depleting chemicals—by the general public and by environmental scientists result in vary different sets of priorities.

Mark Sagoff is director of the Institute for Philosophy and Public Policy at the University of Maryland. He has published numerous articles describing his views on the proper place of economics and risk-benefit analyses in decision making about environmental issues in philosophy of law and environmental journals. He is also the author of *The Economy of the Earth: Philosophy, Law and the Environment* (Cambridge University Press, 1988).

As indicated in the following selection from "At the Shrine of Our Lady of Fatima *or* Why Political Questions Are Not All Economic," *Arizona Law Review* (vol. 23. 1981), Sagoff believes that environmental problems are complex and require more comprehensive assessment than can be made on the basis of economic analyses alone. He maintains that it is impossible to place economic values on aesthetic satisfaction or physical well-being. More profoundly, he challenges the assumption that the true value of all things can be judged by what people are willing to pay for them in the marketplace.

Key Concept: the limitations of economics in making environmental policy

Lewiston, New York, a well-to-do community near Buffalo, is the site of the Lake Ontario Ordinance Works, where years ago the federal government disposed of the residues of the Manhattan Project. These radioactive wastes are buried but are not forgotten by the residents who say that when the wind is southerly, radon gas blows through the town. Several parents at a recent Lewiston conference I attended described their terror on learning that cases of leukemia had been found among area children. They feared for their own lives as well. On the otherside of the table, officials from New York State and from local corporations replied that these fears were ungrounded. People who smoke, they said, take greater risks than people who live close to waste disposal sites. One speaker talked in terms of "rational methodologies of decisionmaking." This aggravated the parents' rage and frustration.

The speaker suggested that the townspeople, were they to make their decision in a free market and if they knew the scientific facts, would choose to live near the hazardous waste facility. He told me later they were irrational —"neurotic"—because they refused to recognize or to act upon their own interests. The residents of Lewiston were unimpressed with his analysis of their "willingness to pay" to avoid this risk or that. They did not see what risk-benefit analysis had to do with the issues they raised.

If you take the Military Highway (as I did) from Buffalo to Lewiston, you will pass through a formidable wasteland. Landfills stretch in all directions and enormous trucks—tiny in that landscape—incessantly deposit sludge which great bulldozers then push into the ground. These machines are the only signs of life, for in the miasma that hangs in the air, no birds, not even scavengers, are seen. Along colossal power lines which criss-cross this dismal land, the dynamos at Niagra send electric power south, where factories have fled, leaving their remains to decay. To drive along this road is to feel, oddly, the mystery and awe one experiences in the presence of so much power and decadence....

At the Shrine of Our Lady of Fatima, on a plateau north of the Military Highway, a larger than life sculpture of Mary looks into the chemical air. The original of this shrine stands in central Portugal where in May, 1917, three children said they saw a Lady, brighter than the sun, raised on a cloud in an evergreen tree. Five months later, on a wet and chilly October day, the Lady again appeared, this time before a large crowd. Some who were skeptical did not see the miracle. Others in the crowd reported, however, that "the sun appeared and seemed to tremble, rotate violently and fall, dancing over the heads of the throng. . . ."

The Shrine was empty when I visited it. The cult of Our Lady of Fatima, I imagine, has only a few devotees. The cult of Pareto Optimality, however, has many. Where some people see only environmental devastation, its devotees perceive efficiency, utility, and the maximization of wealth. They see the satisfaction of wants. They envision the good life. As I looked over the smudged and ruined terrain I tried to share that vision. I hoped that Our Lady of Fatima, worker of miracles, might serve, at least for the moment, as the Patroness of cost-benefit analysis. I thought of all the wants and needs that are satisfied in a landscape of honeymoon cottages, commercial strips, and dumps for hazardous waste. I saw the miracle of efficiency. The prospect, however, looked only darker in that light.

POLITICAL AND ECONOMIC DECISIONMAKING

This essay concerns the economic decisions we make about the environment. It also concerns our political decisions about the environment. Some people have suggested that ideally these should be the same, that all environmental problems are problems in distribution. According to this view, there is an environmental problem only when some resource is not allocated in equitable and efficient ways.

This approach to environmental policy is pitched entirely at the level of the consumer. It is his or her values that count, and the measure of these values is the individual's willingness to pay. The problem of justice or fairness in society becomes, then, the problem of distributing goods and services so that more people get more of what they want to buy: a condo on the beach, a snowmobile for the mountains, a tank full of gas, a day of labor. The only values we have, according to this view, are those that a market can price.

How much do you value open space, a stand of trees, an "unspoiled" landscape? Fifty dollars? A hundred? A thousand? This is one way to measure value. You could compare the amount consumers would pay for a townhouse or coal or a landfill to the amount they would pay to preserve an area in its "natural" state. If users would pay more for the land with the house, the coal mine, or the landfill, then without—less construction and other costs of development—than the efficient thing to do is to improve the land and thus increase its value. That is why we have so many tract developments, pizza stands, and gas stations. How much did you spend last year to preserve open space? How much for pizza and gas? "In principle, the ultimate measure of environmental quality," as one basic text assures us, "is the value people place on these . . . services or their *willingness to pay*."[1]

Willingness to pay: what is wrong with that? The rub is this: not all of us think of ourselves simply as *consumers*. Many of us regard ourselves *as citizens* as well. We act as consumers to get what we want *for ourselves.* We act as citizens to achieve what we think is right or best *for the community.* The question arises, then, whether what we want for ourselves individually as consumers is consistent with the goals we would set for ourselves collectively as citizens. Would I vote for the sort of things I shop for? Are my preferences as a consumer consistent with my judgments as a citizen?

They are not. I am schizophrenic. Last year, I fixed a couple of tickets and was happy to do so since I saved fifty dollars. Yet, at election time, I helped to vote the corrupt judge out of office. I speed on the highway; yet I want the police to enforce laws against speeding. I used to buy mixers in returnable bottles—but who can bother to return them? I buy only disposables now, but to soothe my conscience, I urge my state senator to outlaw one-way containers. I love my car; I hate the bus. Yet I vote for candidates who promise to tax gasoline to pay for public transportation. And of course I applaud the Endangered Species Act, although I have no earthly use for the Colorado squawfish or the Indiana bat. I support almost any political cause that I think will defeat my consumer interests. This is because I have contempt for—although I act upon—those interests. I have an "Ecology Now" sticker on a car that leaks oil everywhere it's parked.

The distinction between consumer and citizen preferences has long vexed the theory of public finance. Should the public economy serve the same goals as the household economy? May it serve, instead, goals emerging from our association as citizens? The question asks if we may collectively strive for and achieve only those items we individually compete for and consume. Should we aspire, instead, to public goals we may legislate as a nation? . . .

COST-BENEFIT ANALYSIS VS. REGULATION

On February 19, 1981, President Reagan published Executive Order 12,291 requiring all administrative agencies and departments to support every new major regulation with a cost-benefit analysis establishing that the benefits of the regulation to society outweigh its costs. The order directs the Office of Management and Budget (OMB) to review every such regulation on the basis of the adequacy of the cost-benefit analysis supporting it. This is a departure from tradition. Historically, regulations have been reviewed not by OMB but by the courts on the basis of the relation of the regulation to authorizing legislation, not to cost-benefit analysis.

A month earlier, in January, 1981, the Supreme Court heard lawyers for the American Textile Manufacturers Institute argue against a proposed Occupational Safety and Health Administration (OSHA) regulation which would have severely restricted the acceptable levels of cotton dust in textile plants. The lawyers for industry argued that the benefits of the regulation would not equal the costs. The lawyers for the government contended that the law required the tough standard. OSHA, acting consistently with Executive Order 12,291, asked the Court not to decide the cotton dust case in order to give the agency time to complete the cost-benefit analysis required by the textile industry. The Court declined to accept OSHA's request and handed down its opinion in *American Textile Manufacturers v Donovan* on June 17, 1981.

The Supreme Court, in a 5–3 decision, found that the actions of regulatory agencies which conform to the OSHA law need not be supported by cost-benefit analysis. In addition, the Court asserted that Congress, in writing a statute, rather than the agencies in applying it, has the primary responsibility for balancing benefits and costs. The Court said:

> When Congress passed the Occupational Health and Safety Act in 1970, it chose to place pre-eminent value on assuring employees a safe and healthful working environment, limited only by the feasibility of achieving such an environment. We must measure the validity of the Secretary's actions against the requirements of that Act.

The opinion upheld the finding of the District of Columbia Court of Appeals that "Congress itself struck the balance between costs and benefits in the mandate to the agency."

The Appeals Court opinion in *American Textile Manufacturers v. Donovan* supports the principle that legislatures are not necessarily bound to a particular conception of regulatory policy. Agencies that apply the law therefore may not need to justify on cost-benefit grounds the standards they set. These standards may conflict with the goal of efficiency and still express our political will as a nation. That is, they may reflect not the personal choices of self-interested individuals, but the collective judgments we make on historical, cultural, aesthetic, moral, and ideological grounds.

The appeal of the Reagan Administration to cost-benefit analysis, however, may arise more from political than economic considerations. The intention seen in the most favorable light, may not be to replace political or ideological goals with economic ones, but to make economic goals more apparent in regulation. This is not to say that Congress should function to reveal a collective willingness-to-pay just as markets reveal an individual willingness-to-pay. It is to suggest that Congress should do more to balance economic with ideological, aesthetic, and moral goals. To think that environmental or worker safety policy can be based exclusively on aspiration for a "natural" and "safe" world is as foolish as to hold that environmental law can be reduced to cost-benefit accounting. The more we move to one extreme, as I found in Lewiston, the more likely we are to hear from the other.

SUBSTITUTING EFFICIENCY FOR SAFETY

The labor unions won an important political victory when Congress passed the Occupational Safety and Health Act of 1970. That Act, among other things, severely restricts worker exposure to toxic substances. It instructs the Secretary of Labor to set "the standard which most adequately assures, to the extent feasible . . . that no employee will suffer material impairment of health or functional capacity even if such employee has regular exposure to the hazard... for the period of his working life."

Pursuant to this law, the Secretary of Labor in 1977 reduced from ten to one part per million (ppm) the permissable ambient exposure level for benzene, a carcinongen for which no safe threshold is known. The American Petroleum Institute thereupon challenged the new standard in court. It argued, with much evidence in its favor, that the benefits (to workers) of the one ppm standard did not equal the costs (to industry). The standard therefore did not appear to be a rational response to a market failure in that it did not strike an efficient balance between the interests of workers in safety and the interests of industry and consumers in keeping prices down.

The Secretary of Labor defended the tough safety standard on the ground that the law demanded it. An efficient standard might have required safety until it cost industry more to prevent a risk than it cost workers to accept it. Had Congress adopted this vision of public policy—one which can be found in many economics texts—it would have treated workers not as ends-in-themselves but as means for the production of overall utility. This, as the Secretary saw it, was what Congress refused to do.

The United States Court of Appeals for the Fifth Circuit agreed with the American Petroleum Institute and invalidated the one ppm benzene standard. On July 2, 1980, the Supreme Court affirmed the decision in *American Petroleum Institute v. Marshall* and remanded the benzene standard back to OSHA for revision. The narrowly based Supreme Court decision was divided over the role economic considerations should play in judicial review. Justice Marshall, joined in dissent by three other justices, argued that the Court had undone on the basis of its own theory of regulatory policy an act of Congress inconsistent with that theory. He concluded that the plurality decision of the Court "requires the American worker to return to the political arena to win a victory that he won before in 1970."

The decision of the Supreme Court is important not because of its consequences, which are likely to be minimal, but because of the fascinating questions it raises. Shall the courts uphold only those political decisions that can be defended on economic grounds? Shall we allow democracy only to the extent that it can be construed either as a rational response to a market failure or as an attempt to redistribute wealth? Should the courts say that a regulation is not "feasible" or "reasonable"—terms that occur in the OSHA law—unless it is supported by a cost-benefit analysis?

The problem is this: An efficiency criterion, as it is used to evaluate public policy, assumes that the goals of our society are contained in the preferences individuals reveal or would reveal in markets. Such an approach may appear attractive, even just, because it treats everyone as equal, at least theoretically, by according to each person's preferences the same respect and concern. To treat a person with respect, however, is also to listen and to respond intelligently to his or her views and opinions. This is not the same thing as to ask how much he or she is willing to pay for them. The cost-benefit analyst does not ask economists how much they are willing to pay for what they believe, that is, that the workplace and the environment should be made efficient. Why, then, does the analyst ask workers, environmentalists, and others how much they are willing to pay for what they believe is right? Are economists the only ones who can back their ideas with reasons while the rest of us can only pay a price? The cost-benefit approach treats people as of equal worth because it treats them as of no worth, but only as places or channels at which willingness to pay is found.

LIBERTY: ANCIENT AND MODERN

When efficiency is the criterion of public safety and health, one tends to conceive of social relations on the model of a market, ignoring competing visions of what we as a society should be like. Yet it is obvious that there are competing conceptions of what we should be as a society. There are some who believe on principle that worker safety and environmental quality ought to be protected only insofar as the benefits of protection balance the costs. On the other hand, people argue—also on principle—that neither worker safety nor environmental quality should be treated merely as a commodity to be traded at the margin for other commodities, but rather each should be valued for its own sake. The conflict between these two principles is logical or moral, to be resolved by argument or debate. The question whether cost-benefit analysis should play a decisive role in policy-making is not to be decided by cost-benefit analysis. A contradiction between principles—between contending visions of the good society—cannot be settled by asking how much partisans are willing to pay for their beliefs.

The role of the *legislator,* the political role, may be more important to the individual than the role of *consumer.* The person, in other words, is not to be treated merely as a bundle of preferences to be juggled in cost-benefit analyses. The individual is to be respected as an advocate of ideas which are to be judged according to the reasons for them. If health and environmental statutes reflect a vision of society as something other than a market by requiring protections beyond what are efficient, then this may express not legislative ineptitude but legislative responsiveness to public values. To deny this vision because it is economically inefficient is simply to replace it with another vision. It is to insist that the ideas of the citizen be sacrificed to the psychology of the consumer.

We hear on all sides that government is routinized, mechanical, entranched, and bureaucratized; the jargon alone is enough to dissuade the most mettlesome meddler. Who can make a difference? It is plain that for many of us the idea of a national political community has an abstract and suppositious quality. We have only our private conceptions of the good, if no way exists to arrive at a public one. This is only to note the continuation, in our time, of the trend Benjamin Constant described in the essay *De La Liberte des Anciens Comparee a Celle des Modernes.* Constant observes that the modern world, as opposed to the ancient, emphasizes civil over political liberties, the rights of privacy and property over those of community and participation. "Lost in the multitude," Constant writes, "the individual rarely perceives the influence that he exercises," and, therefore, must be content with "the peaceful enjoyment of private independence."[2] The individual asks only to be protected by laws common to all in his pursuit of his own self-interest. The citizen has been replaced by the consumer; the tradition of Rousseau has been supplanted by that of Locke and Mill.

Nowhere are the rights of the moderns, particularly the rights of privacy and property, less helpful than in the area of the natural environment. Here the values we wish to protect—cultural, historical, aesthetic, and moral—are public values. They depend not so much upon what each person wants individually as upon what he or she thinks is right for the community. We refuse to regard worker health and safety as commodities; we regulate hazards as a matter of right. Likewise, we refuse to treat environmental resources simply as public goods in the economist's sense. Instead, we prevent significant deterioration of air quality not only as a matter of individual self-interest but also as a matter of collective self-respect. How shall we balance efficiency against moral, cultural, and aesthetic values in policy for the workplace and the environment? No better way has been devised to do this than by legislative debate ending in a vote. This is very different from a cost-benefit analysis terminating in a bottom line.

VALUES ARE NOT SUBJECTIVE

It is the characteristic of cost-benefit analysis that it treats all value judgments other than those made on its behalf as nothing but statements of preference, attitude, or emotion, insofar as they are value judgments. The cost-benefit analyst regards as true the judgment that we should maximize efficiency or wealth. The analyst believes that this view can be backed by reasons, but does not regard it as a preference or want for which he or she must be willing to pay. The cost-benefit analyst tends to treat all other normative views and recommendations as if they were nothing but subjective reports of mental states. The analyst supposes in all such cases that "this is right" and "this is what we ought to do" are equivalent to "I want this" and "this is what I prefer." Value judgments are beyond criticism if, indeed, they are nothing but expressions of personal preference; they are incorrigible since every person is in the best position to know what he or she wants. All valuation, according to this approach, happens *in foro interno;* debate *in foro publico* has no point. With this approach, the reasons that people give for their views do not count; what does count is how much they are willing to pay to satisfy their wants. Those who are willing to pay the most, for all intents and purposes, have the right view; theirs is the more informed opinion, the better aesthetic judgment, and the deeper moral insight. . . .

Economists . . . argue that their role as policy-makers is legitimate because they are neutral among competing values in the client society. The political economist, according to James Buchanan, "is or should be ethically neutral: the indicated results are influenced by his own value scale only insofar as this reflects his membership in a larger group." The economist might be most confident of the impartiality of his or her policy recommendations if he or she could derive them formally or mathematically from individual preferences. If theoretical difficulties make such a social welfare function impossible, however, the next best thing, to preserve neutrality, is to let markets function to transform individual preference orderings into a collective ordering of social states. The analyst is able then to base policy on preferences that exist in society and are not necessarily his own.

Economists have used this impartial approach to offer solutions to many significant social problems, for example, the controversy over abortion. An economist argues that "there is an optimal number of abortions, just as there is an optimal level of pollution, or purity.... Those who oppose abortion could eliminate it entirely, if their intensity of feeling were so strong as to lead to payments that were greater at the margin than the price anyone would pay to have an abortion."[3] Likewise, economists, in order to determine whether the war in Vietnam was justified, have estimated the willingness to pay of those who demonstrated against it. Following the same line of reasoning, it should be possible to decide whether Creationism should be taught in the public schools, whether black and white people should be segregated, whether the death penalty should be enforced, and whether the square root of six is three. All of these questions arguably depend upon how much people are willing to pay for their subjective preferences or wants. This is the beauty of cost-benefit analysis: no matter how relevant or irrelevant, wise or stupid, informed or uninformed, responsible or silly, defensible or indefensible wants may be, the analyst is able to derive a policy from them—a policy which is legitimate because, in theory, it treats all of these preferences as equally valid and good.

PREFERENCE OR PRINCIPLE?

In contrast, consider a Kantian conception of value. The individual, for Kant, is a judge of values, not a mere haver of wants, and the individual judges not for himself or herself merely, but as a member of a relevant community or group. The central idea in a Kantian approach to ethics is that some values are more reasonable than others and therefore have a better claim upon the assent of members of the community as such. The world of obligation, like the world of mathematics or the world of empirical fact, is public not private, and objective standards of argument and criticism apply. Kant recognized that values, like beliefs, are subjective states of mind which have an objective content as well. Therefore, both values and beliefs are either correct or mistaken. A value judgment is like an empirical or theoretical judgment in that it claims to be *true* not merely to be *felt*.

We have, then, two approaches to public policy before us. The first, the approach associated with normative versions of welfare economics, asserts that the only policy recommendation that can or need be defended on objective grounds is efficiency or wealth-maximization. The Kantian approach, on the other hand, assumes that many policy recommendations may be justified or refuted on objective grounds. It would concede that the approach of welfare economics applies adequately to some questions, for example, those which ordinary consumer markets typically settle. How many yo-yos should be produced as compared to how many frisbees? Shall pens have black ink or blue? Matters such as these are so trivial it is plain that markets should handle them.

It does not follow, however, that we should adopt a market or quasi-market approach to every public question.

A market or quasi-market approach to arithmetic, for example, is plainly inadequate. No matter how much people are willing to pay, three will never be the square root of six. Similarly, segregation is a national curse and the fact that we are willing to pay for it does not make it better, but only us worse. The case for abortion must stand on the merits; it cannot be priced at the margin. Our failures to make the right decisions in these matters are failures in arithmetic, failures in wisdom, failures in taste, failures in morality—but not market failures. There are no relevant markets which have failed.

What separates these questions from those for which markets are appropriate is that they involve matters of knowledge, wisdom, morality, and taste that admit of better or worse, right or wrong, true or false, and not mere economic optimality. Surely environmental questions—the protection of wilderness, habitats, water, land, and air as well as policy toward environmental safety and health—involve moral and aesthetic principles and not just economic ones. This is consistent, of course, with cost-effectiveness and with a sensible recognition of economic constraints.

The neutrality of the economist is legitimate if private preferences or subjective wants are the only values in question. A person should be left free to choose the color of his or her necktie or necklace, but we cannot justify a theory of public policy . . . on that basis. . . .

The neutrality of economics is not a basis for its legitimacy. We recognize it as an indifference toward value—an indifference so deep, so studied, and so assured that at first one hesitates to call it by its right name.

THE CITIZEN AS JOSEPH K.

The residents of Lewiston at the conference I attended demanded to know the truth about the dangers that confronted them and the reasons for those dangers. They wanted to be convinced that the sacrifice asked of them was legitimate even if it served interests other than their own. One official from a large chemical company dumping wastes in the area told them in reply that corporations were people and that people could talk to people about their feelings, interests, and needs. This sent a shiver through the audience. Like Joseph K. in *The Trial,*[4] the residents of Lewiston asked for an explanation, justice, and truth, and they were told that their wants would be taken care of. They demanded to know the reasons for what was continually happening to them. They were given a personalized response instead.

This response, that corporations are "just people serving people," is consistent with a particular view of power. This is the view that identifies power with the ability to get what one wants as an individual, that is, to satisfy one's personal preferences. When people in official positions in corporations or in the government put aside their personal interests, it would follow that they put aside their power as well. Their neutrality then justifies them in directing the resources of society in ways they determine to be best. This managerial role serves not their own interests but those of their clients. Cost-benefit analysis may be seen as a pervasive form of this paternalism. Behind this paternalism, as William Simon observes of the lawyer-client relationship, lies a theory of value that tends to personalize power. "It resists understanding power as a product of class, property, or institutions and collapses power into the personal needs and dispositions of the individuals who command and obey." . . .

In the 1980s, the citizens of Lewiston, surrounded by dynamos, high tension lines, and nuclear wastes, are powerless. They do not know how to criticize power, resist power, or justify power—for to do so depends on making distinctions between good and evil, right and wrong, innocence and guilt, justice and injustice, truth and lies. These distinctions cannot be made out and have no significance within an emotive or psychological theory of value. To adopt this theory is to imagine society as a market in which individuals trade voluntarily and without coercion. No individual, no belief, no faith has authority over them. To have power to act as a nation we must be able to act, at least at times, on a public philosophy, conviction, or faith. We cannot abandon the moral function of public law. The antinomianism of cost-benefit analysis is not enough.

NOTES

1. A. Freeman, R. Haveman, A. Kneese, *The Economics of Environmental Policy* (1973).
2. *Oeuvres Politiques de Benjamin Constant*, 269 (C. Louandre, ed. 1874), *quoted in* S. Wolin, Politics and Vision 281 (1960).
3. H. Macaulay & B. Yandle, Environmental Use and the Market 120–21 (1978).
4. F. Kafka, The Trial (rev. ed. trans. 1957).

15.2 ROBERT D. BULLARD

Environmentalism and Social Justice

The environmental movement has often been described as reflecting the idealist aspirations of white middle- and upper-income people. Indeed, poor people and minority groups were not well represented among those who gathered for teach-ins and other events organized to celebrate the first Earth Day in April 1970. The planners and leaders of that event were all white and relatively affluent. The speeches they made and the goals they set rarely took into account the special needs of the poor or the possible relationship between racial oppression and the burdens of pollution.

The involvement of poor and minority people in grassroots environmental organizing has been growing dramatically for more than a decade. The movement for environmental justice was triggered in 1982 by demonstrations protesting the decision to locate a poorly planned PCB disposal site adjacent to impoverished African American and Native American communities in Warren County, North Carolina. It has since grown to encompass local, regional, and national groups organized to protest what they claim is systematic discrimination in the setting of environmental goals and the siting of polluting industries and waste disposal facilities in their backyards. At the 1992 UN Conference on Environment and Development in Rio de Janeiro, a set of "Principles of Environmental Justice" was widely circulated and discussed. In 1993 the U.S. Environmental Protection Agency, opened an Office of Environmental Equity with plans for cleaning up sites in several poor communities. On February 11, 1994, President Bill Clinton made environmental equity a national priority by issuing a sweeping Executive Order on Environmental Justice.

Sociology professor Robert D. Bullard has emerged as one of the most influential leaders of the environmental justice movement. He has authored two books and numerous articles on the inequitable treatment of African Americans and other minorities in environmental planning and decision making, and he has taken an active political role in working with both political leaders and grassroots organizations to combat environmental discrimination. In 1992 he was appointed to President Clinton's transition team.

In the following selection from his book *Dumping in Dixie: Race, Class, and Environmental Quality* (Westview Press, 1990), Bullard defines environmental discrimination and justifies his claim that the environmental inequities experienced by racial minorities cannot be explained as simply a consequence of the general discrimination experienced by the poor.

Key Concept: environmental discrimination against racial minorities

The environmental movement in the United States emerged with agendas that focused on such areas as wilderness and wildlife preservation, resource conservation, pollution abatement, and population control. It was supported primarily by middle- and upper-middle-class whites. Although concern about the environment cuts across racial and class lines, environmental activism has been most pronounced among individuals who have above-average education, greater access to economic resources, and a greater sense of personal efficacy.

Mainstream environmental organizations were late in broadening their base of support to include blacks and other minorities, the poor, and working-class persons. The "energy crisis" in the 1970s provided a major impetus for the many environmentalists to embrace equity issues confronting the poor in this country and in the countries of the Third World. Over the years, environmentalism has shifted from a "participatory" to a "power" strategy, where the "core of active environmental movement is focused on litigation, political lobbying, and technical evaluation rather than on mass mobilization for protest marches."

An abundance of documentation shows blacks, lower-income groups, and working-class persons are subjected to a disproportionately large amount of pollution and other environmental stressors in their neighborhoods as well as in their workplaces. However, these groups have only been marginally involved in the nation's environmental movement. Problems facing the black community have been topics of much discussion in recent years. (Here, we use sociologist James Blackwell's definition of the black community, "a highly diversified set of interrelated structures and aggregates of people who are held together by forces of white oppression and racism.") Race has not been eliminated as a factor in the allocation of community amenities.

Research on environmental quality in black communities has been minimal. Attention has been focused on such problems as crime, drugs, poverty, unemployment, and family crisis. Nevertheless, pollution is exacting a heavy toll (in health and environmental costs) on black communities across the nation. There are few studies that document, for example, the way blacks cope with environmental stressors such as municipal solid-waste facilities, hazardous-waste landfills, toxic-waste dumps, chemical emissions from industrial plants, and on-the-job hazards that pose extreme risks to their health. Coping in this case is seen as a response to stress and is defined as "efforts, both action-oriented and intrapsychic, to manage, i.e., master, tolerate, reduce, minimize, environmental and internal demands, conflicts among them, which tax or exceed a person's resources." Coping strategies employed by individuals confronted with a stressor are of two general types: *problem-focused coping* (e.g., individual and/or group efforts to directly address the problem) and *emotion-focused coping* (e.g., efforts to control one's psychological response to the stressor). The decision to take direct action or to tolerate a stressor often depends on how individuals perceive their ability to do something about or have an impact on the stressful situation. Personal efficacy, therefore, is seen as a factor that affects environmental and political activism.

Much research has been devoted to analyzing social movements in the United States. For example, hundreds of volumes have been written in the past several years on the environmental, labor, antiwar, and civil rights movements. Despite this wide coverage, there is a dearth of material on the convergence (and the divergence, for that matter) of environmentalism and social justice advocacy. This appears to be the case in and out of academia. Moreover, few social scientists have studied environmentalism among blacks and other ethnic minorities. This oversight is rooted in historical and ideological factors and in the composition of the core environmental movement and its largely white middle-class profile.

Many of the interactions that emerged among core environmentalists, the poor, and blacks can be traced to distributional equity questions. How are the benefits and burdens of environmental reform distributed? Who gets what, where, and why? Are environmental inequities a result of racism or class barriers or a combination of both? After more than two decades of modern environmentalism, the equity issues have not been resolved. There has been, however, some change in the way environmental problems are presented by mainstream environmental organizations. More important, environmental equity has now become a major item on the local (grassroots) as well as national civil rights agenda. . . .

Social justice and the elimination of institutionalized discrimination were the major goals of the civil rights movement. Many of the HBCUs [historically black colleges and universities] are located in some of the most environmentally polluted communities in the nation. These institutions and their students, thus, have a vested interest in seeing that improvements are made in local environmental quality. Unlike their move to challenge other forms of inequity, black student-activists have been conspicuously silent and relatively inactive on environmental problems. Moreover, the resources and talents of the faculties at these institutions have also been underutilized in assisting affected communities in their struggle against polluters, including government and private industries.

The problem of polluted black communities is not a new phenomenon. Historically, toxic dumping and the location of locally unwanted land uses (LULUs) have followed the "path of least resistance," meaning black and poor communities have been disproportionately burdened with these types of externalities. However, organized black resistance to toxic dumping, municipal waste facility siting, and discriminatory environmental and land-use decisions is a relatively recent phenomenon. Black environmental concern has been present but too often has not been followed up with action.

Ecological concern has remained moderately high across nearly all segments of the population. Social equity and concern about distributive impacts, however, have not fared so well over the years. Low-income and minority communities have had few advocates and lobbyists at the national level and within the mainstream environmental movement. Things are changing as environmental problems become more "potent political issues [and] become increasingly viewed as threatening public health."[1]

The environmental movement of the 1960s and 1970s, dominated by the middle class, built an impressive political base for environmental reform and regulatory relief. Many environmental problems of the 1980s and 1990s, however, have social impacts that differ somewhat from earlier ones. Specifically, environmental problems have had serious regressive impacts. These impacts have been widely publicized in the media, as in the case of the hazardous-waste problems at Love Canal and Times Beach. The plight of polluted minority communities is not as well known as the New York and Missouri tragedies. Nevertheless, a disproportionate burden of pollution is carried by the urban poor and minorities.

Few environmentalists realized the sociological implications of the not-in-my-backyard (NIMBY) phenomenon. Given the political climate of the times, the hazardous wastes, garbage dumps, and polluting industries were likely to end up in somebody's backyard. But whose backyard? More often than not, these LULUs ended up in poor, powerless, black communities rather than in affluent suburbs. This pattern has proven to be the rule, even though the benefits derived from industrial waste production are directly related to affluence. Public officials and private industry have in many cases responded to the NIMBY phenomenon using the place-in-blacks'-backyard (PIBBY) principle.

Social activists have begun to move environmentalism to the left in an effort to address some of the distributional impact and equity issues. Documentation of civil rights violations has strengthened the move to make environmental quality a basic right of all individuals. Rising energy costs and a continued erosion of the economy's ability to provide jobs (but not promises) are factors that favor blending the objectives of labor, minorities, and other "underdogs" with those of middle-class environmentalists. Although ecological sustainability and socioeconomic equality have not been fully achieved, there is clear evidence that the 1980s ushered in a new era of cooperation between environmental and social justice groups. While there is by no means a consensus on complex environmental problems, the converging points of view represent the notion that "environmental problems and ... material problems have common roots."

When analyzing the convergence of these groups, it is important to note the relative emphasis that environmental and social justice organizations give to "instrumental" versus "expressive" activities. Environmental organizations have relied heavily on environmentally oriented expressive activities (outdoor recreation, field trips, social functions, etc.), while the social justice movements have made greater use of goal-oriented instrumental activities (protest demonstrations, mass rallies, sit-ins, boycotts, etc.) in their effort to produce social change.

The push for environmental equity in the black community has much in common with the development of the modern civil rights movement that began in the South. That is, protest against discrimination has evolved from "organizing efforts of activists functioning through a well-developed indigenous base."[2] Indigenous black institutions, organizations, leaders, and networks are coming together against polluting industries and discriminatory environmental policies. This book addresses this new uniting of blacks against institutional barriers of racism and classism.

RACE VERSUS CLASS IN SPATIAL LOCATION

Social scientists agree that a multidimensional web of factors operate in sorting out stratification hierarchies. These factors include occupation, education, value of dwellings, source and amount of income, type of dwelling structures, government and private industry policies, and racial and ethnic makeup of residents. Unfortunately, American society has not reached a color-blind state. What role does race play in sorting out land uses? Race continues to be a potent variable in explaining the spatial layout of urban areas, including housing patterns, street and highway configurations, commercial development, and industrial facility siting.

Houston, Texas, the nation's fourth largest city, is a classic example of an area where race has played an integral part in land-use outcomes and municipal service delivery. As late as 1982, there were neighborhoods in Houston that still did not have paved streets, gas and sewer connections, running water, regular garbage service, and street markers. Black and Hispanic neighborhoods were far more likely to have service deficiencies than their white counterparts. One of the neighborhoods (Bordersville) was part of the land annexed for the bustling Houston Intercontinental Airport. Another area, Riceville, was a stable black community located in the city's sprawling southwest corridor, a mostly white sector that accounted for nearly one-half of Houston's housing construction in the 1970s.

The city's breakneck annexation policy stretched municipal services thin. Newly annexed unincorporated areas, composed of mostly whites, often gained at the expense of older minority areas. How does one explain the service disparities in this modern Sunbelt city? After studying the Houston phenomenon for nearly a decade, I have failed to turn up a single case of a white neighborhood (low- or middle-income) in the city that was systematically denied basic municipal services. The significance of race may have declined, but racism has not disappeared when it comes to allocating scarce resources.

Do middle-income blacks have the same mobility options that are available to their white counterparts? The answer to this question is no. Blacks have made tremendous economic and political gains in the past three decades with the passage of equal opportunity initiatives at the federal level. Despite legislation, court orders, and federal mandates, institutional racism and discrimination continue to influence the quality of life in many of the nation's black communities.

The differential residential amenities and land uses assigned to black and white residential areas cannot be explained by class alone. For example, poor whites and poor blacks do not have the same opportunities to "vote with their feet." Racial barriers to education, employment, and housing reduce mobility options available to the black underclass and the black middle-class.

Housing is a classic example of this persistent problem. Residential options available to blacks have been shaped largely by (1) federal housing policies, (2) institutional and individual discrimination in housing markets, (3) geographic changes that have taken place in the nation's urban centers, and (4) limited incomes. Federal policies, for example, played a key role in the development of spatially differentiated metropolitan areas where blacks and other visible minorities are segregated from whites, and the poor from the more affluent citizens. Government housing policies fueled the white exodus to the suburbs and accelerated the abandonment of central cities. Federal tax dollars funded the construction of freeway and interstate highway systems. Many of these construction projects cut paths through minority neighborhoods, physically isolated residents from their institutions, and disrupted once-stable communities.... The result of the nation's apartheid-type policies has been limited mobility, reduced housing options and residential packages, and decreased environmental choices for black households.

Environmental degradation takes an especially heavy toll on inner-city neighborhoods because the "poor or nearpoor are the ones most vulnerable to the assaults of air and water pollution, and the stress and tension of noise and squalor."[3] A high correlation has been discovered between characteristics associated with disadvantage (i.e., poverty, occupations below management and professional levels, low rent, and a high concentration of black residents [due to residential segregation and discriminatory housing practices]) and poor air quality. Individuals that are in close proximity to health-threatening problems (i.e., industrial pollution, congestion, and busy freeways) are living in endangered environs. The price that these individuals pay is in the form of higher risks of emphysema, chronic bronchitis, and other chronic pulmonary diseases.

Blacks and other economically disadvantaged groups are often concentrated in areas that expose them to high levels of toxic pollution: namely, urban industrial communities with elevated air and water pollution problems or rural areas with high levels of exposure to farm pesticides....

Air pollution in inner-city neighborhoods can be up to five times greater than in suburban areas. Urban areas, in general, have "dirtier air and drinking water, more wastewater and solid-waste problems, and greater exposure to lead and other heavy metals than nonurban areas."...

All Americans, white or black, rich or poor, are entitled to equal protection under the law. Just as this is true for such areas as education, employment, and housing, it also applies to one's physical environment. Environmental discrimination is a fact of life. Here, environmental discrimination is defined as disparate treatment of a group or community based on race, class, or some other distinguishing characteristic. The struggle for social justice by black Americans has been and continues to be rooted in white racism. White racism is a factor in the impoverishment of black communities and has made it easier for black residential areas to become the dumping grounds for all types of health-threatening toxins and industrial pollution.

Government and private industry in general have followed the "path of least resistance" in addressing externalities as pollution discharges, waste disposal, and nonresidential activities that may pose a health threat to nearby communities. Middle- and upper-class households can often shut out the fumes, noise, and odors with their air conditioning, dispose of their garbage to keep out the rats and roaches, and buy bottled water for drinking. Many lower-income households (black or white) cannot afford such "luxury" items; they are subsequently forced to adapt to a lower-quality physical environment.

Minority and low-income residential areas (and their inhabitants) are often adversely affected by unregulated growth, ineffective regulation of industrial toxins, and public policy decisions authorizing locally unwanted land uses that favor those with political and economic clout. Zoning is probably the most widely applied mechanism to regulate land use in the United States. Externalities such as pollution discharges to the air and water, noise, vibrations, and aesthetic problems are often segregated from residential areas for the "public good." Negative effects of nonresidential activities generally decrease with distance from the source. Land-use zoning, thus, is designed as a "protectionist device" to insure a "place for everything and everything in its place."[4] Zoning is ultimately intended to influence and shape land use in accordance with long-range local needs.

Zoning, deed restrictions, and other protectionist land-use mechanisms have failed to effectively protect minority communities, especially low-income minority communities. Logan and Molotch, in their book *Urban Fortunes: The Political Economy of Place,* contend that the various social classes, with or without land-use controls, are "unequally able to protect their environmental interests." In their quest for quality neighborhoods, individuals often find themselves competing for desirable neighborhood amenities (i.e., good schools, police and fire protection, quality health care, and parks and recreational facilities) and resisting negative characteristics (i.e., landfills, polluting industries, freeways, public housing projects, drug-treatment facilities, halfway houses, etc.). . . .

Why has this happened and what have blacks done to resist these practices? In order to understand the causes of the environmental dilemma that many black and low-income communities find themselves in, the theoretical foundation of environmentalism needs to be explored.

THE THEORETICAL BASIS OF ENVIRONMENTAL CONFLICT

Environmentalism in the United States grew out of the progressive conservation movement that began in the 1890s. The modern environmental movement, however, has its roots in the civil rights and antiwar movements of the late 1960s. The more radical student activists splintered off from the civil rights and antiwar movements to form the core of the environmental movement in the early 1970s. The student environmental activists affected by the 1970 Earth Day enthusiasm in colleges and universities across the nation had hopes of bringing environmental reforms to the urban poor. They saw their role as environmental advocates for the poor since the poor had not taken action on their own. They were, however, met with resistance and suspicion. Poor and minority residents saw environmentalism as a disguise for oppression and as another "elitist" movement.

Environmental elitism has been grouped into three categories: (1) *compositional elitism* implies that environmentalists come from privileged class strata, (2) *ideological elitism* implies that environmental reforms are a subterfuge for distributing the benefits to environmentalists and costs to nonenvironmentalists, and (3) *impact elitism* implies that environmental reforms have regressive distributional impacts.

Impact elitism has been the major sore point between environmentalists and advocates for social justice who see some reform proposals creating, exacerbating, and sustaining social inequities. Conflict centered largely on the "jobs versus environment" argument. Imbedded in this argument are three competing advocacy groups: (1) *environmentalists* are concerned about leisure and recreation, wildlife and wilderness preservation, resource conservation, pollution abatement, and industry regulation, (2) *social justice advocates'* major concerns include basic civil rights, social equity, expanded opportunity, economic mobility, and institutional discrimination, and (3) *economic boosters* have as their chief concerns maximizing profits, industrial expansion, economic stability, laissez-faire operation, and deregulation.

Economic boosters and pro-growth advocates convinced minority leaders that environmental regulations were bad for business, even when locational decisions had adverse impacts on the less advantaged. Pro-growth advocates used a number of strategies to advance their goals, including public relations campaigns, lobbying public officials, evoking police powers of government, paying off or co-opting dissidents, and granting small concessions when plans could be modified. Environmental reform proposals were presented as prescriptions for plant closures, layoffs, and economic dislocation. Kazis and Grossman referred to this practice as "job blackmail." They insisted that by "threatening their employees with a 'choice' between their jobs and their health, employers seek to make the public believe there are no alternatives to 'business as usual.' "

Pro-growth advocates have claimed the workplace is an arena in which unavoidable trade-offs must be made between jobs and hazards: If workers want to keep their jobs, they must work under conditions that may be hazardous to them, their families, and their community. Black workers are especially vulnerable to job blackmail because of the threat of unemployment and their concentration in certain types of occupations. The black workforce remains overrepresented in low-paying, low-skill, high-risk blue collar and service occupations where there is more than an adequate supply of replacement labor. Black workers are twice as likely to be unemployed as their white counterparts. Fear of unemployment acts as a potent incentive for many blacks to stay in and accept jobs they know are health threatening.

There is inherent conflict between the interests of capital and of labor. Employers have the power to move jobs (and sometimes hazards) as they wish. For example, firms may choose to move their operations from the Northeast and Midwest to the South and Sunbelt, or they may move the jobs to Third World countries where labor is cheaper and there are fewer health and environmental regulations. Moreover, labor unions may feel it necessary to scale down their demands for improved work safety conditions in a depressed economy for fear of layoffs, plant closings, or relocation of industries (e.g., moving to right-to-work states that proliferate in the South). The conflicts, fears, and anxieties manifested by workers are usually built on the false assumption that environmental regulations are automatically linked to job loss.

The offer of a job (any job) to an unemployed worker appears to have served a more immediate need than the promise of a clean environment. There is evidence that new jobs have been created as a direct result of environmental reforms. Who got these new jobs? The newly created jobs are often taken by people who already have jobs or by migrants who possess skills greater than the indigenous workforce. . . .

Minority residents can point to a steady stream of industrial jobs leaving their communities. Moreover, social justice advocates take note of the miserable track record that environmentalists and preservationists have on improving environmental quality in the nation's racially segregated inner cities and hazardous industrial workplaces, and on providing housing for low-income groups. Decent and affordable housing, for example, is a top environmental problem for inner-city blacks. On the other hand, environmentalists' continued emphasis on wilderness and wildlife preservation appeal to a population that can afford leisure time and travel to these distant locations. This does not mean that poor people and people of color are not interested in leisure or outdoor activities. Many wilderness areas and national parks remain inaccessible to the typical inner-city resident because of inadequate transportation. Physical isolation, thus, serves as a major impediment to black activism in the mainstream conservation and resource management activities.

NOTES

1. Riley E. Dunlap, "Public Opinion on the Environment in the Reagan Era: Polls, Pollution, and Politics Revisited," *Environment, 29* (July/Aug 1987): 6–11, 32–37.
2. Aldon D. Morris, *The Origins of the Civil Rights Movement: Black Communities Organizing for Change* (New York: Free Press, 1984), p. xii.
3. Daniel Zwendling, "Poverty and Pollution," *Progressive, 37* (January 1973): 25–29.
4. See Constance Perrin, *Everything in Its Place: Social Order and Land Use in America,* (Princeton, NJ: Princeton University Press, 1977).

15.3 JANET N. ABRAMOVITZ

Valuing Nature's Services

Many of the policies that have been recommended by practitioners of the relatively new field of environmental economics have generated controversy. For example, proposals to replace economic measures of a nation's productivity, such as the gross national product (GNP), have failed to win the support of establishment economists and many environmentalists. Recently, environmental economists have focused attention on the fact that many goods and services provided by nature support human society. In addition, these goods and services are free. Examples include the cycling of nutrients in order to produce renewable resources such as fish and forest products, the pollination of flowering plants, and the regulation of the climate. Many environmental economists argue that because nature provides these services at no cost, little attention is given to prevent activities that may compromise nature's ability to sustain human life. In response to this concern, a group of economists, ecologists, and other scientists made the first comprehensive effort to assess the dollar value of these ecosystem services. By reporting their result in the May 15, 1997, issue of the prominent British science journal *Nature,* they hoped to draw attention to the enormous importance of environmental protection. They concluded that the estimated value for the entire biosphere is $33 trillion per year. This report received much media attention and has generated both supportive and deprecatory responses.

Janet N. Abramovitz is a senior researcher at the Worldwatch Institute, a private, nonprofit research organization devoted to the analysis of global economic and environmental issues. Her fields of expertise include biodiversity, natural resources management, human development, and social equity. In the following selection taken from her chapter "Valuing Nature's Services," in Lester R. Brown et al., *State of the World 1997* (W. W. Norton, 1997), Abramovitz discusses the political importance of attaching economic values to nature's "free services." She acknowledges and supports the criticism of those who have pointed out that much of nature's value cannot have a price tag attached to it. She also stresses the need to take into account nature's spiritual, cultural, religious, and aesthetic values.

Key Concept: the enormous value of nature

Nature's "free" services form the invisible foundation that supports our societies and economies. We rely on the oceans to provide abundant fish, on forests for wood and new medicines, on insects to pollinate our crops, on birds and frogs to keep pests in check, and on rivers to supply clean water. We expect that when we need timber we can harvest it, that when we need new crops we can find them in nature, that when we drill a well we will find water, that the wastes we generate will disappear, that clean air will blow in to refresh our cities, and that the climate will be stable and predictable. Nature's services have always been there free for the taking, and our expectations—and economies—are based on the premise that they always will be.

Yet economies unwittingly provide incentives to misuse and destroy nature by underappreciating and undervaluing its services. Nature in turn is increasingly less able to supply the services that the earth's expanding population and economy demand. It is not an exaggeration to state that the continued loss of nature's services threatens not only today's human enterprise, but ultimately the prospects for our continued existence.

How did this unhealthy dynamic come into being? Nature is viewed as a boundless and inexhaustible resource and sink. Human impact is seen as insignificant or beneficial. The very tools used to gauge the economic health and progress of a nation can reinforce and encourage these attitudes. The gross domestic product (GDP), for example, supposedly measures the value of the goods and services produced in a nation. But the most valuable goods and services—the ones provided by nature, on which all else rests—are measured poorly or not at all. The unhealthy dynamic is compounded by the fact that activities that pollute or deplete natural capital are counted as contributions to economic well-being.

In the human economy, it is easy to distinguish between goods and services. In nature's economy, such a distinction is less useful as goods and services are highly integrated in subtle and complex ways. Indeed, treating nature as a box filled with unrelated objects that we can remove or replace at will, with little or no effect, is ultimately counterproductive.

Nature's ecosystem services include producing raw materials, purifying and regulating water, absorbing and decomposing wastes, cycling nutrients, creating and maintaining soils, providing pollination and pest control, and regulating local and global climates. (See Table.) Forests, for example, do much more than supply timber. They provide habitat for birds and insects that pollinate crops and control disease-carrying and agricultural pests. Their canopies break the force of the winds and reduce rainfall's impact on the ground, which lessens soil erosion. Their roots hold soil in place, further reducing erosion. A forest's watershed protection values alone can exceed the value of its timber. Forests also act as effective water pumping and recycling machinery, helping to stabilize the local climate. And through photosynthesis, plants generate life-giving oxygen and hold vast amounts of carbon in storage, which stabilizes the global climate.

Around the world, the conversion, degradation, fragmentation, and simplification of ecosystems has been extensive. In many countries, including some of the largest, more than half the territory has been converted from natural habitat to other uses, much of it unsustainably and irreversibly. In countries that stayed relatively undisturbed until the eighties, significant portions of remaining ecosystems have been lost in the last decade. These trends have been

TABLE 1
Nature's Services

Raw Materials Production (food, fisheries, timber and building materials, non-timber forest products, fodder, genetic resources, medicines, dyes)
Pollination
Biological Control of Pests and Diseases
Habitat and Refuge
Water Supply and Regulation
Waste Recycling and Pollution Control
Nutrient Cycling
Soil Building and Maintenance
Disturbance Regulation
Climate Regulation
Atmospheric Regulation
Recreation
Cultural
Educational/Scientific

Source: Worldwatch Institute.

accelerating everywhere. Lost with these natural ecosystems are the valuable services they provide.

Nature's living library—the genes, species, populations, communities, and ecosystems in existence today—represent a wealth of options for future generations and for change in the biosphere. Unfortunately, we are running a "biodiversity deficit," destroying species and ecosystems faster than nature can create new ones. Species alone are now vanishing 100 to 1,000 times faster than natural extinction rates as a result of human actions.

By reducing the number of species and the size and integrity of ecosystems, we are also reducing nature's capacity to evolve and create new life. In just a few centuries we have gone from living off nature's interest to spending down the capital that has accumulated over millions of years of evolution, as well as diminishing the capacity of nature to create new capital. Humans are only one part of the evolutionary product. Yet we have taken on a major role in shaping its future production course and potential. We are pulling out the threads of nature's safety net even as we need it to support the world's expanding human population and economy....

APPRECIATING SERVICE PROVIDERS

Nature nurtures innumerable species that are not harvested directly but that provide important "free" services. These creatures pollinate crops, keep potentially harmful organisms in check, build and maintain soils, and decompose dead matter so it can used to build new life. Nature's "service providers"—the birds and bees, insects, worms, and microorganisms—show how small and seemingly insignificant things can have disproportionate value. Unfortunately, their services are in increasingly short supply because chemicals, disease, hunting, and habitat fragmentation and destruction have drastically reduced their numbers and ability to function. In the words of Stephen Buchmann and Gary Paul Nabhan, coauthors of a recent book on pollinators, "nature's most productive workers [are] slowly being put out of business."

Pollinators, for example, are of enormous value to agriculture and the functioning of natural ecosystems. Without them, plants cannot produce the seeds that ensure their survival—and ours. . . .

Eighty percent of the world's 1,330 cultivated crop species (including fruits, vegetables, beans and legumes, coffee and tea, cocoa, and spices) are pollinated by wild and semiwild pollinators. Without these services, crops yield less and wild plants produce few seeds—with large economic and ecological consequences. One third of U.S. agricultural output is from insect-pollinated plants (the remainder is from wind-pollinated grain plants such as wheat, rice, and corn). Honeybee pollination services are 60–100 times more valuable than the honey they produce. The value of wild blueberry bees is so great, with each one pollinating 15–19 liters of blueberries in its life, that they are viewed by farmers as "flying $50 bills."

In the United States, more than half the honeybee colonies have been lost in the last 50 years, with 25 percent lost within the last 5 years alone. Widespread threats to pollinators include habitat fragmentation and disturbance, loss of nesting and overwintering sites, intense exposure of pollinators to pesticides and of nectar plants to herbicides, breakdown of "nectar corridors" that provide food sources to pollinators during migration, new diseases, competition from exotic species, and excessive hunting. A "forgotten pollinators" campaign was recently launched to raise awareness of the importance and plight of these service providers.

Ironically, many modern agricultural practices actually limit the productivity of crops by reducing pollination. The high levels of pesticides used on cotton are estimated to reduce annual yields by 20 percent (worth $400 million) in the United States alone by killing bees and other insect pollinators. . . .

Many of the disturbances that have harmed pollinators are also hurting creatures that provide other beneficial services, such as biological control of pests and disease. Much of the wild and semiwild habitat inhabited by beneficial predators like birds has been eliminated. Chemicals have killed beneficial insects along with the pests. The "pest control services" that nature provides are significant. Bat colonies in Texas can eat 250 tons of insects each night. Without birds, leaf-eating insects are more abundant and can slow the growth of trees or damage crops.

The loss of nature's pest control services has also contributed to the rise of vector-borne diseases and to increased reliance on harmful chemicals to try to control ever-more-abundant pests. . . .

How can we encourage nature's service providers? By protecting them, their habitat, and the relationships among them. No-till farming methods can reduce soil erosion and allow nature's underground economy to flourish. Substantial reductions in the use of agricultural chemicals that harm service providers (and people) can help. Across the whole landscape, migratory routes and nectar corridors need protection to ensure the survival of nature's pollinators and pest control agents. . . .

CYCLING AND RECYCLING

Many of nature's services arise from its ability to regulate and recycle water, nutrients, and waste. But human disruptions have impaired this ability to filter and regulate water, to recharge groundwater supplies, and to move nutrients and sediments—indeed, to support life.

One of the most basic aspects of the cycling and recycling service is that water falls as precipitation, running across the landscape to streams and rivers and ultimately to the sea. Human actions have even changed that fundamental force of nature by removing natural plant cover, plowing fields, draining wetlands, separating rivers from their floodplains, and paving over land. In many places, water now races across the landscape much too quickly, causing flooding and droughts. In the Mississippi basin, for example, conversion of forests, prairies, wetlands, and floodplains to agriculture over the past 150 years has dramatically accelerated soil erosion, flood damage, and loss of native species. These changes have also reduced the water-holding capacity of the soils by as much as 70 percent. . . .

One way to picture the value of nature's free services is to estimate what it costs society to replace them. The huge expenditures for building and maintaining infrastructure such as water treatment, irrigation, levees, and flood control illustrate a few of the economic costs. The value of mangroves for flood control alone, for instance, has been calculated at $300,000 per kilometer in Malaysia—the cost of the rock walls that would be needed to replace them. New York City has always relied on the natural filtering capacity of its rural watersheds to cleanse the water that serves 10 million people each day. Rather than spend $7 billion to build water treatment facilities, the city will pay one tenth that amount helping upstream counties protect the watersheds around its drinking-water reservoirs. . . .

NATURE'S STABILITY AND RESILIENCE

A fundamental service provided by nature is ensuring that ecosystems and the entire biosphere are relatively stable and resilient. The ability to withstand disturbances and bounce back from regular "shocks" is essential to keeping the life-support system operating. Maintaining the integrity of the web of species, functions, and processes within a system and the webs that connect different systems is critical for ensuring stability and resilience. As systems are simplified and their webs become disconnected, they become more brittle and vulnerable to catastrophic, irreversible decline. From global climate change and the breakdown of the ozone layer to the biodiversity deficit, the collapse of fisheries, frequent outbreaks of red tides, and increasingly severe floods and droughts, there is now ample evidence that the biosphere is becoming less resilient.

Unfortunately, much of the economy is based on practices that convert natural systems into something simpler for ease of management or to maximize the production of a desired commodity (trees or wheat or minerals, for example). But simplified systems lack the resilience that allows them to survive short-term adversities (like disease or pest outbreaks, forest fires, or pollution) or long-term alterations such as climate change. A tree plantation or fish farm may provide some of the things we need, but it cannot supply the array of goods and services over a range of conditions that natural diverse systems do.

The results of decisions to maximize agricultural productivity while eliminating diversity show the consequences of making systems uniform and brittle. Today, fewer than 100 species provide most of the world's food supply, and within these species, genetic diversity has been drastically reduced. In eight major crops in the United States, for instance, fewer than nine varieties accounted for 50–75 percent of the total. . . .

The cumulative effects of local land use changes have global implications as well. One of the planet's first ecosystem services was the production of oxygen over billions of years of photosynthetic activity, which allowed oxygen-breathing organisms such as ourselves to exist. Our future existence will continue to depend on ecosystems maintaining the proper balance of atmospheric gases like oxygen and carbon dioxide. There is no technological substitute for this vital service.

Humans have begun to impair this basic service by generating too much carbon dioxide and other greenhouse gases, and reducing the ability of ecosystems to absorb carbon dioxide. The benefits of intact forests for global carbon sequestration alone have been estimated at several hundred to several thousand dollars per hectare. The consequences of this service disruption are beginning to be evident in the form of global climate change. Maintaining nature's ability to regulate local and global climates will be even more valuable under the predicted climate change scenario. . . .

LETTING NATURE DO ITS WORK

What can be done to stop the unraveling of nature's life-support system and ensure that it can continue to supply the services on which we depend? First, our understanding of the true extent and value of nature's services, and the tools and processes we use to make decisions, need to be redirected toward ensuring the sustainability of the planet's life-support system. Understanding and valuing nature's services, and ensuring that they are used equitably and within nature's limits, are essential to that sustainable path.

We need to understand the interconnected web of life that we are part of and that supports us. Public and policy education as well as more fundamental ecological research about nature's services and cycles, and the true extent of our reliance on them, are needed. Realizing the cumulative impact of our activities and learning how we can conduct the human enterprise within nature's regenerative capacity are essential.

Economies and societies often use faulty signals that encourage people to make decisions that run counter to their own long-range interests—and those of society and future generations. Economic calculations grossly underestimate the current and future value of nature. While a fraction of nature's goods are counted when they enter the marketplace, many of the goods are not measured. And nature's services—the life-support systems—are not counted at all.

When nature's goods are considered free and therefore valued at zero, the market signals that they are only economically valuable when converted into something else. For example, the profit from deforesting land is counted as a plus on a nation's ledger sheet, but the depletions of the timber stock, watershed, and fisheries are not subtracted. Clearly the costs of environmental degradation and lost ecosystem services are external to economic calculations: the damage from a massive oil spill is not subtracted from a nation's GDP, but the amounts spent on cleanup and health impacts are counted as additions to the national economy....

Another faulty economic signal is that financial benefits from resource use are realized by private individuals or entities, but the costs of any loss are distributed across society. Economists call this "socializing costs." Stated simply, private costs and benefits are counted, but social and environmental costs to current and especially future generations are "external" to the calculations. The people who get the benefits are different from the ones who pay the costs. Thus, there is little economic incentive for those exploiting a resource to use it judiciously or in a manner that maximizes public good....

Fixing a more accurate price for nature by better "internalizing" the economic "externalities" is an important step, but it is also necessary to acknowledge that everything does not have a price. Much of nature's value is quite literally beyond measure. Evolution and resilience, for instance, are priceless. Assessing the total value, economic and otherwise, of an ecosystem requires looking at more than the amount of money that can be made from a piece of land. The nonmonetary benefits of nature's services and the costs of its absence must also be recognized. Beyond looking at the economic value of exploiting nature for a particular commodity, for example, it is also important to examine how many people—and which people—use the resources, and what portion of their livelihoods and larders nature provides. Community stability, self-sufficiency, and livelihood flexibility and security are extremely valuable, even though a monetary value cannot be assigned to them.

There are spiritual, cultural, religious, and aesthetic values of nature that are immeasurable yet powerful incentives for good stewardship. The desire to pass on a healthy environment to future generations is a part of most cultures. Many traditional resource management strategies and religious rituals and taboos may have been developed in response to the need to use biological resources sustainably. For example, the sacred groves of India and Ghana as well as Indonesia's water temples are rooted in religious tradition, but they are now recognized as highly evolved forms of wise resource management as well.

... Charting a more sustainable path will require using better economic tools to measure and value nature's services, economic performance, and human well-being. Standard indicators have not been good at measuring environmental or human well-being.

As the authors of a new Genuine Progress Indicator (GPI) put it, "The GDP makes no distinction between economic transactions that add to well-being and those which diminish it.... As a result, the GDP masks the breakdown of social structure and natural habitat; and worse, it portrays this breakdown as economic gain." While global GDP has been rising in recent decades, for example, the world's population living in poverty has been increasing, the distribution of income has become less equitable, the biodiversity deficit is growing, and the loss of nature's services has worsened. The values of nature's unmarketed goods and ecosystem services as well as the unpaid labor in households and communities must be incorporated into economic calculations and performance indicators.

New performance indicators are being developed to give more accurate assessments of well-being and development. The GPI, for example, has been proposed by a group called Redefining Progress. It expands on the landmark methodology developed for the Index of Sustainable Economic Welfare by Herman Daly and John Cobb in 1989. The GPI counts the positive contributions of household and community work and subtracts for depletion of natural habitat, pollution costs, income distribution, and crime. (It does not, however, reflect the value of nature's services except when they are lost.) In contrast to rising GDP, the GPI in the United States has been declining since the seventies.

The new discipline of ecological economics is providing much-needed guidance on charting a more sustainable path. Revision of the GDP-based system of national accounts is now being taken seriously. And indicators will continue being refined to include estimates of the values of natural capital and ecosystem services. But it is not necessary to wait for a global consensus on these changes to begin taking action. Governments, lenders, businesses, and others can begin today by using the newer indicators and fuller valuation techniques that are already available. They can also end destructive subsidies, apply appropriate economic incentives, and link receipt of benefits to payment of costs. They can start by incorporating social and environmental indicators and nonmonetary criteria as fundamental parts of their decision-making processes....

In project and local planning, most current cost-benefit analyses and environmental impact assessments are not up to measuring the true costs of depleting nature's capital and the real ecological and socioeconomic benefits of nature's services. Thus decisions made based on these measures (as with the GDP) will be flawed. Local entities, banks, private lenders, and businesses can use an expanded set of social and environmental indicators to judge the relative merits of development alternatives.

When measuring the potential environmental impacts of planned developments (from dams and highways to irrigation and land use), investors and developers need to consider lost environmental benefits as well as social, economic, and ecological costs. For example, displacing communities in the name of progress tears the social fabric and the social safety net, frequently leading people to become environmental refugees and aggravating poverty and environmental destruction. . . .

Laws and policies should be designed to protect nature's systems and services, not just individual elements in those systems. Management decisions should cover larger geographic areas and longer time horizons, and should shift from managing for individual elements (such as one species or use) to managing for ecosystem health and processes. . . .

We can reduce and reverse the destructive impact of our activities by consuming less and by placing fewer demands on nature's services—for example, by increasing water and energy use efficiency. It is not necessary to undermine ecosystems services in order to produce the food and materials that people require. Food can still be produced using no-till farming methods that dramatically reduce soil erosion, fertilizer and chemical runoff into waters, and air pollution. This approach can also improve the organic matter content of soil and make it more hospitable to valuable microorganisms.

The continued conversion, simplification, and degradation of ecosystems needs to be reversed. And degraded habitat should be restored so it can perform critical services. Examples of steps needed in this direction include using artificial wetlands for flood control and nitrogen abatement, and promoting reforestation for watershed protection and carbon sequestration. Policies such as "no net loss of wetlands" can fail to protect the myriad values and services of wetlands. The goal should be no more loss of natural wetlands and restoration of degraded wetlands.

Finally, we can no longer assume that nature's services will always be there free for the taking. We must become more cautious and forward-thinking before taking any actions that disrupt natural systems and services and limit the options of future generations. We have already seen that the loss of ecosystem services can have severe economic, social, and ecological costs, even though we can only measure a fraction of them. We can rarely determine the full impact of our actions. The consequences for nature are often unforeseen and unpredictable. The loss of individual species and habitat, and the degradation and simplification of ecosystems, can impair nature's ability to provide the services we need. Many of these losses are irreversible, and much of what is lost is simply irreplaceable.

Maintaining nature's services requires looking beyond the needs of this generation, with the goal of ensuring sustainability for many generations to come. We must act under the assumption that future generations will need at least the same level of nature's services as we have today—the safe minimum standard. Thus reason and equity dictate that we operate under the precautionary principle. We can neither practically nor ethically decide what future generations will need and what they can survive without.

CHAPTER 16 Environmental Ethics and Worldviews

16.1 WORLD COMMISSION ON ENVIRONMENT AND DEVELOPMENT

Towards Sustainable Development

In 1983 the United Nations secretary general, responding to a call by the UN General Assembly, established a World Commission on Environment and Development, which was charged with the task of producing a "global agenda for change." Gro Harlem Brundtland, currently Director-General of the World Health Organization, was chosen to organize and chair this commission. Brundtland, a former Norwegian prime minister and minister of environment, is the author of numerous articles on political, environmental, and developmental issues. She responded with vigor to the UN mandate, which called for proposals of long-term environmental strategies for achieving sustainable development by the year 2000 and beyond and for ways in which countries at different stages of economic development could cooperate in dealing with global environmental concerns.

After three years of intense work the commission produced the report *Our Common Future* (Oxford University Press, 1987). The report included a proposal to hold an international conference on environment and development as a follow-up to the 1972 UN Conference on the Human Environment. The UN Conference on Environment and Development was indeed held—in Rio de Janeiro in June 1992—and it resulted in several preliminary political agreements and an ambitious agenda that is presently guiding the international effort to address global environmental issues while promoting equitable economic development.

"Sustainable development" has recently become something of a catchphrase that appears in most discussions of long-term environmental issues. However, it frequently conveys different meanings when used by industrial developers rather than by environmental advocates. The following selection is an excerpt from chapter 2 of *Our Common Future.* It provides an environmentally sound definition of a global sustainable development policy and some suggestions about what needs to be done to begin its implementation.

Key Concept: sustainable development

Sustainable development is development that meets the needs of the present without compromising the ability of future generations to meet their own needs. It contains within it two key concepts:

- the concept of 'needs', in particular the essential needs of the world's poor, to which overriding priority should be given; and
- the idea of limitations imposed by the state of technology and social organization on the environment's ability to meet present and future needs.

Thus the goals of economic and social development must be defined in terms of sustainability in all countries—developed or developing, market-oriented or centrally planned. Interpretations will vary, but must share certain general features and must flow from a consensus on the basic concept of sustainable development and on a broad strategic framework for achieving it.

Development involves a progressive transformation of economy and society. A development path that is sustainable in a physical sense could theoretically be pursued even in a rigid social and political setting. But physical sustainability cannot be secured unless development policies pay attention to such considerations as changes in access to resources and in the distribution of costs and benefits. Even the narrow notion of physical sustainability implies a concern for social equity between generations, a concern that must logically be extended to equity within each generation.

THE CONCEPT OF SUSTAINABLE DEVELOPMENT

The satisfaction of human needs and aspirations is the major objective of development. The essential needs of vast numbers of people in developing countries —for food, clothing, shelter, jobs—are not being met, and beyond their basic needs these people have legitimate aspirations for an improved quality of life. A world in which poverty and inequity are endemic will always be prone to ecological and other crises. Sustainable development requires meeting the basic needs of all and extending to all the opportunity to satisfy their aspirations for a better life.

Living standards that go beyond the basic minimum are sustainable only if consumption standards everywhere have regard for long-term sustainability. Yet many of us live beyond the world's ecological means, for instance in our patterns of energy use. Perceived needs are socially and culturally determined, and sustainable development requires the promotion of values that encourage consumption standards that are within the bounds of the ecological possible and to which all can reasonably aspire.

Meeting essential needs depends in part on achieving full growth potential, and sustainable development clearly requires economic growth in places where such needs are not being met. Elsewhere, it can be consistent with economic growth, provided the content of growth reflects the broad principles of sustainability and non-exploitation of others. But growth by itself is not enough. High levels of productive activity and widespread poverty can coexist, and can endanger the environment. Hence sustainable development requires that societies meet human needs both by increasing productive potential and by ensuring equitable opportunities for all.

An expansion in numbers can increase the pressure on resources and slow the rise in living standards in areas where deprivation is widespread. Though the issue is not merely one of population size but of the distribution of resources, sustainable development can only be pursued if demographic developments are in harmony with the changing productive potential of the ecosystem.

A society may in many ways compromise its ability to meet the essential needs of its people in the future—by overexploiting resources, for example. The direction of technological developments may solve some immediate problems but lead to even greater ones. Large sections of the population may be marginalized by ill-considered development.

Settled agriculture, the diversion of watercourses, the extraction of minerals, the emission of heat and noxious gases into the atmosphere, commercial forests, and genetic manipulation are all examples of human intervention in natural systems during the course of development. Until recently, such interventions were small in scale and their impact limited. Today's interventions are more drastic in scale and impact, and more threatening to life-support systems both locally and globally. This need not happen. At a minimum, sustainable development must not endanger the natural systems that support life on Earth: the atmosphere, the waters, the soils, and the living beings.

Growth has no set limits in terms of population or resource use beyond which lies ecological disaster. Different limits hold for the use of energy, materials, water, and land. Many of these will manifest themselves in the form of rising costs and diminishing returns, rather than in the form of any sudden loss of a resource base. The accumulation of knowledge and the development of technology can enhance the carrying capacity of the resource base. But ultimate limits there are, and sustainability requires that long before these are reached, the world must ensure equitable access to the constrained resource and reorient technological efforts to relieve the pressure.

Economic growth and development obviously involve changes in the physical ecosystem. Every ecosystem everywhere cannot be preserved intact.

A forest may be depleted in one part of a watershed and extended elsewhere, which is not a bad thing if the exploitation has been planned and the effects on soil erosion rates, water regimes, and genetic losses have been taken into account. In general, renewable resources like forests and fish stocks need not be depleted provided the rate of use is within the limits of regeneration and natural growth. But most renewable resources are part of a complex and interlinked ecosystem, and maximum sustainable yield must be defined after taking into account system-wide effects of exploitation.

As for non-renewable resources, like fossil fuels and minerals, their use reduces the stock available for future generations. But this does not mean that such resources should not be used. In general the rate of depletion should take into account the criticality of that resource, the availability of technologies for minimizing depletion, and the likelihood of substitutes being available. Thus land should not be degraded beyond reasonable recovery. With minerals and fossil fuels, the rate of depletion and the emphasis on recycling and economy of use should be calibrated to ensure that the resource does not run out before acceptable substitutes are available. Sustainable development requires that the rate of depletion of non-renewable resources should foreclose as few future options as possible.

Development tends to simplify ecosystems and to reduce their diversity of species. And species, once extinct, are not renewable. The loss of plant and animal species can greatly limit the options of future generations; so sustainable development requires the conservation of plant and animal species.

So-called free goods like air and water are also resources. The raw materials and energy of production processes are only partly converted to useful products. The rest comes out as wastes. Sustainable development requires that the adverse impacts on the quality of air, water, and other natural elements are minimized so as to sustain the ecosystem's overall integrity.

In essence, sustainable development is a process of change in which the exploitation of resources, the direction of investments, the orientation of technological development, and institutional change are all in harmony and enhance both current and future potential to meet human needs and aspirations.

EQUITY AND THE COMMON INTEREST

Sustainable development has been described here in general terms. How are individuals in the real world to be persuaded or made to act in the common interest? The answer lies partly in education, institutional development, and law enforcement. But many problems of resource depletion and environmental stress arise from disparities in economic and political power. An industry may get away with unacceptable levels of air and water pollution because the people who bear the brunt of it are poor and unable to complain effectively. A forest may be destroyed by excessive felling because the people living there have no alternatives or because timber contractors generally have more influence than forest dwellers.

Ecological interactions do not respect the boundaries of individual ownership and political jurisdiction. Thus:

- In a watershed, the ways in which a farmer up the slope uses land directly affect run-off on farms downstream.
- The irrigation practices, pesticides, and fertilizers used on one farm affect the productivity of neighbouring ones, especially among small farms.
- The efficiency of a factory boiler determines its rate of emission of soot and noxious chemicals and affects all who live and work around it.
- The hot water discharged by a thermal power plant into a river or a local sea affects the catch of all who fish locally.

Traditional social systems recognized some aspects of this interdependence and enforced community control over agricultural practices and traditional rights relating to water, forests, and land. This enforcement of the 'common interest' did not necessarily impede growth and expansion though it may have limited the acceptance and diffusion of technical innovations.

Local interdependence has, if anything, increased because of the technology used in modern agriculture and manufacturing. Yet with this surge of technical progress, the growing 'enclosure' of common lands, the erosion of common rights in forests and other resources, and the spread of commerce and production for the market, the responsibilities for decision making are being taken away from both groups and individuals. This shift is still under way in many developing countries.

It is not that there is one set of villains and another of victims. All would be better off if each person took into account the effect of his or her acts upon others. But each is unwilling to assume that others will behave in this socially desirable fashion, and hence all continue to pursue narrow self-interest. Communities or governments can compensate for this isolation through laws, education, taxes, subsidies, and other methods. Well-enforced laws and strict liability legislation can control harmful side effects. Most important, effective participation in decision-making processes by local communities can help them articulate and effectively enforce their common interest.

Interdependence is not simply a local phenomenon. Rapid growth in production has extended it to the international plane, with both physical and economic manifestations. There are growing global and regional pollution effects, such as in the more than 200 international river basins and the large number of shared seas.

The enforcement of common interest often suffers because areas of political jurisdictions and areas of impact do not coincide. Energy policies in one jurisdiction cause acid precipitation in another. The fishing policies of one state affect the fish catch of another. No supranational authority exists to resolve such issues, and the common interest can only be articulated through international cooperation.

In the same way, the ability of a government to control its national economy is reduced by growing international economic interactions. For example, foreign trade in commodities makes issues of carrying capacities and resource scarcities an international concern. If economic power and the benefits of trade were more equally distributed, common interests would be generally recognized. But the gains from trade are unequally distributed, and patterns of trade in, say, sugar affect not merely a local sugar-producing sector, but the economies and ecologies of the many developing countries that depend heavily on this product.

The search for common interest would be less difficult if all development and environment problems had solutions that would leave everyone better off. This is seldom the case, and there are usually winners and losers. Many problems arise from inequalities in access to resources. An inequitable landownership structure can lead to overexploitation of resources in the smallest holdings, with harmful effects on both environment and development. Internationally, monopolistic control over resources can drive those who do not share in them to excessive exploitation of marginal resources. The differing capacities of exploiters to commandeer 'free' goods—locally, nationally, and internationally—is another manifestation of unequal access to resources. 'Losers' in environment/development conflicts include those who suffer more than their fair share of the health, property, and ecosystem damage costs of pollution.

As a system approaches ecological limits, inequalities sharpen. Thus when a watershed deteriorates, poor farmers suffer more because they cannot afford the same anti-erosion measures as richer farmers. When urban air quality deteriorates, the poor, in their more vulnerable areas, suffer more health damage than the rich, who usually live in more pristine neighbourhoods. When mineral resources become depleted, late-comers to the industrialization process lose the benefits of low-cost supplies. Globally, wealthier nations are better placed financially and technologically to cope with the effects of possible climatic change.

Hence, our inability to promote the common interest in sustainable development is often a product of the relative neglect of economic and social justice within and amongst nations.

STRATEGIC IMPERATIVES . . .

Reorienting Technology and Managing Risk

The fulfilment of all these tasks will require the reorientation of technology—the key link between humans and nature. First, the capacity for technological innovation needs to be greatly enhanced in developing countries so that they can respond more effectively to the challenges of sustainable development. Second, the orientation of technology development must be changed to pay greater attention to environmental factors.

The technologies of industrial countries are not always suited or easily adaptable to the socio-economic and environmental conditions of developing countries. To compound the problem, the bulk of world research and development addresses few of the pressing issues facing these countries, such as arid-land agriculture or the control of tropical diseases. Not enough is being done to adapt recent innovations in materials technology, energy conservation, information technology, and biotechnology to the needs of developing countries. These gaps must be covered by enhancing research, design, development, and extension capabilities in the Third World.

In all countries, the processes of generating alternative technologies, upgrading traditional ones, and selecting and adapting imported technologies should be informed by environmental resource concerns. Most technological research by commercial organizations is devoted to product and process innovations that have market value. Technologies are needed that produce 'social goods', such as improved air quality or increased product life, or that resolve problems normally outside the cost calculus of individual enterprises, such as the external costs of pollution or waste disposal.

The role of public policy is to ensure, through incentives and disincentives, that commercial organizations find it worthwhile to take fuller account of environmental factors in the technologies they develop. Publicly funded research institutions also need such direction, and the objectives of sustainable development and environmental protection must be built into the mandates of the institutions that work in environmentally sensitive areas.

The development of environmentally appropriate technologies is closely related to questions of risk management. Such systems as nuclear reactors, electric and other utility distribution networks, communication systems, and mass transportation are vulnerable if stressed beyond a certain point. The fact that they are connected through networks tends to make them immune to small disturbances but more vulnerable to unexpected disruptions that exceed a finite threshold. Applying sophisticated analyses of vulnerabilities and past failures to technology design, manufacturing standards, and contingency plans in operations can make the consequences of a failure or accident much less catastrophic.

The best vulnerability and risk analysis has not been applied consistently across technologies or systems. A major purpose of large system design should be to make the consequences of failure or sabotage less serious. There is thus a need for new techniques and technologies—as well as legal and institutional mechanisms—for safety design and control, accident prevention, contingency planning, damage mitigation, and provision of relief.

Environmental risks arising from technological and developmental decisions impinge on individuals and areas that have little or no influence on those decisions. Their interests must be taken into account. National and international institutional mechanisms are needed to assess potential impacts of new technologies before they are widely used, in order to ensure that their production, use, and disposal do not overstress environmental resources. Similar arrangements are required for major interventions in natural systems, such as river diversion or forest clearance. In addition, liability for damages from unintended consequences must be strengthened and enforced.

The common theme throughout this strategy for sustainable development is the need to integrate economic and ecological considerations in decision making. They are, after all, integrated in the workings of the real world. This will require a change in attitudes and objectives and in institutional arrangements at every level.

Economic and ecological concerns are not necessarily in opposition. For example, policies that conserve the quality of agricultural land and protect forests improve the long-term prospects for agricultural development. An increase in the efficiency of energy and material use serves ecological purposes but can also reduce costs. But the compatibility of environmental and economic objectives is often lost in the pursuit of individual or group gains, with little regard for the impacts on others, with a blind faith in science's ability to find solutions, and in ignorance of the distant consequences of today's decisions. Institutional rigidities add to this myopia.

One important rigidity is the tendency to deal with one industry or sector in isolation, failing to recognize the importance of intersectoral linkages. Modern agriculture uses substantial amounts of commercially produced energy and large quantities of industrial products. At the same time, the more traditional connection—in which agriculture is a source of raw materials for industry—is being diluted by the widening use of synthetics. The energy–industry connection is also changing, with a strong tendency towards a decline in the energy intensity of industrial production in industrial countries. In the Third World, however, the gradual shift of the industrial base towards the basic material-producing sectors is leading to an increase in the energy intensity of industrial production.

These intersectoral connections create patterns of economic and ecological interdependence rarely reflected in the ways in which policy is made. Sectoral organizations tend to pursue sectoral objectives and to treat their impacts on other sectors as side effects, taken into account only if compelled to do so. Hence impacts on forests rarely worry those involved in guiding public policy or business activities in the fields of energy, industrial development, crop husbandry, or foreign trade. Many of the environment and development problems that confront us have their roots in this sectoral fragmentation of responsibility. Sustainable development requires that such fragmentation be overcome.

Sustainability requires the enforcement of wider responsibilities for the impacts of decisions. This requires changes in the legal and institutional frameworks that will enforce the common interest. Some necessary changes in the legal framework start from the proposition that an environment adequate for health and well-being is essential for all human beings—including future generations. Such a view places the right to use public and private resources in its proper social context and provides a goal for more specific measures.

The law alone cannot enforce the common interest. It principally needs community knowledge and support, which entails greater public participation in the decisions that affect the environment. This is best secured by decentralizing the management of resources upon which local communities depend, and giving these communities an effective say over the use of these resources. It will also require promoting citizens' initiatives, empowering people's organizations, and strengthening local democracy.

Some large-scale projects, however, require participation on a different basis. Public inquiries and hearings on the development and environment impacts can help greatly in drawing attention to different points of view. Free access to relevant information and the availability of alternative sources of technical expertise can provide an informed basis for public discussion. When the environmental impact of a proposed project is particularly high, public scrutiny of the case should be mandatory and, wherever feasible, the decision should be subject to prior public approval, perhaps by referendum.

Changes are also required in the attitudes and procedures of both public and private-sector enterprises. Moreover, environmental regulation must move beyond the usual menu of safety regulations, zoning laws, and pollution control enactments; environmental objectives must be built into taxation, prior approval procedures for investment and technology choice, foreign trade incentives, and all components of development policy.

The integration of economic and ecological factors into the law and into decision-making systems within countries has to be matched at the international level. The growth in fuel and material use dictates that direct physical linkages between ecosystems of different countries will increase. Economic interactions through trade, finance, investment, and travel will also grow and heighten economic and ecological interdependence. Hence in the future, even more so than now, sustainable development requires the unification of economics and ecology in international relations.

CONCLUSION

In its broadest sense, the strategy for sustainable development aims to promote harmony among human beings and between humanity and nature. In the specific context of the development and environment crises of the 1980s, which current national and international political and economic institutions have not and perhaps cannot overcome, the pursuit of sustainable development requires:

- a political system that secures effective citizen participation in decision making,
- an economic system that is able to generate surpluses and technical knowledge on a self-reliant and sustained basis,
- a social system that provides for solutions for the tensions arising from disharmonious development,
- a production system that respects the obligation to preserve the ecological base for development,
- a technological system that can search continuously for new solutions,

- an international system that fosters sustainable patterns of trade and finance, and
- an administrative system that is flexible and has the capacity for self-correction.

These requirements are more in the nature of goals that should underlie national and international action on development. What matters is the sincerity with which these goals are pursued and the effectiveness with which departures from them are corrected.

16.2 PAUL W. TAYLOR

The Ethics of Respect for Nature

Historically, Western philosophy has paid little attention to the environment. The anthropocentric (human-centered) perspective that has dominated philosophical thought accords no intrinsic value to either nonhuman animate or inanimate objects. Thus, from this perspective, land that was unimproved by human labor was considered valueless, and the value of animals and plants was considered exclusively in terms of their satisfaction of human needs or interests.

During the twentieth century, however, the ecological perspective that views the need for human beings to exist in harmony with nature has begun to supersede the notion that nature is a hostile environment that needs to be subdued and dominated. Although anthropocentrism is still a dominant perspective, the biocentric, or ecocentric, views expressed in the writings of such environmentalists as John Muir and Aldo Leopold have grown in popularity. Leopold's proposal of a land ethic based upon the notion of a community that includes nonhuman life and inanimate objects as well as human beings has stimulated philosophers to consider the possibility of developing a general system of ethics based on an ecological perspective.

Of the recent attempts to create and justify a system of environmental ethics, the work of philosophy professor Paul W. Taylor of the City University of New York may be the most complete. The following selection is from "The Ethics of Respect for Nature," *Environmental Ethics* (Fall 1981). He has expanded upon this article for his book *Respect for Nature: A Theory of Environmental Ethics* (Princeton University Press, 1986). In the following selection, Taylor explains the meaning of the "inherent worth" of all living organisms and how this concept coupled with a rejection of human superiority can form the basis for a justifiable system of environmental ethics.

Key Concept: a life-centered system of ethics

HUMAN-CENTERED AND LIFE-CENTERED SYSTEMS OF ENVIRONMENTAL ETHICS

In this paper I show how the taking of a certain ultimate moral attitude toward nature, which I call "respect for nature," has a central place in the foundations of a life-centered system of environmental ethics. I hold that a set of moral norms (both standards of character and rules of conduct) governing human treatment of the natural world is a rationally grounded set if and only if, first, commitment to those norms is a practical entailment of adopting the attitude of respect for nature as an ultimate moral attitude, and second, the adopting of that attitude on the part of all rational agents can itself be justified. When the basic characteristics of the attitude of respect for nature are made clear, it will be seen that a life-centered system of environmental ethics need not be holistic or organicist in its conception of the kinds of entities that are deemed the appropriate objects of moral concern and consideration. Nor does such a system require that the concepts of ecological homeostasis, equilibrium, and integrity provide us with normative principles from which could be derived (with the addition of factual knowledge) our obligations with regard to natural ecosystems. The "balance of nature" is not itself a moral norm, however important may be the role it plays in our general outlook on the natural world that underlies the attitude of respect for nature. I argue that finally it is the good (well-being, welfare) of individual organisms, considered as entities having inherent worth, that determines our moral relations with the Earth's wild communities of life.

In designating the theory to be set forth as life-centered, I intend to contrast it with all anthropocentric views. According to the latter, human actions affecting the natural environment and its nonhuman inhabitants are right (or wrong) by either of two criteria: they have consequences which are favorable (or unfavorable) to human well-being, or they are consistent (or inconsistent) with the system of norms that protect and implement human rights. From this human-centered standpoint it is to humans and only to humans that all duties are ultimately owed. We may have responsibilities *with regard to* the natural ecosystems and biotic communities of our planet, but these responsibilities are in every case based on the contingent fact that our treatment of those ecosystems and communities of life can further the realization of human values and/or human rights. We have no obligation to promote or protect the good of nonhuman living things, independently of this contingent fact.

A life-centered system of environmental ethics is opposed to human-centered ones precisely on this point. From the perspective of a life-centered theory, we have prima facie moral obligations that are owed to wild plants and animals themselves as members of the Earth's biotic community. We are morally bound (other things being equal) to protect or promote their good for *their* sake. Our duties to respect the integrity of natural ecosystems, to preserve endangered species, and to avoid environmental pollution stem from the fact that these are ways in which we can help make it possible for wild species populations to achieve and maintain a healthy existence in a natural state. Such obligations are due those living things out of recognition of their inherent worth. They are entirely additional to and independent of the obligations we owe to our fellow humans. Although many of the actions that fulfill one set of obligations will also fulfill the other, two different grounds of obligation are involved. Their well-being, as well as human well-being, is something to be realized *as an end in itself.*

If we were to accept a life-centered theory of environmental ethics, a profound reordering of our moral universe would take place. We would begin to look at the whole of the Earth's biosphere in a new light. Our duties with respect to the "world" of nature would be seen as making prima facie claims upon us to be balanced against our duties with respect to the "world" of human civilization. We could no longer simply take the human point of view and consider the effects of our actions exclusively from the perspective of our own good.

THE GOOD OF A BEING AND THE CONCEPT OF INHERENT WORTH

What would justify acceptance of a life-centered system of ethical principles? In order to answer this it is first necessary to make clear the fundamental moral attitude that underlies and makes intelligible the commitment to live by such a system. It is then necessary to examine the considerations that would justify any rational agent's adopting that moral attitude.

Two concepts are essential to the taking of a moral attitude of the sort in question. A being which does not "have" these concepts, that is, which is unable to grasp their meaning and conditions of applicability, cannot be said to have the attitude as part of its moral outlook. These concepts are, first, that of the good (well-being, welfare) of a living thing, and second, the idea of an entity possessing inherent worth. I examine each concept in turn.

(1) Every organism, species population, and community of life has a good of its own which moral agents can intentionally further or damage by their actions. To say that an entity has a good of its own is simply to say that, without reference to any *other* entity, it can be benefited or harmed. One can act in its overall interest or contrary to its overall interest, and environmental conditions can be good for it (advantageous to it) or bad for it (disadvantageous to it). What is good for an entity is what "does it good" in the sense of enhancing or preserving its life and well-being. What is bad for an entity is something that is detrimental to its life and well-being.

We can think of the good of an individual nonhuman organism as consisting in the full development of its biological powers. Its good is realized to the extent that it is strong and healthy. It possesses whatever capacities it needs for successfully coping with its environment and so preserving its existence throughout the various stages of the normal life cycle of its species. The good of a population or community of such individuals consists in the population or community maintaining itself from generation to generation as a coherent system of genetically and ecologically related organisms whose average good is at an optimum level for the given environment. (Here *average good* means that the degree of realization of the good of *individual organisms* in the population or community is, on average, greater than would be the case under any other ecologically functioning order of interrelations among those species populations in the given ecosystem.

The idea of a being having a good of its own, as I understand it, does not entail that the being must have interests or take an interest in what affects its life for better or for worse. We can act in a being's interest or contrary to its interest without its being interested in what we are doing to it in the sense of wanting or not wanting us to do it. It may, indeed, be wholly unaware that favorable and unfavorable events are taking place in its life. I take it that trees, for example, have no knowledge or desires or feelings. Yet it is undoubtedly the case that trees can be harmed or benefited by our actions. We can crush their roots by running a bulldozer too close to them. We can see to it that they get adequate nourishment and moisture by fertilizing and watering the soil around them. Thus we can help or hinder them in the realization of their good. It is the good of trees themselves that is thereby affected. We can similarly act so as to further the good of an entire tree population of a certain species (say, all the redwood trees in a California valley) or the good of a whole community of plant life in a given wilderness area, just as we can do harm to such a population or community.

When construed in this way, the concept of a being's good is not coextensive with sentience or the capacity for feeling pain. William Frankena has argued for a general theory of environmental ethics in which the ground of a creature's being worthy of moral consideration is its sentience. I have offered some criticisms of this view elsewhere, but the full refutation of such a position, it seems to me, finally depends on the positive reasons for accepting a life-centered theory of the kind I am defending in this essay.

It should be noted further that I am leaving open the question of whether machines—in particular, those which are not only goal-directed, but also self-regulating—can properly be said to have a good of their own. Since I am concerned only with human treatment of wild organisms, species populations, and communities of life as they occur in our planet's natural ecosystems, it is to those entities along that the concept "having a good of its own" will here be applied. I am not denying that other living things, whose genetic origin and environmental condition have been produced, controlled, and manipulated by humans for human ends, do have a good of their own in the same sense as do wild plants and animals. It is not my purpose in this essay, however, to set out or defend the principles that should guide our conduct with regard to their good. It is only insofar as their production and use by humans have good or ill effects upon natural ecosystems and their wild inhabitants that the ethics of respect for nature comes into play.

(2) The second concept essential to the moral attitude of respect for nature is the idea of inherent worth. We take that attitude toward wild living things (individuals, species populations, or whole biotic communities) when and only when we regard them as entities possessing inherent worth. Indeed, it is only because they are conceived in this way that moral agents can think of themselves as having validly binding duties, obligations, and responsibilities that are *owed* to them as their *due.* I am not at this juncture arguing why they *should* be so regarded; I consider it at length below. But so regarding them is a pre-

supposition of our taking the attitude of respect toward them and accordingly understanding ourselves as bearing certain moral relations to them. This can be shown as follows:

What does it mean to regard an entity that has a good of its own as possessing inherent worth? Two general principles are involved: the principle of moral consideration and the principle of intrinsic value.

According to the principle of moral consideration, wild living things are deserving of the concern and consideration of all moral agents simply in virtue of their being members of the Earth's community of life. From the moral point of view their good must be taken into account whenever it is affected for better or worse by the conduct of rational agents. This holds no matter what species the creature belongs to. The good of each is to be accorded some value and so acknowledged as having some weight in the deliberations of all rational agents. Of course, it may be necessary for such agents to act in ways contrary to the good of this or that particular organism or group or organisms in order to further the good of others, including the good of humans. But the principle of moral consideration prescribes that, with respect to each being an entity having its own good, every individual is deserving of consideration.

The principle of intrinsic value states that, regardless of what kind of entity it is in other respects, if it is a member of the Earth's community of life, the realization of its good is something *intrinsically* valuable. This means that its good is prima facie worthy of being preserved or promoted as an end in itself and for the sake of the entity whose good it is. Insofar as we regard any organism, species population, or life community as an entity having inherent worth, we believe that it must never be treated as if it were a mere object or thing whose entire value lies in being instrumental to the good of some other entity. The well-being of each is judged to have value in and of itself.

Combining these two principles, we can now define what it means for a living thing or group of living things to possess inherent worth. To say that it possesses inherent worth is to say that its good is deserving of the concern and consideration of all moral agents, and that the realization of its good has intrinsic value, to be pursued as an end in itself and for the sake of the entity whose good it is.

The duties owed to wild organisms, species populations, and communities of life in the Earth's natural ecosystems are grounded on their inherent worth. When rational, autonomous agents regard such entities as possessing inherent worth, they place intrinsic value on the realization of their good and so hold themselves responsible for performing actions that will have this effect and for refraining from actions having the contrary effect.

THE ATTITUDE OF RESPECT FOR NATURE

Why should moral agents regard wild living things in the natural world as possessing inherent worth? To answer this question we must first take into account the fact that, when rational, autonomous agents subscribe to the principles of moral consideration and intrinsic value and so conceive of wild living things as having that kind of worth, such agents are *adopting a certain ultimate moral attitude toward the natural world*. This is the attitude I call "respect for nature." It parallels the attitude of respect for persons in human ethics. When we adopt the attitude of respect for persons as the proper (fitting, appropriate) attitude to take toward all persons as persons, we consider the fulfillment of the basic interests of each individual to have intrinsic value. We thereby make a moral commitment to live a certain kind of life in relation to other persons. We place ourselves under the direction of a system of standards and rules that we consider validly binding on all moral agents as such.

Similarly, when we adopt the attitude of respect for nature as an ultimate moral attitude we make a commitment to live by certain normative principles. These principles constitute the rules of conduct and standards of character that are to govern our treatment of the natural world. This is, first, an *ultimate* commitment because it is not derived from any higher norm. The attitude of respect for nature is not grounded on some other, more general, or more fundamental attitude. It sets the total framework for our responsibilities toward the natural world. It can be justified, as I show below, but its justification cannot consist in referring to a more general attitude or a more basic normative principle.

Second, the commitment is a *moral* one because it is understood to be a disinterested matter of principle. It is this feature that distinguishes the attitude of respect for nature from the set of feelings and dispositions that comprise the love of nature. The latter seems from one's personal interest in and response to the natural world. Like the affectionate feelings we have toward certain individual human beings, one's love of nature is nothing more than the particular way one feels about the natural environment and its wild inhabitants. And just as our love for an individual person differs from our respect for all persons as such (whether we happen to love them or not), so love of nature differs from respect for nature. Respect for nature is an attitude we believe all moral agents ought to have simply as moral agents, regardless of whether or not they also love nature. Indeed, we have not truly taken the attitude of respect for nature ourselves unless we believe this. To put it in a Kantian way, to adopt the attitude of respect for nature is to take a stance that one wills it to be a universal law for all rational beings. It is to hold that stance categorically, as being validly applicable to every moral agent without exception, irrespective of whatever personal feelings toward nature such an agent might have or might lack.

Although the attitude of respect for nature is in this sense a disinterested and universalizable attitude, anyone who does adopt it has certain steady, more or less permanent dispositions. These dispositions, which are themselves to be considered disinterested and universalizable, comprise three interlocking sets: dispositions to seek certain ends, dispositions to carry on one's practical reasoning and deliberation in a certain way, and dispositions to have certain feelings. We may accordingly analyze the attitude of respect for nature into the following components. (a) The disposition to aim at, and to take steps to bring about, as final and disinterested ends, the promoting and protecting of the good of organisms, species populations, and life communities in natural ecosystems. (These ends are "final" in not being pursued as means to further ends. They are "disinterested" in being independent of the self-interest of the agent.) (b) The disposition to consider actions that tend to realize those ends to be prima facie obligatory *because* they have that tendency. (c) The disposition to experience positive and negative feelings toward states of affairs in the world *because* they are favorable or unfavorable to the good of organisms, species populations, and life communities in natural ecosystems.

The logical connection between the attitude of respect for nature and the duties of a life-centered system of environmental ethics can now be made clear. Insofar as one sincerely takes that attitude and so has the three sets of dispositions, one will at the same time be disposed to comply with certain rules of duty (such as nonmaleficence and noninterference) and with standards of character (such as fairness and benevolence) that determine the obligations and virtues of moral agents with regard to the Earth's wild living things. We can say that the actions one performs and the character traits one develops in fulfilling these moral requirements are the way one *expresses* or *embodies* the attitude in one's conduct and character. In his famous essay, "Justice as Fairness," John Rawls describes the rules of the duties of human morality (such as fidelity, gratitude, honesty, and justice) as "forms of conduct in which recognition of others as persons is manifested." I hold that the rules of duty governing our treatment of the natural world and its inhabitants are forms of conduct in which the attitude of respect for nature is manifested.

THE JUSTIFIABILITY OF THE ATTITUDE OF RESPECT FOR NATURE

I return to the question posed earlier, which has not yet been answered: why *should* moral agents regard wild living things as possessing inherent worth? I now argue that the only way we can answer this question is by showing how adopting the attitude of respect for nature is justified for all moral agents. Let us suppose that we were able to establish that there are good reasons for adopting the attitude, reasons which are intersubjectively valid for every rational agent. If there are such reasons, they would justify anyone's having the three sets of dispositions mentioned above as constituting what it means to have the attitude. Since these include the disposition to promote or protect the good of wild living things as a disinterested and ultimate end, as well as the disposition to perform actions for the reason that they tend to realize that end, we see that such dispositions commit a person to the principles of moral consideration and intrinsic value. To be disposed to further, as an end in itself, the good of any entity in nature just because it is that kind of entity, is to be disposed to give consideration to *every* such entity and to place intrinsic value on the realization of its good. Insofar as we subscribe to these two principles we regard living things as possessing inherent worth. Subscribing to the principles is what is *means* to so regard them. To justify the attitude of respect for nature, then, is to justify commitment to these principles and thereby to justify regarding wild creatures as possessing inherent worth.

We must keep in mind that inherent worth is not some mysterious sort of objective property belonging to living things that can be discovered by empirical observation or scientific investigation. To ascribe inherent worth to an entity is not to describe it by citing some feature discernible by sense perception or inferable by inductive reasoning. Nor is there a logically necessary connection between the concept of a being having a good of its own and the concept of inherent worth. We do not contradict ourselves by asserting that an entity that has a good of its own lacks inherent worth. In order to show that such an entity "has" inherent worth we must give good reasons for ascribing that kind of value to it (placing that kind of value upon it, conceiving of it to be valuable in that way). Although it is humans (persons, valuers) who must do the valuing, for the ethics of respect for nature, the value so ascribed is not a human value. That is to say, it is not a value derived from considerations regarding human well-being or human rights. It is a value that is ascribed to nonhuman animals and plants themselves, independently of their relationship to what humans judge to be conducive to their own good.

Whatever reasons, then, justify our taking the attitude of respect for nature as defined above are also reasons that show why we *should* regard the living things of the natural world as possessing inherent worth. We saw earlier that, since the attitude is an ultimate one, it cannot be derived from a more fundamental attitude nor shown to be a special case of a more general one. On what sort of grounds, then, can it be established?

The attitude we take toward living things in the natural world depends on the way we look at them, on what kind of beings we conceive them to be, and on how we understand the relations we bear to them. Underlying and supporting our attitude is a certain *belief system* that constitutes a particular world view or outlook on nature and the place of human life in it. To give good reasons for adopting the attitude of respect for nature, then, we must first articulate the belief system which underlies and supports that attitude. If it appears that the belief system is internally coherent and well-ordered, and if, as far as we can now tell, it is consistent with all known scientific truths relevant to our knowledge of the object of the attitude (which in this case includes the whole set of the Earth's natural ecosystems and their communities of life), then there remains the task of indicating why scientifically informed and rational thinkers with a developed capacity of reality awareness can find it acceptable as a way of conceiving of the natural world and our place in it. To the extent we can do this we provide at least a reasonable argument for accepting the belief system and the ultimate moral attitude it supports.

I do not hold that such a belief system can be *proven* to be true, either inductively or deductively. As we shall see, not all of its components can be stated in the form of empirically verifiable propositions. Nor is its internal order governed by purely logical relationships. But the system as a whole, I contend, constitutes a coherent, unified, and rationally acceptable "picture" or "map" of a total world. By examining each of its main components and seeing how they fit together, we obtain a scientifically informed and well-ordered conception of nature and the place of humans in it.

The belief system underlying the attitude of respect for nature I call (for want of a better name) "the biocentric outlook on nature." Since it is not wholly analyzable into empirically confirmable assertions, it should not be thought of as simply a compendium of the biological sciences concerning our planet's ecosystems. It might best be described as a philosophical world view, to distinguish it from a scientific theory or explanatory system. However, one of its major tenets is the great lesson we have learned from the science of ecology: the interdependence of all living things in an organically unified order whose balance and stability are necessary conditions for the realization of the good of its constituent biotic communities.

Before turning to an account of the main components of the biocentric outlook, it is convenient here to set forth the overall structure of my theory of environmental ethics as it has now emerged. The ethics of respect for nature is made up of three basic elements: a belief system, an ultimate moral attitude, and a set of rules of duty and standards of character. These elements are connected with each other in the following manner. The belief system provides a certain outlook on nature which supports and makes intelligible an autonomous agent's adopting, as an ultimate moral attitude, the attitude of respect for nature. It supports and makes intelligible the attitude in the sense that, when an autonomous agent understands its moral relations to the natural world in terms of this outlook, it recognizes the attitude of respect to be the only *suitable* or *fitting* attitude to take toward all wild forms of life in the Earth's biosphere. Living things are now viewed as *the appropriate objects of the attitude of respect* and are accordingly regarded as entities possessing inherent worth. One then places intrinsic value on the promotion and protection of their good. As a consequence of this, one makes a moral commitment to abide by a set of rules of duty and to fulfill (as far as one can by one's own efforts) certain standards of good character. Given one's adoption of the attitude of respect, one makes that moral commitment because one considers those rules and standards to be validly binding on all moral agents. They are seen as embodying forms of conduct and character structures in which the attitude of respect for nature is manifested.

This three-part complex which internally orders the ethics of respect for nature is symmetrical with a theory of human ethics grounded on respect for persons. Such a theory includes, first, a conception of oneself and others as persons, that is, as centers of autonomous choice. Second, there is the attitude of respect for persons as persons. When this is adopted as an ultimate moral attitude it involves the disposition to treat every person as having inherent worth or "human dignity." Every human being, just in virtue of her or his humanity, is understood to be worthy of moral consideration, and intrinsic value is placed on the autonomy and well-being of each. This is what Kant meant by conceiving of persons as ends in themselves. Third, there is an ethical system of duties which are acknowledged to be owed by everyone to everyone. These duties are forms of conduct in which public recognition is given to each individual's inherent worth as a person.

This structural framework for a theory of human ethics is meant to leave open the issue of consequentialism (utilitarianism) versus nonconsequentialism (deontology). That issue concerns the particular kind of system of rules defining the duties of moral agents toward persons. Similarly, I am leaving open in this paper the question of what particular kind of system of rules defines our duties with respect to the natural world.

THE BIOCENTRIC OUTLOOK ON NATURE

The biocentric outlook on nature has four main components. (1) Humans are thought of as members of the Earth's community of life, holding that membership on the same terms as apply to all the nonhuman members. (2) The Earth's natural ecosystems as a totality are seen as a complex web of interconnected elements, with the sound biological functioning of each being dependent on the sound biological functioning of the others. (This is the component referred to above as the great lesson that the science of ecology has taught us). (3) Each individual organism is conceived as a teleological center of life, pursuing its own good in its own way. (4) Whether we are concerned with standards of merit or with the concept of inherent worth, the claim that humans by their very nature are superior to other species in a groundless claim and, in the light of elements (1), (2), and (3) above, must be rejected as nothing more than an irrational bias in our own favor.

The conjunction of these four ideas constitutes the biocentric outlook on nature.

16.3 DAVE FOREMAN

In Time of Crisis

A relatively small but active subdivision of the environmental movement, whose members espouse a more radical biocentric egalitarianism, has emerged over the past two decades. The Norwegian philosopher Arne Naess coined the term "deep ecology" to describe the belief system of this group. Naess and Sierra College philosophy professor George Sessions have attempted to codify the basic principles subscribed to by most deep ecologists. In addition to asserting the equal intrinsic worth of all living things, the principles include a prohibition against human-initiated reduction in the diversity of life except to satisfy vital needs, the need for a significant reduction in global human population, and a change in human ideology to that of appreciating life quality rather than aspiring to a higher standard of living.

Deep ecologists reject the policies of the mainstream environmental organizations, which they see as being restricted by their commitment to work within the confines of existing political structures. They have developed a more radical, activist program to try to achieve their ends. The best known leader of this program is Dave Foreman, cofounder of the Earth First! organization, which has the motto "No compromise in defense of Mother Earth." Earth First! activists have gained notoriety by chaining themselves to redwood trees targeted to be cut down by lumbermen and by sabotaging equipment to be used to develop wild areas. Foreman rejects the assertion that Earth First! is a violent organization, and he points out that members only destroy "machines or property that are used to destroy the natural world."

The following selection is from *Confessions of an Eco-Warrior* (Harmony Books, 1991). In it, Foreman explains the goals, principles, and practices of his ecological warriors.

Key Concept: deep ecology and the campaign to implement its principles

In wildness is the preservation of the world.

—Henry David Thoreau

We are living now in the most critical moment in the three-and-a-half-billion-year history of life on Earth. For this unimaginably long time, life has been developing, expanding, blossoming, and diversifying, filling every available niche with different manifestations of itself, intertwined in complex, globe-girdling relationships. But today this diversity of perhaps 30 million species faces radical and unprecedented change. Never before—not even during the mass extinctions of the dinosaurs at the end of the Cretaceous era, 65 million years ago—has there been such a high rate of extinction as we are now witnessing, such a drastic reduction in the planet's biological diversity.

Over the last three or four hundred years, human civilization has declared war on large mammals, leading some respected ecologists to assert that the only large mammals living twenty years from now will be those we humans choose to allow to live. Other prominent biologists, looking aghast on the wholesale devastation of tropical rain forests and temperate-zone old-growth forests, rapidly accelerating desertification, rapacious commercial fishing, and wasting of high-profile large mammals like whales, elephants, and Tigers ("charismatic megafauna") owing to habitat destruction and poaching, say that Earth could lose one-quarter to one-third of *all* species within forty years.

Not only is this blitzkrieg against the natural world destroying ecosystems and their associated species, but our activities are now beginning to have fundamental, systemic effects upon the entire life-support apparatus of the planet: upsetting the world's climate; poisoning the oceans; destroying the atmospheric ozone layer that protects us from excessive ultraviolet radiation; changing the CO_2 ratio in the atmosphere and causing the "greenhouse effect"; and spreading acid rain, radioactive fallout, pesticides, and industrial contamination throughout the biosphere. Indeed, Professor Michael Soulé, founder of the Society for Conservation Biology, recently warned that vertebrate evolution may be at an end due to the activities of industrial humans.

Clearly, in such a time of crisis, the conservation battle is not one of merely protecting outdoor recreation opportunities, or a matter of aesthetics, or "wise management and use" of natural resources. It is a battle for life itself, for the continued flow of evolution. We—this generation of humans—are at our most important juncture since we came out of the trees six million years ago. It is our decision, ours today, whether Earth continues to be a marvelously living, diverse oasis in the blackness of space, or whether the "charismatic megafauna" of the future will consist of Norway Rats and cockroaches.

How have we arrived at this state, at this threshold of biotic terror? Is it because we have forgotten our "place in nature," as the Native American activist Russell Means says?

If there is one thing upon which the nation states of the world today can agree, one thing at which the United States and the Soviet Union, Israel and Iran, South Africa and Angola, Britain and Argentina, China and India, Japan and Malaysia nod in unison, it is that human beings are the measure of all value. As Gifford Pinchot, founder of the United States Forest Service, said, there are only two things on Earth: human beings and natural resources. Humanism is the philosophy that runs the business engines of the modern world.

The picture that most humans have of the natural world is that of a smorgasbord table, continually replenished by a magic kitchen hidden somewhere in the background. While most people perceive that there are gross and immoral inequities in the sizes of the plates handed out and in the number of times some are allowed to belly up to the bar, few of us question whether the items arrayed are there for their sole use, nor do they imagine that the table will ever become empty.

There is another way to think about man's relationship to the natural world, an insight pioneered by the nineteenth-century conservationist and mountaineer John Muir and later by the science of ecology. This is the idea that all things are connected, interrelated, that human beings are merely one of the millions of species that have been shaped by the process of evolution for three and a half billion years. According to this view, all living beings have the same right to be here. This is how I see the world.

With that understanding, we can answer the question, "Why wilderness?"

Is it because wilderness makes pretty picture postcards? Because it protects watersheds for downstream use by agriculture, industry, and homes? Because it's a good place to clean the cobwebs out of our heads after a long week in the auto factory or over the video display terminal? Because it preserves resource-extraction opportunities for future generations of humans? Because some unknown plant living in the wilds may hold a cure for cancer?

No—the answer is, because wilderness *is.* Because it is the real world, the flow of life, the process of evolution, the repository of that three and a half billion years of shared travel.

A Grizzly Bear snuffling along Pelican Creek in Yellowstone National Park with her two cubs has just as much right to life as any human has, and is far more important ecologically. All things have intrinsic value, inherent worth. Their value is not determined by what they will ring up on the cash register of the gross national product, or by whether or not they are *good.* They are good because they exist.

Even more important than the individual wild creature is the wild community—the wilderness, the stream of life unimpeded by human manipulation.

We, as human beings, as members of industrial civilization, have no divine mandate to pave, conquer, control, develop, or use every square inch of this planet. As Edward Abbey, author of *Desert Solitaire* and *The Monkey Wrench Gang,* said, we have a right to be here, yes, but not everywhere, all at once.

The preservation of wilderness is not simply a question of balancing competing special-interest groups, arriving at a proper mix of uses on our public lands, and resolving conflicts between different outdoor recreation preferences. It is an ethical and moral matter. A religious mandate. Human beings have stepped beyond the bounds; we are destroying the very process of life.

The forest ranger and wilderness proponent Aldo Leopold perhaps stated this ethic best:

A thing is right when it tends to preserve the integrity, stability, and beauty of the biotic community. It is wrong when it tends otherwise.

The crisis we now face calls for *passion*. When I worked as a conservation lobbyist in Washington, D.C., I was told to put my heart in a safe deposit box and replace my brain with a pocket calculator. I was told to be rational, not emotional, to use facts and figures, to quote economists and scientists. I would lose credibility, I was told, if I let my emotions show.

But, damn it, I am an animal. A living being of flesh and blood, storm and fury. The oceans of the Earth course through my veins, the winds of the sky fill my lungs, the very bedrock of the planet makes my bones. I am alive! I am not a machine, a mindless automaton, a cog in the industrial world, some New Age android. When a chain saw slices into the heartwood of a two-thousand-year-old Coast Redwood, it's slicing into my guts. When a bulldozer rips through the Amazon rain forest, it's ripping into my side. When a Japanese whaler fires an exploding harpoon into a great whale, my heart is blown to smithereens. I am the land, the land is me.

Why shouldn't I be emotional, angry, passionate? Madmen and madwomen are wrecking this beautiful, blue-green, living Earth. Fiends who hold nothing of value but a greasy dollar bill are tearing down the pillars of evolution a-building for nearly four thousand million years.

In this world ruled by MBAs, we are taught to use only a fraction of our minds: the left hemisphere of the brain, the rational, calculating part. That portion of our brain is valuable and necessary, but it is not the sole seat of our consciousness. We must get back in touch with the emotional, intuitive right hemisphere of our brain, with our reptilian cortex, with our entire body. Then we must go beyond that to think with the whole Earth. David Brower, onetime executive director of the Sierra Club, has pointed out that you cannot imprison a California Condor in the San Diego Zoo and still have a condor. The being of a condor does not end at the tips of the black feathers on its wings. The condor is *place* as well; it is the thermals rising over the Coast Range, the outcroppings on which it lays its eggs, the carrion on which it feeds.

Society has lobotomized us. Our social environment today can work as a drug, like *soma* in *Brave New World,* to keep us in line, to sedate us, to remove our capacity for passion. Robots do not ask questions. Free men and women do. Wild animals cannot be ruled; they can be domesticated, yes, they can be broken, but then they are no longer free, no longer wild.

We must break out of society's freeze on our passions, we must become animals again. We must feel the tug of the full moon, hear goose music overhead. We must love Earth and rage against her destroyers. We must open ourselves to relationships with one another, with the land; we must dare to love, to feel for something—some*one*—else. And when that final kiss of life—death—comes, we mustn't hide, but rather go *joyously* into that good night. When I die, I don't want to be pickled and put away in a lead box. Place me out in the wilderness, let me revel at rejoining the food chain, at being recycled into weasel, vulture, worm, and mold.

Breaking free from the gilded chains of civilized banality is not easy. One cannot achieve a state of wilderness grace through books, through intellectualization, through rational argument. Our passion comes from our connection to the Earth and it is only through direct interaction with the wilderness that we can unite our minds and our bodies with the land, realizing that there is no separation.

Along with passion, we need *vision.* Why should we content ourselves with the world the way it is handed to us by Louisiana-Pacific, Mitsubishi, the Pentagon, and Exxon? Why should we be constrained by the narrow alternatives presented us by Congress and the Forest Service in discussing protection of the land?

We are told that the Gray Wolf and Grizzly Bear are gone from most of the West and can never be restored, that the Elk and Bison and Panther are but shades in the East and will not come back, that Glen Canyon and Hetch Hetchy are beneath dead reservoir water and we shall never see them again, that the Tall Grass Prairie and Eastern Deciduous Forest are only memories and that we can never have big wilderness east of the Rockies again.

Bunk! Why should we be bound by past mistakes? It is up to us to challenge the government and the people with a vision of Big Wilderness, a vision of humans living modestly in a community that also includes bears and rattlesnakes and salmon and oaks and sage-brush and mosquitoes and algae and streams and rocks and clouds.

We should demand that roads be closed and clearcuts rehabilitated, that dams be torn down, that wolves, Grizzlies, Cougars, River Otters, Bison, Elk, Pronghorn, Bighorn Sheep, Caribou, and other extirpated species be reintroduced to their native habitats. We must envision and propose the restoration of biological wildernesses of several million acres in all of America's ecosystems, with corridors between them for the transmission of genetic variability. Wilderness is the arena for evolution, and there must be enough of it for natural forces to have free rein.

John Seed, the Australian founder of the Rainforest Information Centre, tells of a meeting he had with a group of Australian Aborigines in Sydney. After the meeting, they stepped outside into the night air. The great city spread out before them. One of the Abos asked, "What do you see? What do you see out *there*?"

John looked at the pulsating freeways, towers of anodized glass and steel, ships in the harbor, and replied, "I see a city. Lights, pavement, skyscrapers . . . "

The Abo said quietly, "We still see the land. Beneath the concrete we know where the forest grows, where the kangaroos graze. We see where the Platypus digs her den, where the streams flow. That city there . . . it's just a scab. The land remains alive beneath it."

So it is in North America. In the scrub forests of New England, the spirits of 220-foot-tall White Pines still stand. In the feedlots and cornfields of the Great Plains, ghost hooves of Bison and howls of wolves echo back from a century ago. On San Francisco shores, phantom Grizzlies feed on the beached carcasses of whales.

The genocide against these wilderness nations waged around the world by civilized humans has been going on for only an instant in evolutionary time. Some species are gone forever, some ecosystems are hopelessly muddled, but in most cases the land, the wild land, is still alive beneath the scab of concrete. Do we have the sight to see?

Passion and vision are essential, but without *action* they are empty. It is easy to be immobilized by the sheer magnitude of the problems facing Earth, by tasks calling for Hercules when we know we are puny mortals. We feel daunted about demanding changes when we know that our lives are not pure, that we share the lifestyle that is ravaging the planet. We feel powerless in confronting the vast, immobile gray bureaucracy of government and industry.

"It's too much," we whimper, and surrender. "Better not to fight than to be defeated. Besides, where does one person start? I'm not an expert or a leader. Why don't *they* do something?"

We are frozen because the problems are too big. It's easier to turn on the TV, to plunge into the modern game (whoever dies with the most toys, wins!), to dull our expectations and our passions with drink or with lines of white powder.

The Earth is crying. Do we hear? Martin Luther King, Jr., once said that if one has nothing worth dying for, one has nothing worth living for.

It is a time for courage.

There are many forms of courage. It takes courage to not allow your children to become addicted to television. It takes courage to tell the conservation group to which you belong, *No more compromise!* It takes courage to say no more growth in your community. It takes courage to say that the wild is more important than jobs. It takes courage to write letters to your local newspaper. It takes courage to stand up at a public hearing and speak. It takes courage to live a lower-impact life.

And it takes courage to put your body between the machine and the wilderness, to stand before the chain saw or the bulldozer.

In 1848, Henry David Thoreau went to jail for refusing, as a protest against the Mexican War, to pay his poll tax. When Ralph Waldo Emerson came to bail him out, Emerson said, "Henry, what are you doing in there?"

Thoreau quietly replied, "Ralph, what are you doing out there?"

In this insane world where short-term greed rules over long-term life, those of us with a land ethic, with vision and passion, must face the mad machine. We must stand before it as 19-year-old Oregon Earth First! activist Valerie Wade did when she climbed eighty feet up into an ancient Douglas-fir to keep it from being cut down; as Wyoming guide and outfitter and Earth First! founder Howie Wolke did when he pulled up survey stakes along a proposed gas-exploration road in prime Elk habitat. Both put their lives in jeopardy, both went to jail. Both were proud of what they did. Both are heroes of the Earth, as are hundreds of others who have demonstrated courage in defense of the wild.

This defense is not an arrogant defense, an attitude of Lord Man protecting something less than himself. Rather, it is a humble joining with Earth, becoming the rain forest, the desert, the mountain, the wilderness in defense of yourself. It is through becoming part of the wild that we find courage far greater than ourselves, a union that gives us boldness to stand against hostile humanism, against the machine, against the dollar, against jail, against extinction for what is sacred and right: the Great Dance of Life.

Eighty years ago, Aldo Leopold graduated from the Yale School of Forestry and went to work for the newly-created United States Forest Service in the territories of Arizona and New Mexico. He was put to work inventorying potential timber resources in the high, wild White Mountains of eastern Arizona, which were a great roadless area then. One day Leopold stopped for lunch with his crew on a rimrock overlooking a turbulent stream. As they ate, they saw a large animal ford the *rillito.* They thought at first it was a doe, but as a rolling bunch of pups came out of the willows to greet their mother, they realized it was a wolf. In those days, a wolf you saw was a wolf you shot. Leopold and his men hurriedly pulled their .30-30s from the scabbards on their horses and began to blast away. The wolf dropped, a pup dragged a shattered leg into the rocks, and Leopold rode down to finish the job. He later wrote:

> We reached the old wolf in time to watch a fierce green fire dying in her eyes. I realized then, and have known ever since, that there was something new to me in those eyes—something known only to her and to the mountain. I was young then, and full of trigger itch; I thought that because fewer wolves meant more deer, that no wolves would mean hunters' paradise. But after seeing the green fire die, I sensed that neither the wolf nor the mountain agreed with such a view.

Green fire. We need it in the eyes of the wolf. We need it in the land. And we need it in our own eyes.

16.4 VANDANA SHIVA

Women's Indigenous Knowledge and Biodiversity Conservation

Ecologically focused feminists who call themselves *ecofeminists* see important connections between the domination of women and the domination of nature under the patriarchal social and political framework that characterizes most of the world's human cultures. As with most groups that are defined by certain shared ideological perspectives, ecofeminists differ among themselves about the set of principles and beliefs that distinguish their perspective from that of other environmentalists. For example, not all of them agree that women have an intrinsic feminine perspective on the relationship between humans and nature that is distinct from that of males. Most ecofeminists, however, believe that their experiences as women in male-dominated societies provides them with a different way of knowing and thinking about environmental issues.

Vandana Shiva is a physicist, philosopher, and feminist. She directs the Research Foundation for Science Technology and Natural Resource Policy in Dehradun, India. Shiva is an environmental activist as well as a prolific writer and lecturer. Her ecological perspective is informed by the domination and exploitation of Third World peoples by Western industrial nations as well as by the oppression of women in patriarchal societies.

The following selection is from "Women's Indigenous Knowledge and Biodiversity Conservation," which was published in Maria Mies and Vandana Shiva, *Ecofeminism* (Zed Books, 1993). In it, Shiva argues that women's traditional roles in society make them particularly resistant to the economically driven policies that threaten to destroy biodiversity.

Key Concept: women's knowledge as central to the preservation of biodiversity

Gender and diversity are linked in many ways. The construction of women as the 'second sex' is linked to the same inability to cope with difference as is the development paradigm that leads to the displacement and extinction of diversity in the biological world. The patriarchal world view sees man as the measure of all value, with no space for diversity, only for hierarchy. Woman, being different, is treated as unequal and inferior. Nature's diversity is seen as not intrinsically valuable in itself, its value is conferred only through economic exploitation for commercial gain. This criterion of commercial value thus reduces diversity to a problem, a deficiency. Destruction of diversity and the creation of monocultures becomes an imperative for capitalist patriarchy.

The marginalization of women and the destruction of biodiversity go hand in hand. Loss of diversity is the price paid in the patriarchal model of progress which pushes inexorably towards monocultures, uniformity and homogeneity. In this perverted logic of progress, even conservation suffers. Agricultural 'development' continues to work towards erasing diversity, while the same global interests that destroy biodiversity urge the Third World to conserve it. This separation of production and consumption, with 'production' based on uniformity and 'conservation' desperately attempting to preserve diversity militates against protecting biodiversity. It can be protected only by making diversity the basis, foundation and logic of the technology and economics of production.

The logic of diversity is best derived from biodiversity and from women's links to it. It helps look at dominant structures from below, from the ground of diversity, which reveal monocultures to be unproductive and the knowledge that produces them as primitive rather than sophisticated.

Diversity is, in many ways, the basis of women's politics and the politics of ecology; gender politics is largely a politics of difference. Eco-politics, too, is based on nature's variety and difference, as opposed to industrial commodities and processes which are uniform and homogeneous.

These two politics of diversity converge when women and biodiversity meet in fields and forest, in arid regions and wetlands.

DIVERSITY AS WOMEN'S EXPERTISE

Diversity is the principle of women's work and knowledge. This is why they have been discounted in the patriarchal calculus. Yet it is also the matrix from which an alternative calculus of 'productivity' and 'skills' can be built that respects, not destroys, diversity.

The economies of many Third World communities depend on biological resources for their sustenance and well-being. In these societies, biodiversity is simultaneously a means of production, and an object of consumption. The survival and sustainability of livelihoods is ultimately connected to the conservation and sustainable use of biological resources in all their diversity. Tribal and peasant societies' biodiversity-based technologies, however, are seen as backward and primitive and are, therefore, displaced by 'progressive' technologies that destroy both diversity and people's livelihoods.

There is a general misconception that diversity-based production systems are low-productivity systems. However, the high productivity of uniform and homogenous systems is a contextual and theoretically constructed category, based on taking into account only one-dimensional yields and outputs. The alleged low productivity of the one against the alleged high productivity of the other is, therefore, not a neutral, scientific measure but biased towards commercial interests for whom maximizing the one-dimensional output is an economic imperative.

Crop uniformity, however, undermines the diversity of biological systems which form the production system as well as the livelihoods of people whose work is associated with diverse and multiple-use systems of forestry, agriculture and animal husbandry. For example, in the state of Kerala in India (its name derives from the coconut palm), coconut is cultivated in a multi-layered, high-intensity cropping system, along with betel and pepper vines, bananas, tapioca, drumstick, papaya, jackfruit, mango and vegetables. The annual labour requirement in a monoculture of coconut palm is 157 man-days per ha, while in a mixed cropping system, it is 960 man-days per ha. In the dry-land farming systems of the Deccan, the shift from mixed cropping millets, pulses and oilseeds to eucalyptus monocultures led to an annual loss of employment of 250 man-days per ha.

When labour is scarce and costly, labour displacing technologies are productive and efficient, but when labour is abundant, labour displacement is unproductive because it leads to poverty, dispossession and destruction of livelihoods. In Third World situations, sustainability has therefore to be achieved at two levels simultaneously: sustainability of natural resources and sustainability of livelihoods. Consequently, biodiversity conservation must be linked to conservation of livelihoods derived from biodiversity.

Women's work and knowledge is central to biodiversity conservation and utilization both because they work between 'sectors' and because they perform multiple tasks. Women, as farmers, have remained invisible despite their contribution. Economists tend to discount women's work as 'production' because it falls outside the so-called 'production boundary'. These omissions arise not because too few women work, but too many women do too much work of too many different kinds.

Statisticians and researchers suffer a conceptual inability to define women's work inside and outside the house—and farming is usually part of both. This recognition of what is and is not labour is exacerbated by the great volume and variety of work that women do. It is also related to the fact that although women work to sustain their families and communities, most of what they do is not measured in wages. Their work is also invisible because they are concentrated outside market-related or remunerated work, and they are normally engaged in multiple tasks.

Time allocation studies, which do not depend on an a priori definition of work, reflect more closely the multiplicity of tasks undertaken, and the seasonal, even daily movement in and out of the conventional labour force which characterize most rural women's livelihood strategy. Gender studies now being published, confirm that women in India are major producers of food in terms of value, volume and hours worked.

In the production and preparation of plant foods, women need skills and knowledge. To prepare seeds they need to know about seed preparation, germination requirements and soil choice. Seed preparation requires visual discrimination, fine motor co-ordination, sensitivity to humidity levels and weather conditions. To sow and strike seeds demands knowledge of seasons, climate, plant requirements, weather conditions, micro-climatic factors and soil-enrichment; sowing seeds requires physical dexterity and strength. To properly nurture plants calls for information about the nature of plant diseases, pruning, staking, water supplies, companion planting, predators, sequences, growing seasons and soil maintenance. Persistence and patience, physical strength and attention to plant needs are essential. Harvesting requires judgements in relation to weather, labour and grading; and knowledge of preserving, immediate use and propagation.

Women's knowledge has been the mainstay of the indigenous dairy industry. Dairying, as managed by women in rural India, embodies practices and logic rather different from those taught in dairy science at institutions of formal education in India, since the latter is essentially an import from Europe and North America. Women have been experts in the breeding and feeding of farm animals, including not only cows and buffaloes but also pigs, chickens, ducks and goats.

In forestry too, women's knowledge is crucial to the use of biomass for feed and fertilizer. Knowledge of the feed value of different fodder species, the fuel value of firewood types, and of food products and species is essential to agriculture-related forestry in which women are predominately active. In low input agriculture, fertility is transferred from forest and farm trees to the field by women's work either directly or via animals.

Women's work and knowledge in agriculture is uniquely found in the spaces 'in between' the interstices of 'sectors', the invisible ecological flows between sectors, and it is through these linkages that ecological stability, sustainability and productivity under resource-scarce conditions are maintained. The invisibility of women's work and knowledge arises from the gender bias which has a blind spot for realistic assessment of women's contributions. It is also rooted in the sectoral, fragmented and reductionist approach to development which treats forests, livestock and crops as independent of each other.

The focus of the 'green revolution' has been increasing grain yields of rice and wheat by techniques such as dwarfing, monocultures and multicropping. For an Indian woman farmer, rice is not only food, but also a source of cattle fodder and straw for thatch. High yield varieties (HYVs) can increase women's work; the shift from local varieties and indigenous crop-improvement strategies can also take away women's control over seeds and genetic resources. Women have been seed custodians since time immemorial, and their knowledge and skills should be the basis of all crop-improvement strategies.

WOMEN: CUSTODIANS OF BIODIVERSITY

In most cultures women have been the custodians of biodiversity. They produce, reproduce, consume and conserve biodiversity in agriculture. However, in common with all other aspects of women's work and knowledge, their role in the development and conservation of biodiversity has been rendered as non-work and non-knowledge. Their labour and expertise has been defined into nature, even though it is based on sophisticated cultural and scientific practises. But women's biodiversity conservation differs from the dominant patriarchal notion of biodiversity conservation.

Recent concern with biodiversity at the global level has grown as a result of the erosion of diversity due to the expansion of large-scale monoculture-based agricultural production and its associated vulnerability. Nevertheless, the fragmentation of farming systems linked to the spread of monocultures continues to be the guiding paradigm for biodiversity conservation. Each element of the farm eco-system is viewed in isolation, and conservation of diversity is seen as an arithmetical exercise of collecting varieties.

In contrast, in the traditional Indian setting, biodiversity is a relational category in which each element acquires its characteristics and value through its relationships with other elements. Biodiversity is ecologically and culturally embedded. Diversity is reproduced and conserved through the reproduction and conservation of culture, in festivals and rituals which not only celebrate the renewal of life, but also provide a platform for subtle tests for seed selection and propagation. The dominant world-view does not regard these tests as scientific because they do not emerge from the laboratory and the experimental plot, but are integral to the total world-view and lifestyle of people and are carried out, not by men in white coats, but by village woman. But because it is thus that the rich biological diversity in agriculture has been preserved they are systematically reliable.

When women conserve seed, they conserve diversity and therefore conserve balance and harmony. *Navdanya* or nine seeds are the symbol of this renewal of diversity and balance, not only of the plant world, but of the planet and of the social world. This complex relationship web gives meaning to biodiversity in Indian culture and has been the basis of its conservation over millennia....

BIOTECHNOLOGY AND THE DESTRUCTION OF BIODIVERSITY

There are a number of crucial ways in which the Third World women's relationship to biodiversity differs from corporate men's relationship to biodiversity. Women produce through biodiversity, whereas corporate scientists produce through uniformity.

For women farmers, biodiversity has intrinsic value—for global seed and agribusiness corporations, biodiversity derives its value only as 'raw material' for the biotechnology industry. For women farmers the essence of the seed is the continuity of life. For multinational corporations, the value of the seed lies in the discontinuity of its life. Seed corporations deliberately breed seeds that cannot give rise to future generations so that farmers are transformed from seed custodians into seed consumers. Hybrid seeds are 'biologically patented' in that the offspring cannot be used as seeds as farmers must go back to corporations to buy seed every year. Where hybrids do not force the farmers back to the market, legal patents and intellectual property rights' are used to prevent farmers from saving seed. Seed patents basically imply that corporations treat seed as their 'creation.' Patents prevent others from 'making' the patented product, hence patented seed cannot be used for making seed. Royalties have to be paid to the company that gets the patent.

The claim of 'creation' of life by corporate scientists is totally unjustified, it is in fact an interruption in the life flow of creation. It is also unjustified because nature and Third World farmers have made the seed that corporations are attempting to own as their innovation and their private property. Patents on seeds are thus a twenty-first century form of piracy, through which the shared heritage and custody of Third World women peasants is robbed and depleted by multinational corporations, helped by global institutions like GATT [General Agreement on Tariffs and Trade].

Patents and biotechnology contribute to a two-way theft. From Third World producers they steal biodiversity. From consumers everywhere they steal safe and healthy food.

Genetic engineering is being offered as a 'green' technology worldwide. President Bush ruled in May 1992 that genetically engineered foods should be treated as 'natural' and hence safe. However, genetic engineering is neither natural nor safe.

A number of risks associated with genetically engineered foods have been listed recently by the Food and Drug Administration of the US:

- New toxicants may be added to genetically engineered food.
- Nutritional quality of engineered food may be diminished.
- New substances may significantly alter the composition of food.
- New proteins that cause allergic reactions may enter the food supply.
- Antibiotic resistant genes may diminish the effectiveness of some antibiotics to human and domestic animal diseases.
- The deletion of genes may have harmful side effects.
- Genetic engineering may produce 'counterfeit freshness'.
- Engineered food may pose risks to domestic animals.
- Genetically engineered food crops may harm wildlife and change habitats.

When we are being asked to trust genetically engineered foods, we are being asked to trust the same companies that gave us pesticides in our food. Monsanto, which is now selling itself as Green was telling us that 'without chemicals, millions more would go hungry'. Today, when Bhopal has changed the image of these poisons, we are being told by the Monsantos, Ciba-Geigys, Duponts, ICIs and Dows that they will now give us Green products. However, as Jack Kloppenberg has recently said, 'Having been recognized as wolves, the industrial semoticians want to redefine themselves as sheep, and green sheep at that.'

16.5 MARK HERTSGAARD

Ambassadors from Another Time

In recent years many scientists have issued warnings about the continued decline of global ecological health. Readers of these warnings should be aware that no matter how much research is involved in coming to these conclusions, most are based on secondhand knowledge rather than the actual experience of the writer.

Mark Hertsgaard, a journalist and media critic, is the author of several books as well as numerous articles in such publications as *The New Yorker, The Atlantic Monthly, Vanity Fair,* and *The Nation.* The global environment has long been one of his principal concerns. By 1991 he was seriously troubled by conflicting reports about whether or not present developmental practices are likely to bring about a worldwide ecological crisis. During that year he embarked on a six-year journey around the world. He traveled to 19 countries in order to obtain a global perspective on issues concerning the environment.

Earth Odyssey: Around the World in Search of Our Environmental Future (Broadway Books, 1998), from which the following selection is taken, is Hertzgaard's vivid journalistic record of his global investigation. The report has been lauded by such noted environmentalists as Ralph Nader, Lester R. Brown, and Bill McKibben for its vivid, detailed description of the enormous implications of a rapidly deteriorating environment coupled with a growing population. But, despite the enormity of the task of reversing these trends before disaster overcomes us, Hertsgaard emerges from his quest optimistic that it can be accomplished. He cites the moon race of the 1960s and the New Deal of the 1930s as models of how inspiration and commitment might be combined to bring about the broad cooperative effort and the personal sacrifices needed for the top-to-bottom redesign of our economy and our industries.

Key Concept: firsthand report on our global environmental crisis

Climb the mountains and get their glad tidings. Nature's peace will flow into you as sunshine flows into trees. The winds will blow their own freshness into you, and the storms their energy, while cares drop off like autumn leaves.

—John Muir
American naturalist

Three hours drive east of San Francisco, in the western foothills of California's Sierra Nevada mountains, live some of the largest trees on earth. Local people call them, bluntly, the Big Trees. More properly, they are named giant sequoia redwoods, and in truth they are the largest living organisms of any kind on this planet, and very nearly the oldest as well.

The oldest known giant sequoia is 3,300 years old—it was already ancient when Jesus, Buddha, and Mohammed walked the earth—and 2,000 year lifespans are common. Sequoias live in such exquisite balance with their environment that they do not age. The largest of them stand over three hundred feet tall, and they measure as much as thirty-two feet in diameter near the ground: It takes twenty adults with their arms extended to encircle one. Their limbs resemble the trunks of ordinary trees, their root systems encompass an entire acre of soil. Viewed from a distance through afternoon shadows, giant sequoias appear supernaturally oversized, as if they belong to a separate and mightier reality, which in a sense they do.

The author John Steinbeck called giant sequoias "ambassadors from another time" because, as a species, they have existed since the age of the dinosaurs, some 100 million years ago. Redwoods and dinosaurs, perhaps the largest plants and animals this planet has ever known, were the dominant life forms throughout the northern hemisphere for tens of millions of years. But during the last major climatic change, 65 million years ago, when the earth's temperature cooled dramatically, the dinosaurs failed to adapt and became extinct. Giant sequoias survived. Over time, their geographical reach has shrunk, however, and today they grow naturally only in some seventy-five groves scattered along the western slopes of the Sierra Nevada mountain chain.

It was my good fortune... to spend a year living near one of those seventy-five groves. This was in Calaveras County, an area made famous by Mark Twain's early writings about jumping frogs and gold mining. My friends, the Smiths, had rented me a log cabin that was nestled so deep in the forest it was impossible to find if you didn't know exactly where you were going. At the end of the paved road that led back to civilization, a dirt path led up to an old logging trail that was inaccessible in winter but sprinkled with pale golden pine needles the rest of the year. There was no telephone in the cabin, nor television, but I did have access to my friends' truck for once-a-week grocery runs.

My nearest human neighbors lived over a mile away, so I saw bears on my porches as often as I saw people. One morning, I was sitting on the cabin floor, writing away, when a mother bear and two young cubs started scratching at the kitchen door. Luckily, the mother couldn't see inside the cabin well enough to feel secure about pushing the door down to get in. Instead, the trio trooped around to the clearing in front of the cabin, while I stepped out onto the porch above them. The bears spent the next hour in that clearing, fully aware of my presence five yards away but paying me no real attention. The cubs scratched for bugs inside fallen tree trunks, the mother drowsed in the sunshine, the insects droned. Just another day in paradise. And then, too soon, the bears padded slowly into the forest, momma in the lead, the youngsters behind, leaving me a memory I shall treasure the rest of my life.

The forest was full of such delights. Besides bears, I regularly saw deer, coyotes, hawks, and owls, even the occasional mountain lion. Best of all, though, it took only forty minutes to hike over the ridge and visit the giant sequoias in the North Grove of Big Trees State Park.

Even at the height of springtime's exuberance, when Douglas squirrels squawked a counterpoint to the insistent tapping of whiteheaded woodpeckers and dogwood blossoms splashed color among the awakening undergrowth, the grove was pervaded by a deep sense of calm. The giant sequoias towered skyward, massive yet graceful, dominating their environment without overwhelming it. Besides handsome black oaks and ponderosa pines, the sequoias were joined by incense cedars with furrowed chocolate trunks, sugar pines whose limbs drooped from the weight of cones over a foot long, and white firs whose branches ascended in a symmetry as perfectly triangular as an old church steeple. From the forest floor rose the rich, fragrant aroma of pine needles and duff warmed by the morning sun, while the creek gurgled with new life imparted by the melting snows of the High Sierra. Dappled by shifting sunbeams, the sequoias' bark seemed constantly to change its hue, from cinammon brown to red to shades of violet. The air felt still, but far overhead the foliage tossed and swayed, dancing joyfully in the high breeze, singing a song infinitely soothing yet powerful, like the murmured whispers of countless generations past and future. The sequoias, ageless and vibrant, seemed at once inseparable from yet indifferent to the swirl of nature's elements; they had held this ground forever. As moment followed moment, time dissolved into the eternal Now and the splendor of life on this earth stood revealed as a never-ending miracle.

Native Americans had considered the giant sequoias sacred for thousands of years before the white man arrived in the mid-1900s. Members of the Mono tribe would not even touch the mammoth trees, believing a guardian spirit in the form of an owl would punish any who dared abuse them. Some of the first whites who encountered giant sequoias harbored a similarly respectful attitude, including a hunter by the name of Augustus T. Dowd who literally stumbled upon the trees one day in 1852. This was during the gold rush era, when thousands of people were hurrying to California with dreams of striking it rich. It was Dowd's job to furnish meat for construction crews working in the gold mining operations near the town of Murphys. One day, a grizzly bear he had shot and wounded fled across a meadow and disappeared into the forest. Dowd gave chase. Entering the forest, he suddenly encountered a tree so impossibly large that for a moment he thought his eyes were deceiving him. And the giant was not alone; nearby stood more of the incredible creatures. Forgetting about the grizzly, Dowd spent the rest of the day exploring the grove, the same grove I came to know and love over a century later.

When Dowd returned to Murphys and described what he had seen, the miners dismissed his account as the tallest of tales. But Dowd persuaded them to come have a look, and soon the story was out. Reporters from Stockton and San Francisco arrived to investigate. The stories they wrote were not universally believed—exaggeration was a staple of newspaper writing in those days, and besides, the plain facts about the trees did sound too bizarre to be true—but within weeks, visitors were making their way up to the grove to satisfy their curiosity. Where most saw beauty and wonder, however, a reckless few saw the chance for personal gain.

The danger arose in January 1853, less than a year after Dowd's accidental discovery. Three local capitalists, including Dowd's boss at the construction company, Captain W. H. Hanford, decided they would fell the largest of the big trees, cut a huge slab out of it, and display the slab in a traveling exhibition, like a freak in a circus show. The plan was by no means certain of success; many people objected to the idea of mutilating such ancient, majestic creatures. An official history of the giant sequoias published by the University of California speaks of general "expressions of outrage, dismay, and disgust" toward the plan and notes that Dowd in particular refused to have anything to do with it. But appeals to the businessmen's higher selves fell on deaf ears, and since there was not yet any real law enforcement on the frontier, there was no stopping them.

Hanford and his fellow investors had their minds set on the so-called Discovery Tree, the first tree Dowd had seen. It was by far the largest tree in the grove, measuring thirty-two feet wide at ground level and twenty-four feet wide at eight feet above the ground. There was not a saw in the world long and strong enough to topple such a tree. Instead, Hanford and his cronies hired five miners from the Murphys gold camps who decided to fell the tree by drilling holes in it with long metal augers, as wide as a man's fist. They would direct their attack in a line eight feet above the ground, hoisting the augers on specially built sawhorses. If the operation went as planned, the drilling would weaken the base of the tree enough to cause it to fall of its own accord.

The operation began on June 5, 1853. The first step—removing the bark from the areas to be drilled—proved simple enough: after a few incisions, great chunks of bark were pulled off like the peel from an orange. Having breached the tree's skin, however, the men found its underlying bone and muscle less accommodating. Their augers were simple devices; to propel them into the tree, two men turned a wheel at the far end of the auger round and round, as if closing a valve. It was difficult work.

The spectacle was sufficiently arousing that a group of well-to-do ladies and gentlemen journeyed up from Murphys to witness the final stages of the assault. After twenty days of fierce labor, the men had succeeded in punching an uninterrupted line of shafts around the entire circumference of the great tree. This was the moment the drillers and their backers had been longing for, but it turned out to hold a maddening surprise: the tree refused to fall. The sequoia "was so symmetrical, so well balanced, that it continued to stand upright and gave no sign of weakening," says the University of California's history. As if in revenge, the tree resisted the villains another two and a half days before finally giving up the ghost. Adding insult to injury, it waited until the entire entourage was away at lunch before crashing to the forest floor.

Undeterred, the investors pressed on with their plans, which included developing the site into a luxury resort, complete with a saloon and a bowling alley that were carved into the tree's fallen trunk. The stump itself was converted into a dance floor big enough to accommodate a dozen waltzing couples, an act of blasphemy that inspired the title of John Muir's tract *And the Vandals Danced Upon the Stump.* Another grand old tree, known as the Mother of the Forest, was also killed when its bark was stripped off and shipped to New York and London to entice tourists to visit. An accompanying handbill encouraged the curious to take the daily stagecoach from Sacramento City to see the extraordinary "Vegetable Monsters."

Captain Hanford and his fellow investors soon went bankrupt, but by then the larger damage had been done. "Any fool can destroy trees. They cannot defend themselves or run away," Muir later wrote, adding, "Through all the eventful centuries since Christ's time, and long before that, God has cared for these trees, saved them from drought, disease, avalanches, and a thousand storms; but he cannot save them from sawmills and fools; this is left to the American people."

More than once during my years of working . . . I retreated to my cabin in the forest. The sequoias were good company as I pondered the implications of all the places I had visited and people I had interviewed. Had the Discovery Tree not been cut down in 1853, it would now be the largest living thing on earth. Its destruction, I believe, ranks as one of the great crimes modern man has committed against nature. Yet the beauty of the remaining sequoias, and of the forest surrounding them, has never failed to lift my spirits, and their ancient history encourages me to take the long view of the human enterprise. As poet Edwin Markham wrote of the Big Trees in 1914, "Many of them have seen more than a hundred of our human generations rise, and give out their little clamors and perish. They chide our pettiness, they rebuke our impiety."

Now, . . . after one last visit to the Big Trees, I return once more to the question that sent me off around the world: will the human species survive the many environmental pressures crowding in on it at the end of the twentieth century?

Without reiterating all that I have reported, it seems plain to me that most of the key global environmental trends are moving in the wrong direction, and many are picking up speed. If, as Mostafa Tolba of the UN Environment Program put it, the 1990s were the decisive decade when humanity's environmental decline had to be reversed or it would accelerate beyond our control, it seems clear that we failed the test. Yes, progress has been made, but it has been too incremental, too grudging and slow. [I have] focused above all on the threat of climate change, where the IPCC's call for immediate 50 to 70 percent reductions in greenhouse gas emissions dwarfs what the world's governments (provisionally) agreed to in Kyoto. But similar discrepancies exist in regard to many environmental hazards. The gap between what science demands and what our political structures deliver remains vast, and it is vigilantly patrolled by powerful interests that profit from the existing order. The one important exception is population growth, though even there the good news is not all that encouraging: our numbers may eventually stabilize at a mere eight billion, rather than ten or twelve billion. . . .

By not taking decisive remedial action today, we are in effect gambling that the warnings of our most capable experts will turn out to be false alarms. "The climate system is an angry beast and we are poking it with sticks," Wallace Broecker, a geochemist at Columbia University who was one of the earliest investigators of global warming, has said. If provoked, said Broecker, the climate system could undergo *within a decade* a drastic warming or, worse, a return to Ice Age temperatures; such a shift could so devastate global agriculture that citizens of the wealthy North might suddenly discover they have more in common with their Dinka brethren than they ever imagined.

"The inhabitants of planet earth are quietly conducting a gigantic environmental experiment," warned Broecker. "So vast and sweeping will the consequences [of this experiment] be that, were it brought before any responsible council for approval, it would be firmly rejected." Broecker was referring to the buildup of greenhouse gases, but the point applies to other environmental hazards as well, including the vast array of manmade chemicals that pervade our plastics-saturated societies. In their book *Our Stolen Future*, zoologist Theo Colborn and coauthors Dianne Dumanoski and John Peterson Myers argue that DDT, dioxin, PCBs, and countless other bedrock chemicals of industrial civilization are severely undermining the reproductive health of wildlife and humans the world over by disrupting the hormones that regulate pre- and postnatal development. The most dramatic example is the 50 percent drop in human sperm counts reported in various places around the world during the last two generations. In addition, dozens of species—from Florida alligators to Lake Michigan mink to Baltic Sea fish—have suffered documented cases of shriveled penises, female sterility, and other fertility malfunctions. . . .

So, is there any hope? During my travels, I often encountered despair about our chances of reversing course, even among people who were not particularly involved in environmental affairs. When people learned I was writing . . . about whether the human species was going to survive its environmental problems, the question they invariably asked me was, "Well, will we?" And often, before I could reply, my interrogators would ruefully add words to the effect of, "It doesn't look good, does it?"

No, it doesn't, and there's no sense denying it. But there's no sense being paralyzed by it, either. To be sure, the task at hand is enormous, complex, and difficult, but the answers to it are not obscure; such august bodies as the UN Brundtland Commission and Earth Summit have been outlining the way forward for years now, and dedicated NGOs [nongovernmental organizations] and enlightened government and business leaders have fleshed out many of the concrete details. In the short term, we must accelerate changes already underway in our technologies to make them more efficient and environmentally friendly. Furthermore, these technologies must be diffused throughout the planet, which means in concrete terms that the North must help transfer them to the South. In the medium term, population size must be stabilized both in the South and the North, and the hyper-consumption that is now common in the North and among elites in the South must be cut back. In the medium to long term, capitalism will probably have to be transformed so that the constant expansion in material terms of production, consumption, and waste is no longer a central feature of the system. Development, not growth, must become our motto....

The idea is to renovate human civilization from top to bottom in environmentally sustainable ways. Humans would redesign and renovate everything from our farms to our factories, our garages to our garbage dumps, our schools, shops, houses, offices, and everything inside them, and we would do so in both the wealthy North and the impoverished South. The economic activity such renovating could generate would be enormous. Better yet, it would be labor intensive, and so address the problem of poverty that is the irreducible other half of the environmental challenge.

Restoring our embattled environment could become the biggest stimulus program for jobs and business in history. It is odd, and unfortunate, that our environmental movements have not been more effective in publicizing this possibility; the elements of such a program are scattered throughout their literature. Still, there is no time like the present. A political program that made the economic case for environmental restoration in clear and inspiring language could capture the popular imagination and rally it to victory.

Some environmentalists have suggested that the race to the moon in the 1960s serve as the model for the race now needed to save the earth. It's a good idea, and not simply because that earlier race sent back pictures of this blue planet that revolutionized humanity's understanding of itself and its place in the cosmos. The race to the moon showed how a clear mission and deadline can focus resources and fire public enthusiasm. It also demonstrated something rarely acknowledged these days: that certain overarching public challenges cannot be left up to the workings of the marketplace; government must play a central, leading role.

Besides the moon race, it seems to me that another model worth emulating is the New Deal that President Franklin Roosevelt launched in the 1930s to propel the U.S. economy out of depression. After all, the environmental crisis is as much an economic challenge as anything, and the New Deal helped overcome the gravest economic challenge in modern American history. What's more, the problems afflicting today's global economy are strikingly similar to those the New Deal was created to solve in the 1930s. As analysts from Karl Marx to John Maynard Keynes have noted, capitalist economies have an inherent tendency toward stagnation and inequality. The routine workings of the system generate more wealth than can find profitable investment outlets and too little money at the bottom to generate sufficient overall demand to keep the system churning forward....

The basic function of the New Deal was to restore sufficient demand to the economy by raising what can be called the social wage. New Deal policies raised the economy's collective purchasing power by guaranteeing workers a minimum wage and the right to strike for more; by putting unemployed people to work in government-funded public works projects; by providing direct cash payments to tide over the unemployed until they found work; and by establishing the universal pension plan for the elderly known as Social Security....

Why not revive these New Deal policies but apply them in a green and global fashion? The program could even be called the Global Green Deal. It would rely on market mechanisms to the maximum extent possible, while realizing that government must also establish "rules of the road" that compel markets to respect rather than harm the environment. In particular, governments must reform skewed tax, subsidy, and economic accounting systems so that the market internalizes environmental values. Governments should also increase public investment to help nascent industries like solar power achieve commercial takeoff. Priming the pump with steady purchases by the Pentagon in the 1960s was what got the computer industry up and running, and the Clinton administration did much the same in the 1990s by having the federal bureaucracy shift its purchases from virgin to recycled paper. By requiring that the seven million vehicles the U.S. government buys every year be fuel cell or hybrid powered rather than traditional gasoline powered, for example, Washington could help create market demand for green cars, demand that private capital could then step up and accommodate.

There would be a certain poetic justice in this, for the government's lavish subsidization of conventional automobiles throughout the twentieth century is no small cause of our current problems. Imagine what a similar level of government commitment could accomplish for the Global Green Deal! Mass transit would no longer be the unwanted stepchild of government policy but its proud focus, yielding transportation systems of such excellence that people would *want* to patronize them. Heat waves would no longer kill 800 people a summer in Chicago's inner city because they were too poor to afford air-conditioning; their apartment buildings would be renovated with the advanced energy efficiency methods that make air-conditioning unnecessary, with the renovations providing jobs for the neighborhood residents who now look in vain for work. Water would no longer be wasted in such criminally extravagant volumes; farmers would be encouraged to install superefficient drip irrigation systems. And so on, through all of the many technologies that need environmental reforming. . . .

A Global Green Deal can succeed only if rich and poor countries alike participate, but there is no getting around the fact that the rich will have to take the lead, both logistically and financially. Not only will rich nations have to adopt environmental reforms at home before expecting poor nations to follow suit, they will also have to help pay for most of the environmental reforms of the poor. If they don't, many of these reforms simply will not happen, not only because some Southern leaders are malign despots interested solely in lining their personal pockets, but because poverty makes a more urgent claim on limited public funds in the South than long-term environmental considerations do. . . . If even half of the estimated $500–900 billion in environmentally destructive subsidies now being doled out by the world's governments were pointed in the opposite direction, the Global Green Deal would be off to a roaring start.

None of this will happen without a fight, however. Amory Lovins likes to say that the role of government is to "steer, not row." But it must steer in a fundamentally different direction than at present, and that will upset those who profit from the status quo. . . .

The most difficult sharing will not be of money—the rich have plenty of that—but of environmental space. As Maurice Strong explained at the Earth Summit, the wealthy nations must substantially reduce their production of greenhouse gases and other such harmful activities in order to make space available so that Southern nations fighting poverty can increase *their* greenhouse emissions. Many Americans will resist such sharing, in part because they simply do not realize how lavish their lifestyles are compared with the rest of the world's. In taking for granted such luxuries as unlimited hot water at the turn of a tap, to say nothing of cars bigger than many people's houses, Americans inadvertently exhibit the sort of arrogance and self-centeredness that has made people hate the rich since time immemorial. The irony is that Americans are in fact a generous people, or at least like to think of themselves that way; they are simply oblivious to how wasteful and selfish their lives appear to outsiders. Countless millions of people in the South still live like the malnourished Dinka of Sudan, the unemployed workers of Shenyang, the landless squatters of rural Brazil. In the face of such deep, pervasive poverty, how can Americans and the rest of the world's comfortable class insist that they cannot cut back on their own consumption? . . .

The point is to strike a proper balance, and that requires change from individuals as well as institutions. It may be emotionally gratifying to blame corporations and governments for our environmental problems, and god knows they deserve it, but individuals cannot escape responsibility for their actions. "The problems are our lives," argues the writer Wendell Berry. "In the 'developed' countries, at least, the large [environmental] problems occur because all of us are living either partly wrong or almost entirely wrong. It was not just the greed of corporate shareholders and the hubris of corporate executives that put the fate of Prince William Sound into one ship [the Exxon *Valdez*]; it was also our demand that energy be cheap and plentiful. . . . To fail to see this is to go on dividing the world falsely between guilty producers and innocent consumers."

A Global Green Deal that put people to work restoring our environment would, I believe, yield enormous economic and social benefits to the vast majority of earth's current inhabitants, to say nothing of their descendants. At a time when many sectors and regions of the world economy are in danger of sinking into depression, a Global Green Deal could stimulate enough economic activity to keep the system from crashing down around us. It could also reverse current trends toward widening inequality and environmental overload. Such a fundamental shift in direction will not happen by itself, however. Politics must be committed.

Which returns me to the story of the Big Trees. If John Muir and a great number of other nature-loving citizens had not thrown themselves into defending the giant sequoias after the Discovery Tree was cut down, today's remaining seventy-five sequoia groves would have fallen to the logger's axe long ago. Muir's tireless advocacy helped establish Yosemite, Kings Canyon, General Grant, and numerous other national parks in California. (Even so, 35 percent of the original acreage of giant sequoias was logged from the late 1800s through the 1950s, along with untold amounts of sugar and ponderosa pines.) From the time the North Grove went to public auction in 1877, it took fifty years of activism before the grove was decisively removed from harm's way. On July 5,

1931, Calaveras Big Trees State Park was dedicated and declared public land forever.

The loss of the Discovery Tree can never be made right. In its place, the Big Stump now stands as a grotesque monument to the irreversible, shortsighted idiocy of some humans, just as nuclear waste dumps and other environmental scars will in the future. The rest of the sequoia grove, however, remains intact and spectacularly beautiful—living proof that humans can, with intelligence, cooperation, hard work, and perseverance, learn to live amid the natural bounties of this planet.

The moral of this story? The outlook is uncertain, the hour late, the earth a place of both beauty and despair. The fight for what's right is never ending, but the rewards are immense. Humans may or may not still be able to halt the drift toward ecological disaster, but we will find out only if we rouse ourselves and take common and determined action.

I wish us godspeed.

ACKNOWLEDGMENTS

1.1 From John Muir, *The Mountains of California* (Houghton Mifflin, 1916).

1.2 From Gifford Pinchot, *The Fight for Conservation* (Doubleday, 1910).

1.3 From Aldo Leopold, *A Sand County Almanac: And Sketches Here and There* (Oxford University Press, 1949). Copyright © 1949, 1977 by Oxford University Press, Inc. Reprinted by permission.

2.1 From Lynn White, Jr., "The Historical Roots of Our Ecological Crisis," *Science*, vol. 155, no. 2 (March 10, 1967). Copyright © 1967 by The American Association for the Advancement of Science. Reprinted by permission.

2.2 From Barry Commoner, *The Closing Circle: Nature, Man and Technology* (Bantam Books, 1971). Copyright © 1971 by Barry Commoner. Reprinted by permission of Alfred A. Knopf, Inc.

2.3 From Paul R. Ehrlich, *The Population Bomb* (Ballantine Books, 1968). Copyright © 1968, 1971 by Paul R. Ehrlich. Reprinted by permission of Ballantine Books, a division of Random House, Inc. Notes omitted.

2.4 From Garrett Hardin, "The Tragedy of the Commons," *Science*, vol. 162 (December 13, 1968). Copyright © 1968 by The American Association for the Advancement of Science. Reprinted by permission. References omitted.

2.5 From Donella H. Meadows, Dennis L. Meadows, Jørgen Randers, and William W. Behrens III, *The Limits to Growth: A Report for the Club of Rome's Project on the Predicament of Mankind* (Universe Books, 1972). Copyright © 1972 by Dennis L. Meadows. Reprinted by permission of Potomac Associates, Washington, DC.

2.6 From Peter M. Vitousek, Harold A. Mooney, Jane Lubchenco, and Jerry M. Melillo, "Human Domination of Earth's Ecosystems," *Science* (July 25, 1997). Copyright © 1997 by The American Association for the Advancement of Science. Reprinted by permission.

3.1 From G. Evelyn Hutchinson, "Homage to Santa Rosalia, or Why Are There So Many Kinds of Animals?" *The American Naturalist*, no. 870 (May/June 1959). Copyright © 1959 by University of Chicago Press. Reprinted by permission. References omitted.

3.2 From Eugene P. Odum, "Great Ideas in Ecology for the 1990s," *Bioscience*, vol. 42, no. 7 (July/August 1992). Copyright © 1992 by The American Institute of Biological Sciences. Reprinted by permission.

4.1 From John Teal and Mildred Teal, *Life and Death of the Salt Marsh* (Ballantine Books, 1969). Copyright © 1969 by John and Mildred Teal. Reprinted by permission of Frances Collin, Literary Agent.

4.2 From Orrin H. Pilkey, "Geologists, Engineers, and a Rising Sea Level," *Northeastern Geology*, vol. 3, nos. 3/4 (1981). Copyright © 1981 by Orrin H. Pilkey. Reprinted by permission of The Northeastern Science Foundation.

5.1 From Chancey Juday, "The Annual Energy Budget of an Inland Lake," *Ecology*, vol. 21, no. 4 (October 1940). Copyright © 1940 by The Ecological Society of America. Reprinted by permission.

5.2 From John M. Fowler, *Energy and the Environment* (McGraw-Hill, 1975). Copyright © 1975 by The McGraw-Hill Companies. Reprinted by permission. Notes omitted.

6.1 From Amory B. Lovins, *Soft Energy Paths: Toward a Durable Peace* (Friends of the Earth, 1977). Copyright © 1977 by Amory B. Lovins, Director of Research, Rocky Mountain Institute. Reprinted by permission. Some notes omitted.

6.2 From Christopher Flavin and Seth Dunn, "Reinventing the Energy System," in Lester R. Brown et al., *State of the World 1999: A Worldwatch Institute Report on Progress Toward a Sustainable Society* (W.W. Norton, 1999). Copyright © 1999 by The Worldwatch Institute. Reprinted by permission of W.W. Norton & Company, Inc. Notes omitted.

7.1 From *Sierra Club v. Morton*, 405 U.S. 727 (1972). Notes and some case citations omitted.

7.2 From William Cronon, *Uncommon Ground: Toward Reinventing Nature* (W.W. Norton, 1995). Copyright © 1995 by William Cronon. Reprinted by permission of W.W. Norton & Company, Inc. Notes omitted.

7.3 From John Vandermeer and Ivette Perfecto, "Rethinking Rain Forests: Biodiversity and Social Justice," *Food First Backgrounder* (Summer 1995). Copyright © 1995 by The Institute for Food and Development Policy, Oakland, CA. Reprinted by permission. Notes and references omitted.

8.1 From John Evelyn, *Fumifugium: Or the Inconvenience of the Aer and Smoake of London Dissipated* (1661). Notes omitted.

8.2 From Beverly Paigen, "Controversy at Love Canal," *Hastings Center Report* (June 1982). Copyright © 1982 by The Hastings Center. Reprinted by permission. References omitted.

8.3 From J. W. Maurits la Rivière, "Threats to the World's Water," *Scientific American* (September 1989). Copyright © 1989 by Scientific American, Inc. Reprinted by permission.

9.1 From Mario J. Molina and F. S. Rowland, "Stratospheric Sink for Chlorofluoromethanes: Chlorine Atomc-atalysed Destruction of Ozone," *Nature*, vol. 249 (June 28, 1974), pp. 810–812. Copyright © 1974 by Macmillan Magazines Ltd. Reprinted by permission.

9.2 From Richard Elliot Benedick, *Ozone Diplomacy: New Directions in Safeguarding the Planet* (Harvard University Press, 1991). Copyright © 1991 by The World Wildlife Fund & The Conservation Foundation and The Institute for the Study of Diplomacy, Georgetown University. Reprinted by permission of Harvard University Press. Notes omitted.

9.3 From The Intergovernmental Panel on Climate Change (IPCC), *Climate Change 1995: IPCC Second Assessment Report*, <http://www.ipcc.ch/cc95/cont-95.htm> (April 12, 1999). Copyright © 1999 by The Intergovernmental Panel on Climate Change. Reprinted by permission of the Secretary of the IPCC.

10.1 From E. O. Wilson, "The Current State of Biological Diversity," in E. O. Wilson, ed., *Biodiversity* (National Academy Press, 1988). Copyright © 1988 by The National Academy of Sciences. Reprinted by permission of The National Academy Press, Washington, DC. Some references omitted.

10.2 From Stephen Jay Gould, *Eight Little Piggies: Reflections in Natural History* (W.W. Norton, 1993). Copyright © 1993 by Stephen Jay Gould. Reprinted by permission of W.W. Norton & Company, Inc.

11.1 From Betsy Hartmann, *Reproductive Rights and Wrongs: The Global Politics of Population Control*, rev. ed. (South End Press, 1995). Copyright © 1995 by Betsy Hartmann. Reprinted by permission of South End Press, 116 Saint Botolph Street, Boston, MA 02115. Notes omitted.

11.2 From Julian L. Simon, *The Ultimate Resource* (Princeton University Press, 1981). Copyright © 1981 by Princeton University Press. Reprinted by permission.

11.3 From Joel E. Cohen, *How Many People Can the Earth Support?* (W.W. Norton, 1995). Copyright © 1995 by Joel E. Cohen. Reprinted by permission of W.W. Norton & Company, Inc. Notes omitted.

12.1 From Wendell Berry, *The Unsettling of America: Culture and Agriculture* (Sierra Club Books, 1977). Copyright © 1977 by Wendell Barry. Reprinted by permission of Sierra Club Books.

12.2 From Frances Moore Lappé and Joseph Collins, *Food First: Beyond the Myth of Scarcity* (Ballantine Books, 1977). Copyright © 1977, 1978 by The Institute for Food and Development Policy, Oakland, CA. Reprinted by permission of Ballantine Books, a division of Random House, Inc. Notes omitted.

12.3 From Lester Brown, "Food Scarcity: An Environmental Wakeup Call," *The Futurist* (January/February 1998). Copyright © 1998 by The World Future Society. Reprinted by permission of The World Future Society, 7910 Woodmont Ave., Bethesda, MD 20814.

13.1 From Rachel Carson, *Silent Spring* (Houghton Mifflin, 1962). Copyright © 1962 by Rachel Carson; renewed 1990 by Roger Christie. Reprinted by permission of Houghton Mifflin Company.

13.2 From Robert van den Bosch, *The Pesticide Conspiracy* (Doubleday, 1978). Copyright © 1978 by Robert van den Bosch. Reprinted by permission of Doubleday, a division of Random House, Inc. Notes omitted.

14.1 From Sandra Steingraber, *Living Downstream: An Ecologist Looks at Cancer and the Environment* (Addison-Wesley, 1997). Copyright © 1997 by Sandra Steingraber. Reprinted by permission of Perseus Books Publishers, a member of Perseus Books, L.L.C.

14.2 From Theo Colborn, Dianne Dumanoski, and John Peterson Myers, *Our Stolen Future* (Dutton, 1996). Copyright © 1996 by Theo Colborn, Dianne Dumanoski, and John Peterson Myers. Reprinted by permission of Dutton, a division of Penguin Putnam, Inc.

15.1 From Mark Sagoff, "At the Shrine of Our Lady of Fatima *or* Why Political Questions Are Not All Economic," *Arizona Law Review*, vol. 23 (1981). Adapted from Mark Sagoff, *The Economy of the Earth: Philosophy, Law and the Environment* (Cambridge University Press, 1988). Copyright © 1981 by Mark Sagoff. Reprinted by permission of Cambridge University Press and the author. Some notes omitted.

15.2 From Robert D. Bullard, *Dumping in Dixie: Race, Class, and Environmental Quality* (Westview Press, 1990). Copyright © 1990 by Westview Press. Reprinted by permission of Westview Press, a member of Perseus Books, L.L.C. Some notes omitted.

15.3 From Janet N. Abramovitz, "Valuing Nature's Services," in Lester R. Brown et al., *State of the World 1997* (W.W. Norton, 1997). Copyright © 1997 by The Worldwatch Institute. Reprinted by permission of W.W. Norton & Company, Inc. Notes omitted.

16.1 From World Commission on Environment and Development, *Our Common Future* (Oxford University Press, 1987). Copyright © 1987 by The World Commission on Environment and Development. Reprinted by permission of Oxford University Press. Notes omitted.

16.2 From Paul W. Taylor, "The Ethics of Respect for Nature," *Environmental Ethics*, vol. 3 (Fall 1981). Copyright © 1981 by Environmental Philosophy, Inc. Reprinted by permission. Notes omitted.

16.3 From Dave Foreman, *Confessions of an Eco-Warrior* (Harmony Books, 1991). Copyright © 1991 by David Foreman. Reprinted by permission of Harmony Books, a division of Crown Publishers, Inc. Notes omitted.

16.4 From Vandana Shiva, "Women's Indigenous Knowledge and Biodiversity Conservation," in Maria Mies and Vandana Shiva, *Ecofeminism* (Zed Books, 1993). Copyright © 1993 by Maria Mies and Vandana Shiva. Reprinted by permission of Zed Books Ltd.

16.5 From Mark Hertsgaard, *Earth Odyssey: Around the World in Search of Our Environmental Future* (Broadway Books, 1998). Copyright © 1998 by Mark Hertsgaard. Reprinted by permission of Broadway Books, a division of Random House, Inc.

Index